BIOLOGICAL AND MEDICAL PHYSICS
BIOMEDICAL ENGINEERING

Marion Gurfein originally painted the upper water color of the tiger in 1996. In 2006, four years after the onset of macular degeneration, she revisited her original painting as part of her artistic chronicles of the progression.

Mark S. Humayun
James D. Weiland
Gerald Chader
Elias Greenbaum (Eds.)

Artificial Sight

Basic Research, Biomedical Engineering, and Clinical Advances

 Springer

Mark S. Humayun
James D. Weiland
Gerald Chader
Doheny Eye Institute
Los Angeles, CA 90033
USA
humayun@doheny.org
jweiland@doheny.org
gchader@doheny.org

Elias Greenbaum
Oak Ridge National Laboratory
Oak Ridge, TN 37831
USA
greenbaum@ornl.gov

ISBN-13: 978-0-387-49329-9   e-ISBN-13: 978-0-387-49331-2

Library of Congress Control Number: 2006939422

Printed on acid-free paper.

9 8 7 6 5 4 3 2 1

springer.com

Series Preface

The fields of biological and medical physics and biomedical engineering are broad, multidisciplinary and dynamic. They lie at the crossroads of frontier research in physics, biology, chemistry, and medicine. The Biological & Medical Physics/Biomedical Engineering Series is intended to be comprehensive, covering a broad range of topics important to the study of the physical, chemical and biological sciences. Its goal is to provide scientists and engineers with textbooks, monographs, and reference works to address the growing need for information.

Books in the series emphasize established and emergent areas of science including molecular, membrane, and mathematical biophysics; photosynthetic energy harvesting and conversion; information processing; physical principles of genetics; sensory communications; automata networks, neural networks, and cellular automata. Equally important will be coverage of applied aspects of biological and medical physics and biomedical engineering such as molecular electronic components and devices, biosensors, medicine, imaging, physical principles of renewable energy production, advanced prostheses, and environmental control and engineering.

Elias Greenbaum
Oak Ridge, TN

Preface

For over 50 years the U.S. Department of Energy's Biological and Environmental Research (BER) program has advanced environmental and biomedical knowledge that promotes improved energy production, development, and use; international scientific and technological cooperation; and research that improves the quality of life for all peoples. BER supports these vital missions through competitive and peer-reviewed research at national laboratories, universities, and private institutions. This book, *Artificial Sight: Basic Research, Biomedical Engineering, and Clinical Advances* emerged mostly from the research programs of presenters at the Second DOE International Symposium on Artificial Sight. The book, however, is not a symposium proceedings. The editors encouraged the chapter authors to expand on the vision of their research in this field which lies at the intersection of physics, chemistry, biology and biomedical engineering. The members of the organizing committee for this DOE symposium are M.S. Humayun (chairman), E. Greenbaum (co-chairman), D.A. Cole, R. Iezzi, Y. Tano, M.V. Viola, J.D. Weiland and E. Zrenner. The work of the DOE Artificial Retina Program continues with the collaboration of the National Laboratory system, universities, and private industry. The members of this team have helped create micromachine technology design, mathematical modeling of retinal information processing, microelectrode arrays designed for retinal tissue stimulation, and telemetric communications. We thank Eugenie V. Mielczarek, Professor Emeritus of Physics at George Mason University, for bringing our attention to Marion Gurfein's artwork which is exhibited in the frontispiece.

Mark S. Humayun
James D. Weiland
Gerald Chader
Los Angeles, CA

Elias Greenbaum
Oak Ridge, TN

June 2007

Contents

Biological Response to Stimulation

List of Contributors

G. W. Abrams
Kresge Eye Institute
Wayne State University
Ligon Research Center of Vision
Wayne State University
Detroit, MI
USA

A. Agazaryan
Second Sight Medical Products, Inc.
Sylmar, CA
USA

Hossein Ameri
Doheny Retina Institute
Keck School of Medicine
University of Southern California
Los Angeles, CA
USA

G. W. Auner
SSIM/Biomedical
Engineering/Electrical and Computer
Engineering
Wayne State University
Ligon Research Center of Vision
Detroit, MI
USA

Michael Bach
University Eye Hospital Freiburg
Freiburg, Germany

M. S. Blumenkranz
Department of Ophthalmology
Stanford University
Stanford, CA
USA

A. Butterwick
Hansen Experimental Physics
Laboratory
Stanford University
Stanford, CA
USA

Carlos J. Cela
Department of Electrical and
Computer Engineering
North Carolina State University
Raleigh, NC
USA

Gerald Chader
Doheny Retina Institute
Keck School of Medicine
University of Southern California
Los Angeles, CA
USA

Xinyu Chai
Institute for Laser Medicine
and Bio-Photonics
College of Life Science and
Technology
Shanghai Jiao-Tong University
and C-Sight Group
Shanghai, People's Republic of China

I. Chan
Department of Ophthalmology and
Hansen Experimental Physics
Laboratory
Stanford University
Stanford, CA
USA

E. J. Chichilnisky
The Salk Institute for Biological
Studies
La Jolla, CA
USA

A. Chu
Second Sight Medical Products, Inc.
Sylmar, CA
USA

C-Sight Group
Institute for Laser Medicine
and Bio-Photonics
College of Life Science
and Technology
Shanghai Jiao-Tong University
and C-Sight Group
Shanghai, People's Republic of China

Wladyslaw Dabrowski
Faculty of Physics and Applied
Computer Science
AGH University of Science
and Technology
Krakow, Poland

Gislin Dagnelie
Department of Ophthalmology
Johns Hopkins University School of
Medicine
Baltimore, MD
USA

Wilhelm Durst
Centre for Ophthalmology
University Eye Hospital Tübingen
Tübingen, Germany

Matthias Feucht
Division of Ophthalmology
University Medical Center
Hamburg-Eppendorf,
Hamburg, Germany

Ione Fine
Department of Psychology

University of Washington
Seattle, WA
USA

Shelley I. Fried
Vision Science
University of California - Berkeley
Berkeley, CA
USA

Takashi Fujikado
Department of Applied Visual Science
Osaka University Medical School
Osaka, Japan

Elias Greenbaum
Chemical Sciences Division
Oak Ridge National Laboratory
Oak Ridge, TN
USA

R. Greenberg
Second Sight Medical Products, Inc.
Sylmar, CA
USA

John R. Hetling
Department of Bioengineering
University of Illinois at Chicago
Chicago, IL
USA

Akito Hirakata
Department of Ophthalmology
Kyorin University School of Medicine
Tokyo, Japan

Ralf Hornig
IMI Intelligent Medical Implants
GmbH
Bonn, Germany

Pawel Hottowy
Faculty of Physics and Applied
Computer Science
AGH University of Science
and Technology
Krakow, Poland

Hain-Ann Hsueh
Bioengineering
University of California – Berkeley
Berkeley, CA
USA

Zhiyu Hu
Biosciences Division
Oak Ridge National Laboratory
Oak Ridge, TN
USA

P. Huie
Department of Ophthalmology and
Hansen Experimental Physics
Laboratory
Stanford University
Stanford, CA
USA

Mark S. Humayun
Doheny Retina Institute
Keck School of Medicine
Department of Ophthalmology
University of Southern California
Los Angeles, CA
USA

A. Istomin
Second Sight Medical Products, Inc.
Sylmar, CA
USA

Keiichiro Kagawa
Graduate School of Materials Science
Nara Institute of Science
and Technology
Nara, Japan

Thomas Laube
Division of Ophthalmology
University Hospital Essen
Essen, Germany

Gianluca Lazzi
Department of Electrical
and Computer Engineering

North Carolina State University
Raleigh, NC
USA

Alan M. Litke
Santa Cruz Institute for Particle
Physics
University of California – Santa Cruz
Santa Cruz, CA
USA

J. Little
Second Sight Medical Products, Inc.
Sylmar, CA
USA

Wentai Liu
Department of Electrical Engineering
University of California – Santa Cruz
Santa Cruz, CA
USA

M. F. Marmor
Department of Ophthalmology
Stanford University
Stanford, CA
USA

J. P. McAllister
Department of Neurosurgery
Wayne State University
Ligon Research Center of Vision
Wayne State University
Detroit, MI
USA

Douglas McCreery
Neural Engineering Program
Huntington Medical Research
Institutes
Pasadena, CA
USA

K. Morris
Second Sight Medical Products, Inc.
Sylmar, CA
USA

Evan J. Nagler
Chemical Sciences Division
Oak Ridge National Laboratory
Oak Ridge, TN
USA

Kazuaki Nakauchi
Department of Applied Visual Science
Osaka University Medical School
Osaka, Japan

Jun Ohta
Graduate School of Materials Science
Nara Institute of Science
and Technology
Nara, Japan

Motoki Ozawa
Vision Institute
R&D Division
NIDEK, Co., Ltd.
Japan

D. Palanker
Department of Ophthalmology and
Hansen Experimental Physics
Laboratory
Stanford University
Stanford, CA
USA

Qiushi Ren
Institute for Laser Medicine
and Bio-Photonics
College of Life Science and
Technology
Shanghai Jiao-Tong University
and C-Sight Group
Shanghai, People's Republic of China

Gisbert Richard
Division of Ophthalmology
University Medical Center
Hamburg-Eppendorf
Hamburg, Germany

Charlene A., Sanders
Chemical Sciences Division
Oak Ridge National Laboratory
Oak Ridge, TN
USA

Hajime Sawai
Department of Physiology
Osaka University Medical School
Osaka, Japan

Stefan Schmidt
Department of Electrical
and Computer Engineering
North Carolina State University
Raleigh, NC
USA

Chris Sekirnjak
The Salk Institute for Biological
Studies
La Jolla, CA
USA

Alexander Sher
Santa Cruz Institute for Particle
Physics
University of California – Santa Cruz
Santa Cruz, CA
USA

Vinit Singh
Department of Electrical
and Computer Engineering
North Carolina State University
Raleigh, NC
USA

Mohanasankar Sivaprakasam
Department of Electrical Engineering
University of California – Santa Cruz
Santa Cruz, CA
USA

P. Siy
SSIM/Biomedical
Engineering/Electrical and Computer
Engineering

Wayne State University
Detroit, MI
USA

M. Talukder
SSIM/Biomedical
Engineering/Electrical and Computer
Engineering
Wayne State University
Detroit, MI
USA

Yasuo Tano
Department of Ophthalmology
Osaka University Medical School
Osaka, Japan

Yasuo Terasawa
Vision Institute
R&D Division
NIDEK, Co., Ltd.
Japan

Thomas Thundat
Biosciences Division
Oak Ridge National Laboratory
Oak Ridge, TN
USA

Takashi Tokuda
Graduate School of Materials Science
Nara Institute of Science
and Technology
Nara, Japan

Susanne Trauzettel-Klosinski
Centre for Ophthalmology
University Eye Hospital Tübingen
Tübingen, Germany

A. Vankov
Department of Ophthalmology and
Hansen Experimental Physics
Laboratory
Stanford University
Stanford, CA
USA

Michaela Velikay-Parel
Division of Ophthalmology
University Hospital Graz
Graz, Austria

Matthias Walter
Kirchhoff Institute for Physics
Ruprecht Karls University Heidelberg
Heidelberg, Germany

Guoxing Wang
Department of Electrical Engineering
University of California – Santa Cruz
Santa Cruz, CA
USA

James D. Weiland
Doheny Retina Institute
Keck School of Medicine
Department of Ophthalmology
University of Southern California
Los Angeles, CA
USA

Frank Werblin
Molecular and Cell Biology
University of California – Berkeley
Berkeley, CA
USA

Barbara Wilhelm
Steinbeis Transfer Center for
Biomedical Optics
Ofterdingen, Germany

Robert Wilke
Centre for Ophthalmology
University Eye Hospital Tübingen
Tübingen, Germany

Kaijie Wu
Institute for Laser Medicine
and Bio-Photonics
College of Life Science
and Technology
Shanghai Jiao-Tong University
and C-Sight Group
Shanghai, People's Republic of China

Liancheng Yang
Department of Ophthalmology
Johns Hopkins University School of
Medicine
Baltimore, MD
USA

R. You
SSIM/Biomedical
Engineering/Electrical and Computer
Engineering
Wayne State University
Detroit, MI
USA

Thomas Zehnder
IMI Intelligent Medical Implants AG
Zug, Switzerland

Mingcui Zhou
Department of Electrical Engineering
University of California – Santa Cruz

Santa Cruz, CA
USA

Chuanqing Zhou
Institute for Laser Medicine
and Bio-Photonics
College of Life Science
and Technology
Shanghai Jiao-Tong University
and C-Sight Group
Shanghai, People's Republic of China

David M. Zhou
Second Sight Medical Products, Inc.
Sylmar, CA
USA

Eberhart Zrenner
Centre for Ophthalmology
University Eye Hospital Tübingen
Tübingen, Germany

List of Acronyms

ADL/O	Activities of daily living and orientation
adRP	Autosomal dominant retinitis pigmentosa
AER	Averaged evoked response
AMD	Age-related macular degeneration
ANOVA	Analysis of variance
APB	2-amino-4-phosphonobutyrate
ARVO	Association for Research on Vision and Ophthalmology
ASP	Aspartate
AVD	Artificial vision device
BaGa	Basic grating acuity
BaLM	Basic light and motion
BCC	Biphasic current controller
BDNF	Brain-derived neurotrophic factor
bFGF	Basic fibroblast growth factor
BHE	Bioheat equation
BSI	Brief symptom inventory
CCD	Charge coupled device
CMG	Common mode gain
CMOS	Complementary metal oxide semiconductor
CMRR	Common mode rejection ratio
CNTF	Ciliary neurotrophic factor
CNV	Choroidal neovascular membrane
CT	Computed tomography
CV	Cyclic voltammetry
DAC	Digital-to-analog converter
DACC	Digital to analog current converter
DC	Direct current
DI	Deionized
DMG	Differential mode gain
DPLL	Digital phase-locked loop
DSP	Digital signal processing
ECG	Electrocardiogram
EEG	Electroencephalogram
EEP	Electrically evoked potential
EER	Electrically evoked response
eERG	Electrically elicited ERG

EIS	Electrochemical impedance spectroscopy
EKG	Electrocardiogram
EP	Evoked potential
ERG	Electroretinogram
ETDRS	Early Treatment Diabetic Retinopathy Study
FA	Fluorescein angiogram
FDTD	Finite-Difference Time-Domain
fERG	Focal ERG
fMRI	Functional magnetic resonance imaging
FrACT	Freiburg visual acuity and contrast test
GABA	Glutamate and γ-aminobutyric acid
GCL	Ganglion cell layer
GS	Glutamine synthetase
HMD	Head-mounted display
IC	Integrated circuit
ICMS	Intracortical microstimulation
ID	Identification
IEEE	Institute of Electrical and Electronics Engineers
IGF-1	Insulin-like growth factor-1
IIP	IIP-Technologies gmbh
ILM	Inner limiting membrane
INL	Inner nuclear layer
IOP	Intraocular pressure
IR	Infrared
IrOx	Iridium oxide
ISCEV	International Society for Clinical Electrophysiology of Vision
IT	Inferotemporal cortex
LEP	Light evoked potential
LFP	Local field potential
LGN	Lateral geniculate nucleus
LSI	Large-scale integration
LVES	Low vision enhancement system
mfERG	Multifocal ERG
MOEMS	Micro-optoelectromechanical systems
MRI	Magnetic resonance imaging
NAIST	Nara Institute of Science and Technology
NEDO	New Energy Development Organization
NFL	Nerve fiber layer
NIR	Near-infrared
NMDA	N-methyl-d-aspartic acid
NMOS	n-Channel metal-oxide semiconductor
OCT	Optical coherence tomography
ON	Optic nerve
ONL	Outer nuclear layer
PBS	Phosphate buffered saline

PDA	*cis*-2,3-piperidinecarboxylate
PDE	Partial differential equation
PDMS	Poly-dimethylsiloxane
pERG	Pattern ERG
PFCL	Perfluorodecaline
pfERG	Paired-flash ERG
PFM	Pulse frequency modulation
PIM	Partial inductance method
PLL	Phase-locked loop
PMMA	Polymethyl-methacrylate
PMOS	p-channel metal-oxide semiconductor
PS	Current pulse stimulation
RCCS	Regulated cascade current sink
RCS	Royal College of Surgeons
RF	Radio frequency
RGC	Retinal ganglion cell
RMS	Root mean square
RP	Retinitis pigmentosa
RPE	Retinal pigment epithelium
RS	Retinal stimulator
RSA	Retinal stimulating array
RT-PCR	Reverse transcriptase–polymerase chain reaction
SC	Superior colliculus
SEM	Scanning electron micrograph
SEM	Standard errors of the mean
SHE	Standard hydrogen electrode
SIDNE	Stimulation-induced depression of electrical excitability
SNR	Signal to noise ratio
SOC	System-on-chip
SPC	Serial to parallel converter
SSMP	Second Sight Medical Products
STS	Suprachoroidal transretinal stimulation
TES	Transcorneal electrical stimulation
TFF	T flip-flops
TTX	Tetrodotoxin
ULR	Unit-like response
VEP	Visually evoked potentials
VER	Visually evoked response
VGA	Video graphics array
VPU	Visual processing unit
WIBI	Wireless implantable biodevice interface

1
Biological Considerations for an Intraocular Retinal Prosthesis

Hossein Ameri, James D. Weiland and Mark S. Humayun
Doheny Retina Institute, Doheny Eye Institute,
Keck School of Medicine, University of Southern California

Introduction

The idea of replacing a dysfunctional or missing body part with a prosthesis is perhaps as old as the history of humankind; yet, it still continues to be a fascinating theme. While early prostheses were limited to external organs (e.g. wooden legs), the twentieth century witnessed the implantation of numerous devices inside the human body, some of which have since become routine surgical procedures. For external prostheses, both proper fitting and functionality are needed to make a device acceptable for use; however, internally implanted prostheses have additional challenges, and the implantation of intelligent devices which interact with surrounding tissues is even more difficult.

Despite being one of the most delicate organs of the body, the eye has proven to be a hospitable environment for a variety of materials. Whereas some metals such as iron and copper are highly toxic to the eye, other substances such as silicone, polymethyl-metacrylate (PMMA), and acrylic are well tolerated. Intraocular lenses made of these substances are routinely used to replace the natural lens during cataract surgery, the most commonly performed intraocular surgical procedure. Prior to surgical procedures that could be performed within the back of the eye (vitrectomy), retinal detachment had almost exclusively been treated with scleral buckling, a procedure whereby silicone rubber is placed around the eyeball. Even now, a significant number of retinal detachments are treated this way. Ocular implants are also used in some procedures relating to orbital, keratorefractive, and glaucoma surgeries. Thus, the use of prosthetic devices in an ocular environment is well established and a number of materials are proven to be suitable in the fabrication and use of these devices.

Retinal prostheses are being developed to apply electrical stimulation to the retina in order to restore vision. Several different configurations have been proposed and these will be reviewed in detail later in the chapter. However, common qualities of almost all of the implants are: (1) a light-sensitive device for

1

capturing image data, (2) implanted microelectronic for converting image data into a stimulus pattern, and (3) a microelectrode array interface for delivering the stimulus current to the retina. The implementation of these functions is system dependent; however, the potential for harmful interaction between the device and the eye is considerable in all cases. The following issues must be considered for any of these implants.

- *Toxicity*: The device should either be made of (or hermetically coated with) materials that are not toxic to the retina or any other ocular tissue.
- *Degradation*: Even if the outer coating is not toxic, over time, the degradation of the device coating can expose the retina to toxic materials which are part of the implant electronics.
- *Mechanical damage*: The electrode array should also properly fit over the area of the retina that is being implanted. The retina is a delicate neural tissue, and rigid electrodes and microelectronic chips can easily cut the retina, without careful mechanical design. Additionally, limited space in the eye and orbit will constrain the device size and shape.
- *Electrical damage*: An electrical stimulus beyond safe limits may damage the retina.
- *Thermal damage*: The retinal prosthesis electronics produce heat which can be damaging to the retina and other ocular structures, if beyond a given threshold.
- *Reversibility*: Extreme measures in fabricating a robust prosthesis and preoperative testing of the device reduce the chance of its becoming dysfunctional; however, despite best efforts, some prostheses may break down, and it is important to be able to remove the faulty device.

Background

Anatomy and Physiology of the Eye

The eyeball is classically divided into two segments: the anterior segment, composed of the cornea, iris, lens, and ciliary body; and the posterior segment, composed of the sclera, choroid, retina, vitreous, and optic nerve (Figure 1.1).

The eyeball is located within a bony socket called the orbit (Figure 1.2). In addition to housing the eyeball, the orbital cavity contains extraocular muscles, vessels, nerves, connective tissue, and orbital fat.

Although it may be possible to insert a prosthetic device around the eyeball, the device size and shape are important limiting factors. For instance, a poorly shaped, rigid object inside the orbital cavity may damage vessels and nerves, interfere with eye movement, and directly damage the eyeball. On the other hand, a bulky device may displace the eyeball, jeopardize the blood supply to the eye, or raise the intraocular pressure (IOP), thereby resulting in nerve fiber layer damage. These constraints will obviously limit the capability of an electronic device because of necessary size restrictions.

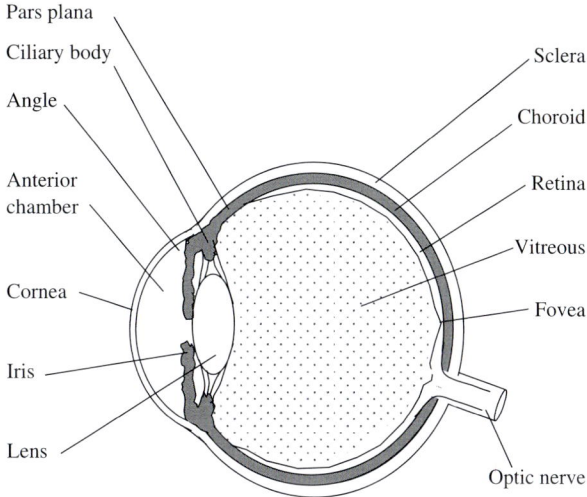

FIGURE 1.1. Schematic cross section of the eyeball as viewed from the side.

Anterior Segment

The iris divides the anterior segment into anterior and posterior chambers and gives the eye its color (Figure 1.3). The opening in the center of the iris is the pupil (Figure 1.3). The pupil regulates light entry to the back of the eye by dilatation or constriction.

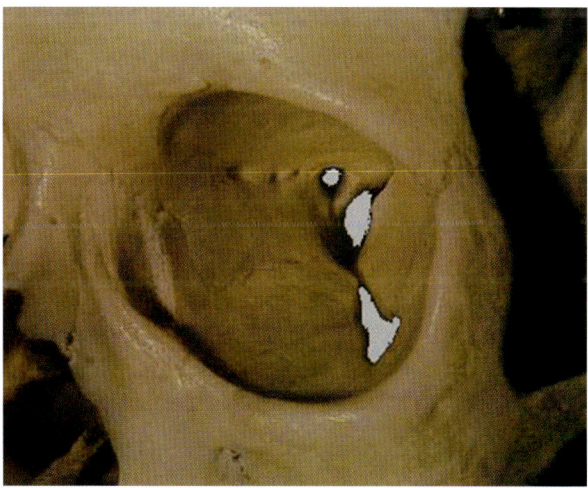

FIGURE 1.2. The bony orbit that accommodates the eyeball, and all surrounding structures (adapted with permission from http://anatome.ncl.ac.uk/tutorials/clinical/eye/page3.html).

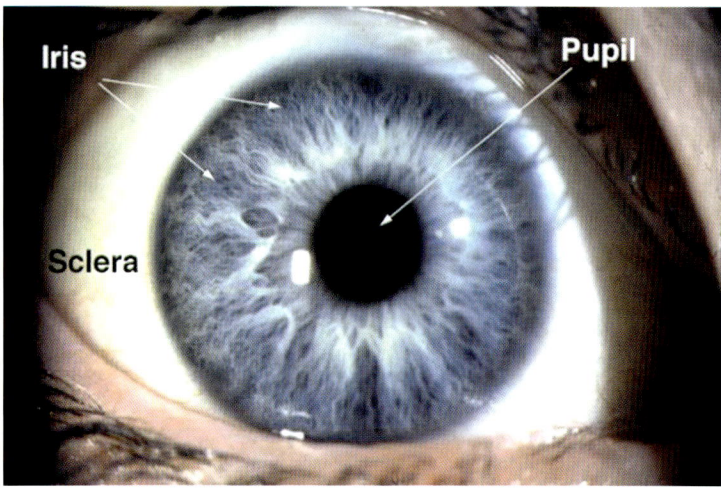

FIGURE 1.3. The front view of the eye; the sclera is covered by a thin transparent membrane known as conjunctiva which contains small vessels (not visible in this photo) (adapted with permission from http://webvision.med.utah.edu/index.html).

Both anterior and posterior chambers are filled with aqueous humor, a fluid produced by the ciliary body in the posterior chamber and drained through trabecular meshwork in the anterior chamber (Figure 1.4).

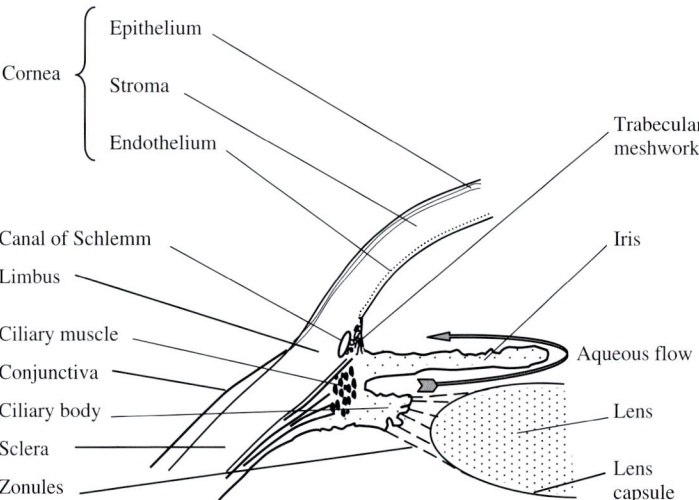

FIGURE 1.4. Structures of the anterior segment of the eye; note that the aqueous humor produced in the posterior chamber passes through the pupil to reach the trabecular meshwork in the anterior chamber.

The rate at which aqueous is produced and drained determines IOP; the overproduction or more often outflow obstruction will result in raised IOP, which may in turn lead to retinal nerve fiber layer damage and loss of peripheral vision – a common eye disease known as glaucoma. Thus, a retinal prosthesis, which has components in the anterior chamber, risks disrupting the fluid flow in the eye.

The lens is surrounded by a very thin capsule and is kept in place by zonules, tiny fibers which attach the capsule to the ciliary body. The lens is normally clear, but may become opaque following injury secondary to trauma or intraocular surgery. The opaque lens is known as cataract and more commonly occurs with age. During cataract surgery, the lens is removed from the capsule and replaced by an artificial lens made of silicone, PMMA, acrylic, etc. (Figure 1.5). If necessary, the capsular bag can also be used to accommodate some parts of the retinal prosthesis.

Posterior Segment

The sclera, or outermost layer of the eye, merges with the cornea and conjunctiva at the limbus (Figures 1.1 and 1.4). Unlike the cornea which is clear, the sclera is white and blocks most light from entering the posterior chamber. The conjunctiva is a partially transparent membrane that contains vessels and covers anterior part of the sclera (Figure 1.4). The sclera and cornea give the eye its shape.

The choroid is a highly vascular tissue located underneath the sclera, and merges anteriorly with the ciliary body. This structure supplies nutrients to the outer third of the retina, including the photoreceptor outer segments which are particularly metabolically demanding. The inner retina is perfused by a separate circulatory pathway, retinal vessels (Figure 1.6). The vitreous, which fills the space between the lens and the retina, is jelly-like in consistency and helps to keep the retina attached to the underlying layer.

Retina

The retina is positioned as the innermost layer of the posterior segment of the eye and functions as both the sensory detector and the early processing for the

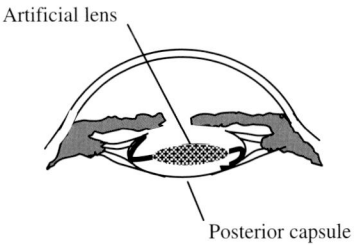

Artificial lens

Posterior capsule

FIGURE 1.5. Schematic cross section of the anterior segment after cataract surgery, showing an artificial intraocular lens inside the capsular bag.

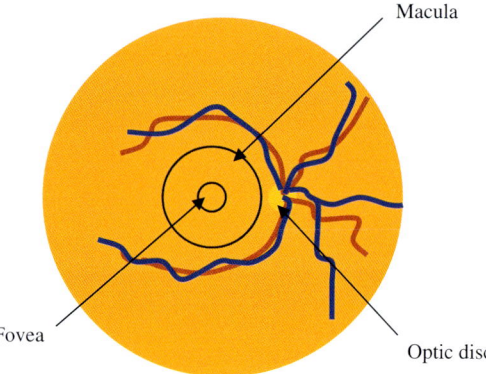

Macula

Fovea

Optic disc

FIGURE 1.6. The fundus as viewed through the pupil; retinal arteries are shown in red and retinal veins in blue.

visual system. After the initial processing of a given image, the retina sends coded signals to the brain via the optic nerve.

The outermost layer of the retina is the retinal pigment epithelium (RPE), a monocellular layer firmly attached to the underlying choroid (Figure 1.7). The remainder of the retina is composed of several layers that collectively form the neurosensory retina. Although the RPE is often considered a part of the retina, the term "retina," in practice and in literature, implies neurosensory retina; in this chapter, retina also implies the neurosensory retina unless stated otherwise.

The retina is generally firmly attached to the RPE but, in a number of pathological conditions, detachment can occur (Figure 1.8). Trauma, age, and disease processes can all lead to separation, often with ensuing photoreceptor cell death.

The integrity of RPE cells is vital for the viability of the photoreceptors. In addition to absorbing stray light, RPE cells also supply nutrients and oxygen to photoreceptors, remove and digest discs shed by photoreceptors, and have a regulatory role in ion and metabolite transport. Photoreceptors detect light and are divided into two main classes, based on wavelength sensitivity and their location in the retina: the cones are highly concentrated at the central part of the retina (macula) and are responsible for both fine and color vision, whereas rods are more prevalent in the periphery and mainly function in peripheral vision, movement detection, and night vision. Color vision is due to the presence of three types of cones: blue, green, and red. Each type is sensitive to a different spectrum of light, but there is a considerable overlapping amongst them. The nuclei of photoreceptors are located within the outer nuclear layer (ONL). Another major group of cells are bipolar cells, located within the inner nuclear layer (INL). Bipolar cells are involved in the processing and transferring of visual signals from the photoreceptors to the ganglion cells. Other cells within the INL include amacrine and horizontal cells, which modify the neural response as it passes through the retina. The bipolar cells connect to ganglion cells; the axons of the ganglion cells form the nerve fiber layer. These axons converge at the optic disc to form the optic nerve; the axons travel via the optic nerve and make

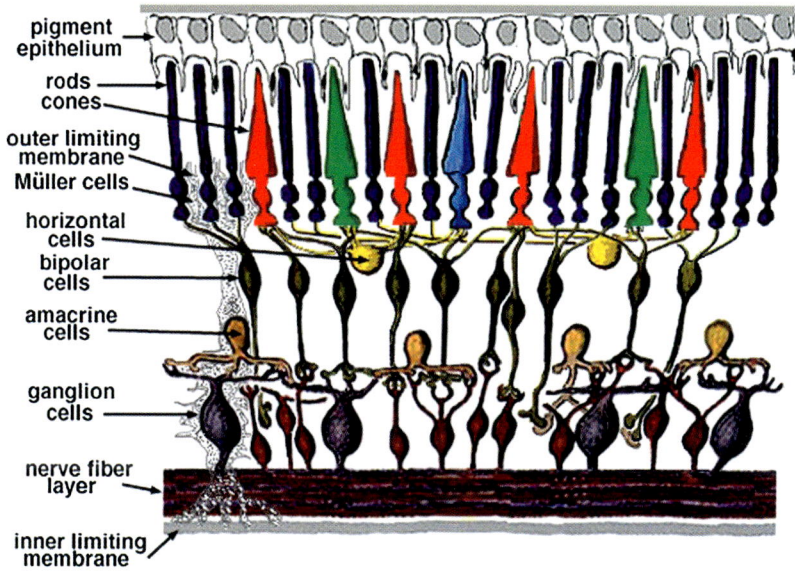

pigment epithelium

rods
cones

outer limiting membrane
Müller cells

horizontal cells
bipolar cells

amacrine cells

ganglion cells

nerve fiber layer

inner limiting membrane

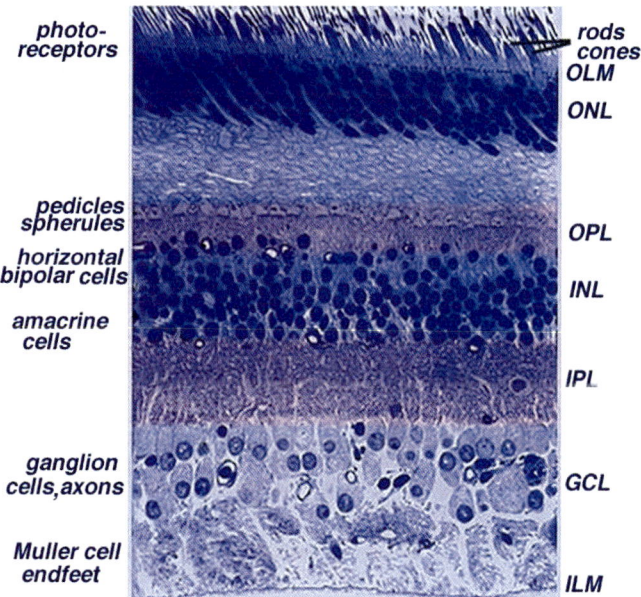

photo-receptors

rods
cones
OLM
ONL

pedicles
spherules

horizontal bipolar cells

OPL

INL

amacrine cells

IPL

ganglion cells, axons

GCL

Muller cell endfeet

ILM

FIGURE 1.7. (a) Simple diagram of the organization of the retina (adapted with permission from http://webvision.med.utah.edu/index.html). (b) Light micrograph of a vertical section through central human retina (adapted with permission from http://webvision.med.utah.edu/index.html).

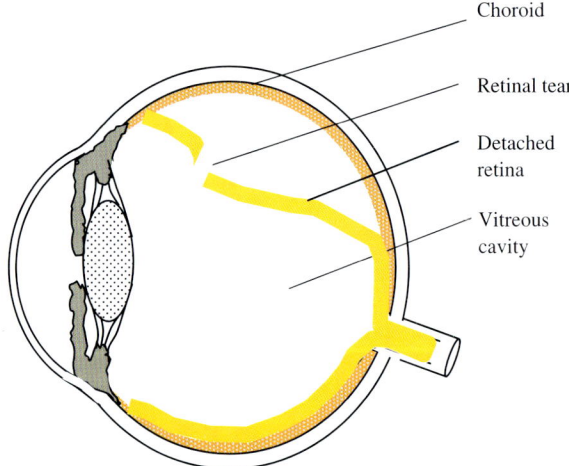

Choroid

Retinal tear

Detached
retina

Vitreous
cavity

FIGURE 1.8. Schematic diagram of a retinal detachment showing separation of the retina from the underlying RPE and choroid (RPE and choroid are shown as a single layer).

connections with the central visual pathways. The outer plexiform layer contains the contacts between photoreceptors and bipolar cells, and the inner plexiform layer contains the contacts between bipolar cells and ganglion cells. Each of the retinal layers are transparent, and light can easily reach the photoreceptors. Muller cells are glia which extend through virtually the entire thickness of the retina and are thought to have a supporting metabolic role. Other retinal glial cells include astrocytes and microglia.

Retinal neurons do not simply transfer signals to a higher level. Indeed, there is a very complex system of signal processing in both the retina and the brain which involves many regulatory cells. Of note, even within one class of neurons (for instance bipolar cells), there may be different reactions to a given stimulus.

Average Ocular Dimensions in Human

- Weight: 7.5 g [1]
- Volume: 6.5 ml [1]
- Axial length (from cornea to retina): approximately 24 mm (between 21 and 26 mm) [1]
- Transverse and vertical diameters of the eye: approximately 24 mm (between 23 and 26 mm) [1]
- The distance between insertion of ocular muscle to limbus: between 5.5 and 7.7 mm [1]
- Cornea [2]:
    - Diameter: 11.7 mm horizontal, 10.6 mm vertical
    - Radius of curvature: 7.8 mm
    - Thickness: $520\,\mu$ in the center, $670\,\mu$ at the periphery

- Depth of anterior chamber: 2.6 mm [2]
- Lens [1]:
  - Weight: from 130 to 25 mg
  - Anteroposterior diameter: from 3.5 to 5 mm
  - Equatorial diameter: 9 mm (after removal of the lens, the capsule can stretch to accommodate objects with larger diameters)
  - Radius of curvature: anterior 10 mm; posterior 6 mm
- Sclera [1]:
  - Diameter: 22 mm
  - Thickness: $800\,\mu$ near limbus; $300\,\mu$ just behind the muscle insertions; $1000\,\mu$ around the optic nerve
- Diameter of optic disc: 1.86 mm × 1.75 mm [3]
- Distance between the center of the fovea and optic disc: 3.4 mm [4]
- One degree of visual angle is equal to $288\,\mu$ on the retina [5] (hence, the field of vision in a patient with a 5 mm × 5 mm array is around 17°)
- Normal field of vision in each eye: vertical 120°; horizontal 150°
- Diameter of macular area: 5 mm
- Retinal thickness: $150\,\mu$ at center of fovea (thinnest part); $400\,\mu$ at foveal rim [6]; $249\,\mu$ mean retinal thickness; $109\,\mu$ mean nerve fiber layer thickness [7]

Common Methods of Evaluating Retinal Anatomy and Function

Retinal prosthesis implantation is an experimental procedure. New devices that become available will require extensive biological testing before they can be declared safe to test in humans. The following diagnostics are standard in ophthalmology and can be used with either human or animals implanted with retinal prosthesis:

Fundus Photo

The fundus (or interior of the back of the eye) can be seen using an ophthalmoscope; it is also possible to photograph it with a fundus camera which reverses the optics of the eye (Figure 1.9).

Fluorescein Angiogram

Fluorescent dye (fluorescein) can be used to assess the integrity of the vessels of the retina and choroid. Within a few seconds after systemic intravenous injection of fluorescein, the dye reaches the retinal vessels where they can be seen and photographed with a fundus camera (Figure 1.10). Fluorescein angiogram (FA) is a useful test for the diagnosis of many retinal conditions, especially those which affect the vessels or RPE.

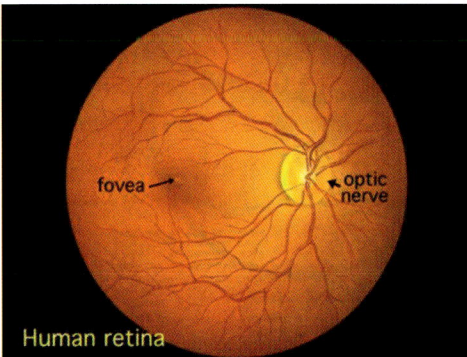

FIGURE 1.9. Front view of the retina as seen during fundus photography (adapted with permission from http://webvision.med.utah.edu/index.html).

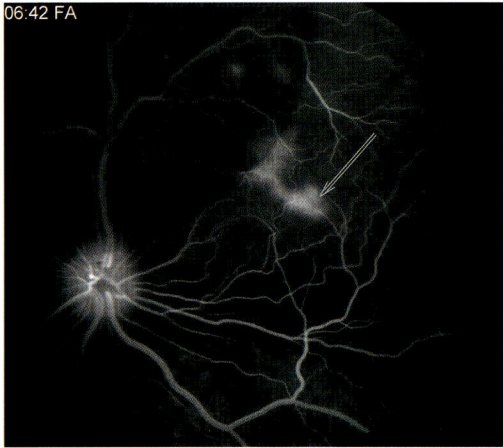

FIGURE 1.10. Fluorescein angiogram of a dog eye; usually the retinal vessels and the optic disc are the only parts that appear as white in a dark background. However, in cases of retinal damage, fluorescein leaks out of the either retinal or choroidal vessels and appears as a white area (arrow).

Optical Coherence Tomography

Optical Coherence Tomography (OCT) measures retinal thickness using visible wavelengths (Figure 1.11). It is a valuable tool for a retinal prosthesis, since it can be used to measure the distance between the retina and the electrode array and can assist in studying the effects of a retinal implant (Figure 1.12).

Electroretinogram

The collective response of retinal cells to light stimulation creates an electrical potential which can be recorded by placing electrodes on the cornea and

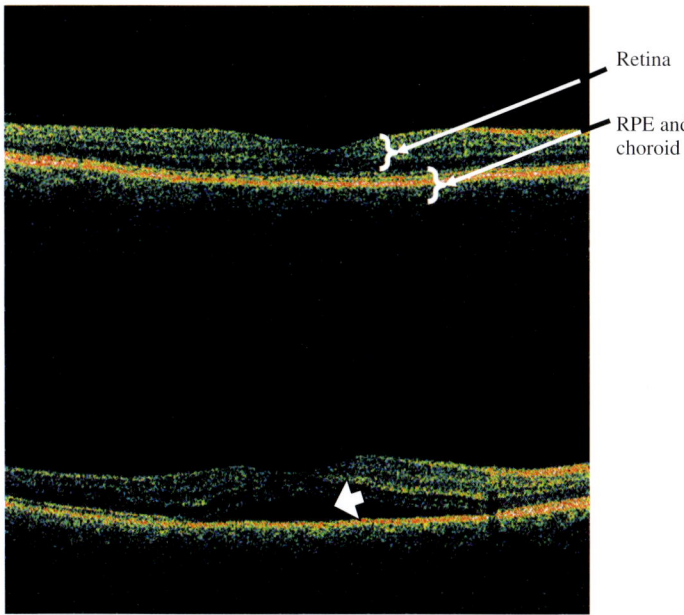

FIGURE 1.11. OCT showing normal fovea (top) and a fovea with subretinal fluid (arrowhead) which has resulted in separation of the retina from the RPE and choroid (bottom).

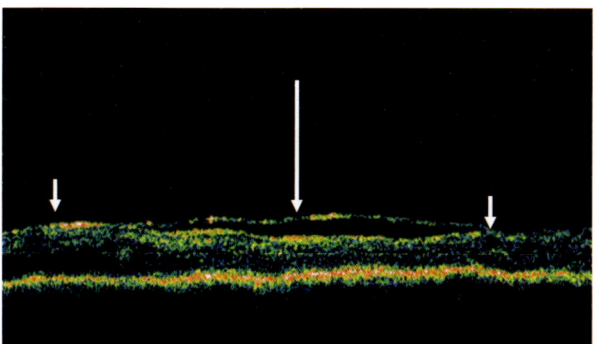

FIGURE 1.12. OCT showing a retinal array in contact with the retina at each end (short arrows) but slightly elevated in the middle (long arrow).

under the eyelids (Figure 1.13). Light stimulation is performed under both bright (photopic) and dark (scotopic) background conditions to induce a retinal response. A diseased retina may alter the shape of the waveform; in some conditions (such as severe retinitis pigmentosa) the *Electroretinogram* (ERG) becomes flat, i.e. an "extinguished" electrical response. Because the ERG is

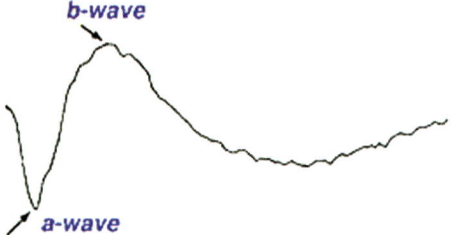

FIGURE 1.13. The basic waveform of the ERG (adapted with permission from http://webvision.med.utah.edu/index.html).

often non-recordable in advanced degenerative retinal disease, it has little utility for assessment of blind patients with retinal implants. The ERG can aid in the development of retinal prostheses by providing a functional measure during preclinical safety testing [8, 9]; however, its lack of spatial sensitivity makes it unable to detect small non-functional areas in the retina. At best, the ERG can serve as a gross measurement that indicates that the retina is not experiencing a large scale, negative reaction to implantation.

Retinal Surgery

Retinal surgery enables the implantation of a device in the vitreous cavity. Due to the nature of the vitreoretinal surgery, intravitreal implants are limited in size and shape. A significant number of retinal diseases are treated surgically. In some cases, surgery can be performed without entering the eyeball. For example, in certain cases of retinal detachment, the retina can be reattached by scleral buckling, i.e. the suturing of a silicone prosthesis over the sclera, which results in indentation of the sclera, choroids, and RPE toward the retina. However, in most conditions, surgery is performed by vitrectomy (i.e. the insertion of surgical instruments into the vitreous cavity and removal of the vitreous). First, the conjunctiva is cut at the limbus and the sclera becomes exposed. The vitreous cavity is then accessed via scleral incisions (sclerotomy) made 3.5–4 mm behind the limbus (Figure 1.14), the area overlying the pars plana (Figure 1.1). Anterior entry to this point may injure the lens and thus cause cataract or damage the ciliary body or iris; on the other hand, entry behind this point may cause retinal detachment or choroidal hemorrhage.

In a standard vitreoretinal surgery case, three scleral incisions are made: one for the infusion terminal, through which fluid constantly flows into the eye to maintain IOP and prevent the collapse of the eye; a second one for the insertion of an intraocular light probe; and a third one is used to cut the vitreous and for intraocular manipulation. The size of a scleral incision is usually approximately 1 mm (between about 0.6 mm and 1.4 mm). Since the vitreous is gelatinous and is attached to some parts of the retina, it is not possible to

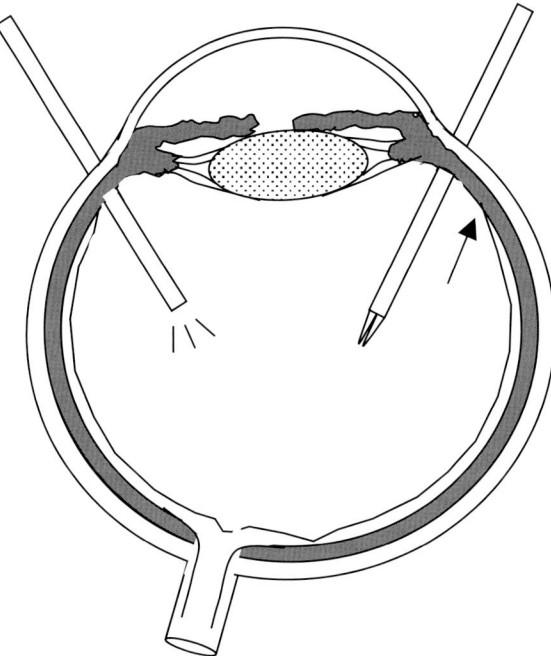

FIGURE 1.14. Schematic diagram of intraocular instruments in vitreoretinal surgery; note that the instruments pass through the pars plana, an area located between the ciliary body and the beginning of the retina (arrow).

remove it with simple aspiration; conversely, vigorous aspiration may also result in retinal tears and subsequent retinal detachment. To overcome these problems, the vitrectomy probe simultaneously cuts the vitreous in a guillotine-like manner and aspirates the small chunks of vitreous, hence reducing traction on the retina. For an epiretinal implant to be placed close to the retina, the vitreous should be completely removed at the site of implantation. At the end of the vitrectomy, the scleral incisions are closed using absorbable sutures, and the conjunctiva is reattached to the limbus to cover the scleral incisions.

Retinal Disease

Common retinal diseases fall into one of two main categories: retinal vascular disease and degenerative retinal disease. In general, vascular disease principally damages the inner retinal layers, while degenerative disease primarily affects outer retinal layers.

Inner Retinal Disease

Diseases of the inner retina are often associated with systemic diseases such as diabetes mellitus and hypertension. In fact, diabetic retinopathy is the leading cause of blindness in the United States in individuals between 20 and 74 years of age [10]. The inner part of the retina is completely dependent upon the retinal vessels for nutrients and oxygen, and any compromise in these vessels may result in the loss of ganglion cells and bipolar cells, and subsequently (in severe cases), atrophy of the retinal nerve fiber layer and optic nerve. Obviously, electrical retinal stimulation is not an option for inner retinal disease, but a visual cortex prosthesis may someday be an alternative.

Outer Retinal Disease

The two most prevalent diseases in this category are age-related macular degeneration (AMD) and retinitis pigmentosa (RP).

Age-related Macular Degeneration

The AMD is the leading cause of blindness in the elderly in developed countries [11–13]. The incidence of AMD increases dramatically from 0.7% in the 65–74-year-old age group to 5.4% in the 75–84-year–old age group, and 18.5% in individuals over 85 years old; more than two-thirds of people over 90 years of age have early AMD, and one-quarter have late AMD [12, 14]. There are two types of AMD: dry AMD and neovascular AMD (also known as exudative or wet) (Figure 1.15). Ninety percent of affected individuals have dry AMD; however, the majority of people having severe visual loss have neovascular AMD [15]. Dry AMD is associated with the gradual degeneration of the outer retina and RPE, and may take years or even decades to progress to geographic atrophy, the final stage of the disease with marked visual loss. However, visual loss in wet AMD may occur within days. In wet AMD, new vessels (or the so-called "choroidal neovascular membrane" [CNV]) grow under the retina or

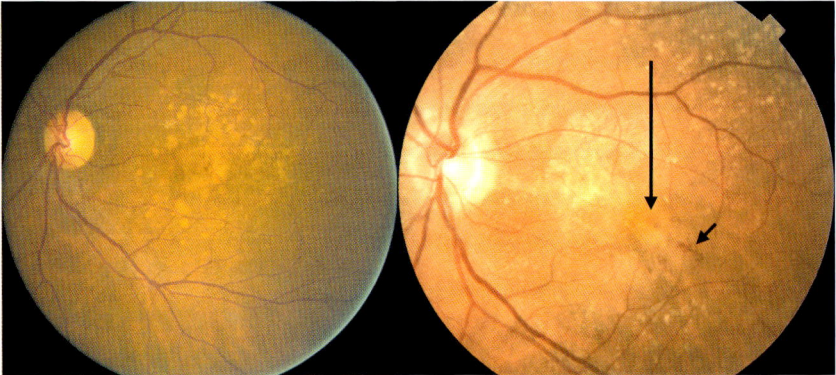

FIGURE 1.15. Retinal appearance in AMD; dry AMD (left), wet AMD (right); small subretinal hemorrhage (short arrow) is due to leakage of the blood from choroidal neovascular membrane (long arrow).

RPE, making the outer retina dysfunctional. In the long term, a scar may form under the retina (known as disciform macular degeneration).

The range of central visual loss in AMD can vary from very mild to very severe. Nevertheless, because the peripheral retina remains relatively normal, affected individuals retain peripheral vision; i.e. they may not be able to see the objects they are directly looking at, but they can see other objects around them.

Current treatments for AMD: Unfortunately there is no effective treatment for dry AMD which comprises 90% of AMD patients. However, progression to advanced disease may be reduced in high-risk patients by daily supplementation of specific vitamins and zinc [16].

Unlike dry AMD, wet AMD often results in severe visual loss, and therefore the majority of research has been concentrated in treating this type of disease. Until the late 1990s, the only proven treatment was laser photo-coagulation [17], but since then a growing number of treatments have entered human clinical trials. The aim of treatment in wet AMD is to destroy the CNV without damaging the retina or RPE. Other treatment modalities for wet AMD include the following: surgical treatments [18, 19], photodynamic therapy [20], transpupillary thermotherapy [21], intravitreal injection of antian-giogenic drugs [22–24], and periocular injection of anecortave acetate [25].

Retinitis Pigmentosa

This term is used for a heterogeneous group of hereditary disorders charac-terized by dysfunction and loss of photoreceptors and RPE, and has a worldwide prevalence of 1:3000 to 1:5000 [26]. Rod photoreceptors are predominantly affected; therefore, clinical symptoms begin with the loss of peripheral and night vision. Depending on the specific gene mutation, clinical symptoms may start in childhood or later in life. The disease is bilateral and progressive, and finally results in loss of central vision in the majority of affected individuals. In extreme cases even light perception may disappear.

The appearance of the fundus in patients with RP is characteristic and displays a narrowing of the vessels, peculiar pigmentation of the peripheral retina, and optic disc pallor (Figure 1.16). The ERG is usually abnormal, even in early stages of the disease when patients still have relatively good vision (Figure 1.17).

Current treatments for retinitis pigmentosa: Like dry AMD, there is no proven treatment for RP. However, in a randomized, controlled clinical trial on 601 patients with RP and with 4–6 years follow-up, it was demonstrated that vitamin A supplementation could slow the rate of decline in the retinal function in these patients. Potential treatments are being investigated in animals, and may some day cure the disease, but are unlikely to be helpful to those who have already lost their sight. These experimental treatments include the following: gene therapy [27], stem cells [28], and RPE and retinal transplantation [29].

Viability of Inner Retinal Cells in Outer Retinal Disease

For a retinal prosthesis to be possible, there must be some part of the retina remaining to which the prosthesis can communicate. Morphometric studies of

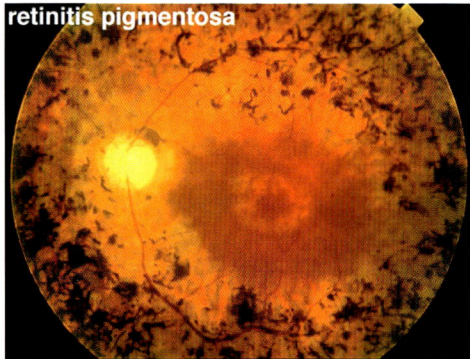

FIGURE 1.16. Retinal appearance in retinitis pigmentosa (adapted with permission from http://webvision.med.utah.edu/index.html).

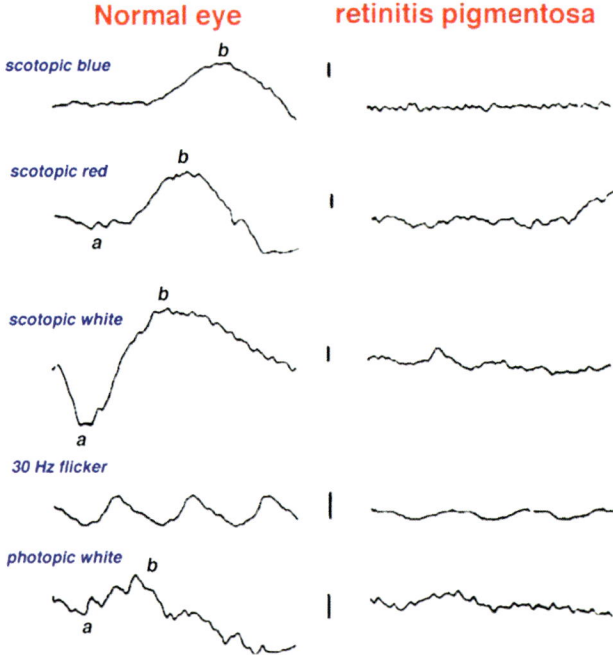

FIGURE 1.17. ERG recording in a normal person and a patient with retinitis pigmentosa (adapted with permission from http://webvision.med.utah.edu/index.html).

enucleated eyes from patients with AMD or RP have demonstrated the partial preservation of inner retinal cells. When comparing the number of cells in 10 eyes with geographic atrophy to the number of cells in 5 normal eyes, one study demonstrated cell loss in all three nuclear layers in areas where RPE was completely absent [30]. Cell loss was most pronounced in the ONL (photoreceptor nuclei), with a 76.9% reduction in the number of cells. Ganglion cells were reduced by 30.7%. However, cell count in the INL (bipolar, horizontal, and amacrine cells) was not significantly different from the control group. In another study, morphometric analysis of six eyes with disciform AMD showed a similar pattern [31].

In RP, cell loss occurs in all layers [32–34]. In both the macula and the extramacular regions, cell loss is most profound in the ONL, followed by the ganglion cell layer (GCL) and then the INL. More cells in the INL and GCL are preserved in the macula, when compared with extramacular regions. In the macula, 78–88% of the inner nuclear cells and 30–48% of ganglion cells are retained, whereas in the extramacular regions, 40% of the inner nuclear cells and 20–30% of ganglion cells are retained. In addition to the reduction of viable cells, retinal degeneration is also associated with a progressive retinal remodeling which involves rewiring and anomalous circuitry within the retinal layers. In the final phase of remodeling, retina displays hypertrophy of Muller cells, formation of microneuromas (tangles of cell processes from different layers), and cellular translocation (migration of amacrine cells into the GCL and movement of ganglion cells into the INL) [35–38].

Electrical Stimulation of the Nerve

Nerve cells are excitable, which means that they can change their transmembrane potential in response to a stimulus. This is accomplished normally through either a physical perturbation for sensory receptor neurons (e.g. photoreceptors) or synaptic transmission where one neuron signals another at a connection called a synapse. In either case, a neuron is excited and the conductance of the cell membrane changes, resulting in ions crossing the membrane at different rates different from equilibrium. This movement of charged particles results in a transient change in the membrane potential, which serves as a signaling mechanism. The signal propagates along the nerve membrane to the synapse, where it can serve to either excite or inhibit activity in other nerve cells.

Nerve cells can be artificially activated by electric fields that depolarize the cell membrane. Electrical stimulation from an extracellular microelectrode has been extensively used in both experimental and clinical settings. Early experiments suffered from an inability to precisely control electrical energy. With the emergence of microelectronic technology, more precise stimulus currents could be delivered to neurons. Concomitant advances in medicine have enabled increasing access to neural structures. The combined efforts in both engineering and medicine have yielded electrically active, microelectronic implants that can stimulate nerves to restore lost sensory function or modulate neural function.

Visual Response to Electrical Stimulation of the Retina in Outer
Retinal Disease

Many studies have documented that electrical stimulation of neurons in the visual pathway produces the perception of light. Fewer have looked at electrical stimulation in the case of outer retinal disease. This chapter will only consider the latter case of experiments.

One of the first experiments in this area did not involve an implant, but instead used an external electrode to stimulate the eye [39–42]. It has been demonstrated that an electrically evoked response (EER) can be recorded through scalp electrodes, during electrical stimulation of the eye through a contact lens electrode. It has also been shown that this stimulation elicits sensations of streaks or dots of light (phosphenes) in individuals with advanced photoreceptor degeneration. The discrepancies between the EER and the visually evoked response (VER) in some patients with RP lead the investigators to believe that the cells that respond to electrical stimulation are probably different from the photoreceptors (which are the cells that respond to light stimulation).

Direct retinal stimulation in five blind subjects (secondary to RP, AMD, or unspecified retinal degeneration from birth) with visual acuity of little or no light perception resulted in perception of phosphenes in all subjects [43]. When the retina was stimulated with intraocular probes held very close to the retina and ranging from 50 to 200 μ in diameter, a typical stimulus pulse of 300–800 μA and 1–4 milliseconds in duration was required to elicit a phosphene. All subjects were able to localize the stimuli and give a detailed description of visual phenomena, except one subject with unspecified retinal degeneration and who had been blind since birth. Shapes of variable size and colors including yellow, white, and yellow-green were perceived in response to electrical stimulations. This study did not show any difference between anodic-first and cathodic-first monopolar stimulation, nor between monopolar and bipolar electrode configurations.

In another experiment involving nine patients with either RP or AMD, the retina was stimulated with multiple electrodes of an electrode array [44]. Patients perceived simple forms in response to a pattern electrical stimulation. In addition, stimulating frequencies between 40 and 50 Hz gave rise to a non-flickering perception. It was also found that the macular region required lower threshold currents than the peripheral retina. Also, patients with less advanced RP or AMD required lower threshold currents than those with more advanced disease. A separate study done on normal and blind humans, and several animal studies have confirmed lower threshold values in healthier retinas [45–48].

In a more recent study on a normal-sighted person and five patients with RP with a visual acuity ranging from hand motion to 20/800, it was found that threshold charge densities in severely blind patients were substantially higher than that in the normal-sighted subject. Retinal stimulation was performed with either a single hand-held electrode of 250 μ in diameter or an array of iridium oxide electrodes ranging between 50 and 400 μ in diameter. In the blind subjects, the threshold charge density at 0.25 millisecond pulse duration was 4.1 mC/cm^2 for the 100 μ electrode and 0.3 mC/cm^2 for the 400 μ electrode. It was also found that the threshold charge density was lower at shorter stimulus durations. On average, three blind patients reported percepts that matched the stimulation pattern 32–48% of the time [49, 50].

The most recent acute retinal stimulation in human involved 20 RP patients with visual acuities ranging from 1/50 to no light perception. Threshold charges needed to generate visual perception with a single electrode were different between individual patients and ranged from 20 to 380 nC. Nineteen of 20 patients perceived colored objects ranging in size from a head of a match to a football as seen from a distance of 1 m [51].

The background on ocular anatomy, ophthalmic surgery, retinal disease, and electrical stimulation support the development of a retinal prosthesis. The ocular anatomy can accommodate artificial implants. Ophthalmic surgery provides access to all parts of the eye, although size limitations for scleral incisions will restrict intravitreal implant sizes. Incurable retinal diseases mandate that we seek cures outside traditional medical practice. The diseases primarily afflict the outer retina, so the inner retina can be artificially activated to produce the perception of light. The remainder of the chapter will review the various implementations of retinal prosthetic systems.

Retinal Implant

Epiretinal Approach

Surgical Limitations of Epiretinal Implant

The surgical approach for implantation of the epiretinal implant depends on the location of the microelectronic part (intraocular vs. extraocular). When the chip and the coil are located extraocularly, the cable attached to the intraocular stimulating array should pass through a sclerotomy to be connected to the extraocular components (Figure 1.18).

Typical surgery involves a standard vitrectomy (see above), followed by the extension of one of the scleral incisions to the required length for insertion of the array. The array, then, is inserted into the eye and is fixed onto the retina with a tack. Thereafter, the scleral incision, through which the cable is passing, is partially closed using absorbable sutures. Care is taken to close the scleral incision as tightly as possible around the cable without damaging it. Subsequently, other sclerotomies and conjunctiva are closed. Other surgical steps would depend on the location of the other components of the prosthesis.

Because of the presence of the cable at the sclerotomy, the wound may not completely heal, and intraocular fluid may leak out of the eye into the periocular space, resulting in lower than normal IOP (hypotony). Hypotony may give rise to choroidal detachment, choroidal folds, and decreased vision. However, there have not been cases of postoperative hypotony reported in animals and humans that have undergone this type of surgery [8, 52, 53]. Although there are no data to suggest what the safe limit for a scleral incision is, it has been our experience that a 5 mm sclerotomy for insertion of the device is well tolerated. Because the cable can be made narrower than the array, the sclerotomy can be partially closed

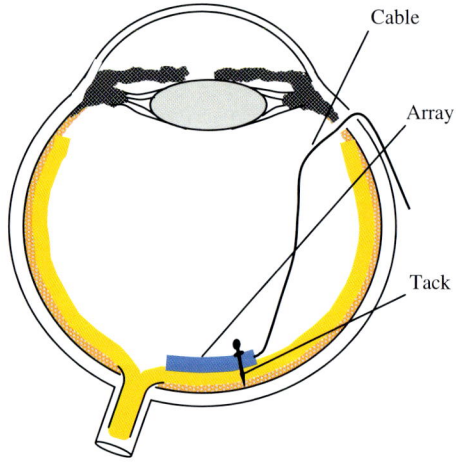

FIGURE 1.18. Schematic diagram of an epiretinal implant showing the array, the cable, and the tack, and their relation to the surrounding ocular structures.

after insertion of the array. The cable should be made as narrow as possible, since the final opening depends on the width of cable.

Attachment of the Array to the Retina

With the epiretinal approach, the electrode array is fixed on the retina by means of a tack. Because the tack penetrates the choroid (which is a highly vascular tissue), there is a risk of intraocular bleeding at the time of tacking. It is possible, however, to reduce this risk by briefly increasing the IOP by raising the infusion line. While massive intravitreal bleeding may be associated with subsequent complications, minimal bleeding is usually absorbed within days or weeks without any consequence.

Another concern with the tack is the risk of gliosis (fibrosis), i.e. proliferative tissue that may grow from the penetration site and may either cover the retinal surface or encapsulate the array and interfere with retinal stimulation. The use of only one tack for fixation of the array reduces this risk. In one study which an epiretinal array was fixed onto the retina by two tacks in four dog eyes, retinal hyper-pigmentation was observed around the site of the tack. In that study, histological examination revealed minimal fibrosis around the site of implantation [8].

Pressure Effect on the Retina

When the array is tacked to the retina, some of its parts may exert unwanted pressure, which may subsequently result in retinal damage. Normal IOP is 10–21 mmHg, and it is well established that raised IOP can damage the nerve fiber layer of the retina. Glaucoma is a common cause of blindness in the elderly and is caused by nerve fiber layer damage secondary to raised IOP. It has been demonstrated that reducing IOP in individuals with higher IOP may delay or prevent the onset of glaucoma [54, 55]. It is believed that high pressure interferes with axoplasmic flow inside the nerve fibers, leading to the death of ganglion cells and optic nerve atrophy. In addition, it is thought that high IOP may compromise optic nerve head circulation and cause nerve fiber layer loss. Because peripheral nerve fiber layers are more sensitive to the effects of pressure, the disease starts with the loss of peripheral vision; if left untreated for months or years, the macular fibers may also be affected and eventually total blindness may ensue. Because of the loss of ganglion cells, retinal prosthesis does not play any role in the treatment of this condition. Compared to glaucoma (in which pressure is applied to all parts of the retina) pressure is only exerted to some areas of the retina underneath the array with an epiretinal prosthesis. Because of the curvature of the inner eye, the point of maximum pressure is likely to be around the tack and at the edges of the array. Other factors affecting the amount of pressure exerted on the retina include the shape of the tack, the stiffness of the array, and improper alignment of the array in relation to the scleral incision during surgery, which may cause rotation of the cable and tilting of the array. Since pressure from a device will be localized, and since elevated IOP applies pressure globally, research into the effect of localized pressure on the retina may reveal a different type of damage other than that caused by raised IOP.

Bioadhesives for Attaching the Array to the Retina

Given the potential problems with the use of a tack, tissue glue may provide a safe alternative for attaching the array to the retina. In addition to being biocompatible, ideal glue should create an attachment which is strong enough to keep the array close to the retina, but not too strong to pull the retina off the RPE layer. Moreover, the attachment should be reversible to allow for the removal of the array, in case of improper functioning, or any other complication. Tissue glues are usually used for creating an adhesion between different parts of a tissue or between two different tissues and usually disintegrate with time, whereas glues used for retinal prosthesis should have an adhesive effect on both the retina and the array, and should remain stable for years.

A study using dissected rabbit retina demonstrated that hydrogels have from 2 to 39 times greater adhesive force than commercial fibrin sealant, autologous fibrin, Cell-Tak, or photocurable glues [56]. Hydrogels liquefied at body temperature anywhere from 3 days to a few months. In terms of safety, the relative strength of adhesion and consistency, the specific hydrogel SS-PEG appeared to be an excellent substance for intraocular use; however, it was short lasting and liquefied after 72 hours under wet conditions.

Implantation of a Microelectronic Device in the Eye (All-Intraocular Retinal Prosthesis)

In this approach, all of the components of a retinal prosthesis are inserted into the eye (Figure 1.19). The inductive transmitter can be accommodated in a spectacle

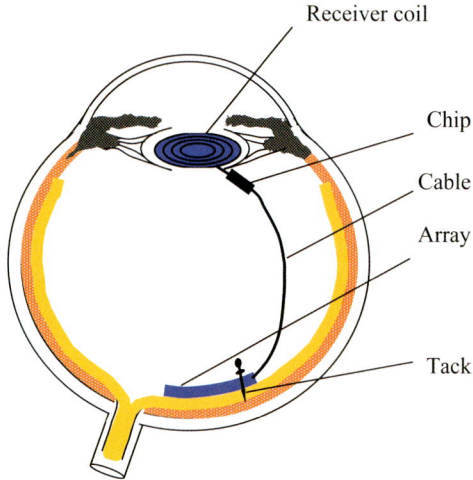

FIGURE 1.19. Schematic diagram of an all-intraocular retinal prosthesis; although the coil is shown inside the capsular bag, it can alternatively be placed in the sulcus, in front of the anterior capsule (the area between the iris and the ciliary body).

frame, and both the receiver coil and the chip inside the eye. Perhaps the best place for the receiver coil is inside the capsular bag, after removal of the natural lens (a place at which, following cataract surgery, the artificial intraocular lens is inserted). The chip can be placed either along the cable or adjacent to the coil.

Surgery involves both anterior and posterior segments of the eye. In a typical surgery, a standard vitrectomy is followed by a corneal incision along the limbus. A circular window from the anterior portion of the lens capsule is then removed and the lens is subsequently expressed out of the eye; finally, the central portion of the posterior capsule is also removed. At this stage, the anterior and posterior segments of the eye are connected to each other, and the array and cable can be inserted into the eye through the corneal incision, and then passed through the remnants of the lens capsule. The coil (which usually has a large diameter) can be placed either between the anterior and the posterior part of the capsule (inside the capsular bag) or in front of the capsule, and sutured to the sclera if necessary. Finally, the array is tacked onto the retina, and both the corneal and the scleral incisions are then closed with sutures. The feasibility of this surgery has been demonstrated on cats [57].

The advantage of the all-intraocular method is that there would be no connection between the inside and the outside of the eye at the conclusion of the surgery. However, there are some other concerns. The heat produced by the receiver coil may be damaging to the eye; the stability of the coil in the capsule is also a concern.

Unlike the current epiretinal prosthesis in which the excess cable can either be folded inside the eye or pulled outside of it, with the all-intraocular design the excess cable may touch and damage the retina. Considering the variability of the axial length of the eye (anteroposterior axis) in different people, one solution could be that the different length cables be used to fit an individual's eye (The axial length of the eye and the depth of the anterior chamber can be measured with an ultrasound probe). However, this may increase the cost of the device.

Subretinal Approach

Surgical Limitations of Implantation in the Subretinal Space

There are two types of subretinal implants: those that rely on incident light (passive) and those with external connections for a power supply (active) [58–61]. The implant without an external connection is a disc comprised of a microphotodiode array ranged from 2 to 2.5 mm in diameter. This device is implanted as follows. After standard vitrectomy, some fluid is injected under the retina to create a small bulla, a small retinal incision is then made at the edge of the bulla, and a microphotodiode array is pushed under the retina (ab-interno technique) (Figure 1.20). To decrease the chance of subsequent retinal detachment, air is injected into the vitreous cavity, thus creating a surface tension which prevents leakage of fluid under the retina. The air is then spontaneously absorbed within a few days.

Subretinal implantation of semiconductor microphotodiode arrays (which were $250\,\mu$ in thickness and ranged from 1.5 to 3 mm in diameter) in rabbits showed a normal retina at locations away from the surgical and implant sites. However, abnormalities were seen in retinal sections taken from areas overlying the implant. While the outer retina displayed a significant damage to photoreceptors, the inner retina demonstrated a decline in inner nuclear and GCL densities and a thinning of the inner plexiform layer [59]. In another study, subretinal implantation of inactive or electrically active semiconductor microphotodiode arrays

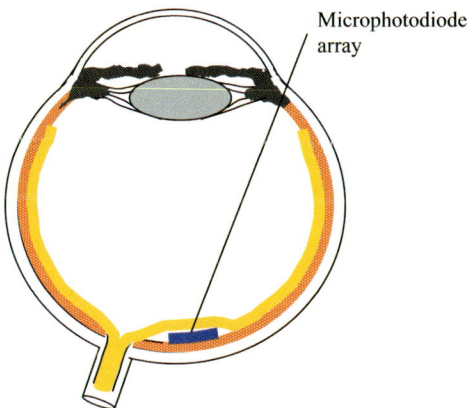

FIGURE 1.20. Schematic diagram of a subretinal implant with no external component placed under the retina.

(which were $50\,\mu$ in thickness and ranged from 2 to 2.5 mm in diameter) in normal cats showed a marked loss of photoreceptors in areas directly overlying the implant. Nevertheless, despite some disorganization of the inner retinal layers, there was not a significant reduction in either INL or GCL densities [60].

The surgical approach for a subretinal prosthesis with external connections is completely different (Figure 1.21). In a technique known as ab-externo, a scleral and choroidal incision is made about 6 mm or more behind the limbus without cutting the retina. The array is next advanced under the retina to the desired position, and then the sclera is sutured around the cable. A partial vitrectomy may be done to reduce the pressure from the eye before cutting the choroid, since the retina is a very delicate tissue and can easily tear either during the choroidal incision or at the time of insertion of the array. As an alternative to a partial vitrectomy, some fluid can be withdrawn from the anterior chamber to soften the eye. Also, either some fluids or viscoelastic material (a jelly-like material used especially during cataract surgery to maintain the shape of the eye and to protect the cornea) can be injected under the retina, via the choroid immediately before or after the choroidal incision to separate the retina from the choroid and reduce the chance of retinal tear.

This technique is very difficult, and despite all preventative measures, the risk of retinal tear and detachment is high. Unlike with the epiretinal approach or ab-interno subretinal implantation, not all vitreoretinal surgeons may be able to perform (nor be willing to take the risk associated with) the ab-externo technique. In addition, the effect of the array rubbing against the retina and the RPE during insertion may further damage the RPE or the already diseased retina.

In a modified technique known as ab-interno-transscleral [62], the surgery begins with a vitrectomy. Viscoelastic material is next injected under the retina at the point where the scleral incision will be made to separate the retina from

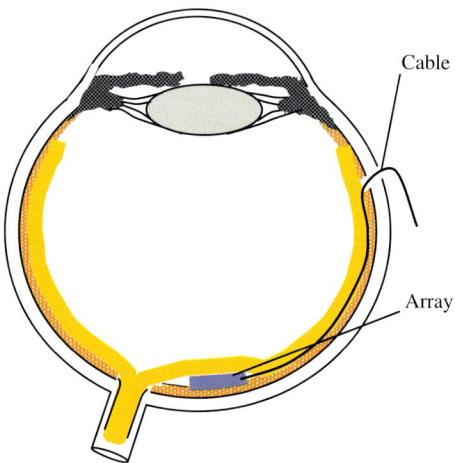

Figure 1.21. Schematic diagram of a subretinal implant with external components showing the array and the cable running under the retina.

the choroid. Then a scleral and choroidal incision is made about 6 mm or more behind the limbus and the array is slipped under the retina. The risk of a retinal tear and retinal damage is reduced by the use of this technique, but there is still considerable risk.

The advantage of the ab-externo or ab-interno-transscleral subretinal approach over the epiretinal approach is that the array is more likely to have a consistent position with the retina, yet without exerting pressure on it. The disadvantage of it is that it may interfere with the ability of the choroid to supply nutrition to the outer retina, although a porous array may address this problem. There are also more limitations on the size of the array when compared to the epiretinal approach, as a larger array may lift and detach the retina. Most important of all, there is a considerable chance of complications during the surgery, including retinal detachment and choroidal hemorrhage.

The advantage of the epiretinal approach is that the surgery is relatively easy and is also associated with a lower rate of complications compared to subretinal surgery. In addition, provided that a given array can conform to the surface of the retina, a large electrode array may be inserted into the eye to increase the field of vision, yet without significantly interfering with the physiology of the retina. Some of the disadvantages, however, include that: (1) there is a risk of gliosis secondary to the use of a tack; (2) an array may rotate if the cable is not properly fixed; (3) the array may not remain uniformly close to the retina; (4) the array may exert too much pressure on the retina; and (5) the tack may fall off of the retina.

Potential problems for both the epiretinal and the subretinal approach include the following risks: (1) retinal detachment; (2) intraocular bleeding during the surgery; (3) infection; (4) exposure of the part of device that is located under the conjunctiva; (5) thermal damage to the retina or other intraocular structures; and (6) possible electrical retinal damage.

Summary

Retinal prostheses have the potential to cure many blinding diseases, but face many difficult hurdles before a useful system can be implemented. Many of these constraints are presented by the complex anatomy and physiology of the eye. A successful retinal prosthesis must be constructed such that it can be implanted with few complications and it will not damage the eye. One specific challenge will be positioning the electrode array to achieve consistent communication between the implant and the retina. At the same time, the implant must be programmed such that it can apply a pattern of stimulus to the retina that results in a usable perception. Topics that have not been covered in this chapter include engineering challenges such as hermetic packaging and microelectronic design. A development process that considers both biology and technology together will be required to produce a treatment for blindness through prosthetic development.

References

1. Hogan MJ, Alvaredo JA, Weddell JE. Histology of the Human Eye. Philadelphia: W. B. SAUNDERS COMPANY, 1971.
2. Maurice DM. Physiology of the Eye. Boston: Little, Brown and Co., 1969; 6,14,5.
3. Straatsma BR, Foos RY, Spencer LM. The retina-topography and clinical correlations. Symposium on the Retina and Retinal Surgery. Trans New Orleans Acad Ophthal 1969:1.
4. Kolb H, Fernandez E, Nelson R, Jones BW. http://webvision.med.utah.edu. Webvision, 2005.
5. Drasdo N, Fowler CW. Non-linear projection of the retinal image in a wide-angle schematic eye. Br J Ophthalmol 1974; 58:709–14.
6. Yamada E. Some structural features of the fovea centralis in the human retina. Arch Ophthalmol 1969; 82:151–9.
7. Alamouti B, Funk J. Retinal thickness decreases with age: an OCT study. Br J Ophthalmol 2003; 87(7):899–901.
8. Majji AB, Humayun MS, Weiland JD, et al. Long-term histological and electrophysiological results of an inactive epiretinal electrode array implantation in dogs. Invest Ophthalmol Vis Sci 1999; 40(9):2073–81.
9. Guven D, Weiland JD, Fujii G, et al. Long-term stimulation by active epiretinal implants in normal and RCD1 dogs. J Neural Eng 2005; 2(1):S65–73.
10. Palmberg P. Diabetic retinopathy. Diabetes 1977; 26:703–9.
11. Klein R, Klein BEK, Linton KLP. Prevalence of age related maculopathy, The Beaver Dam Eye Study. Ophthalmology 1992; 99:933–43.
12. Mitchell P, Smith W, Attebo K, Wang JJ. Prevalence of age-related maculopathy in Australia. Ophthalmology 1995; 102:1450–60.
13. Vingerling JR, Dielemans I, Hoffman A, et al. The prevalence of age related maculopathy in the Rotterdam study. Ophthalmology 1995; 102:205–10.
14. VanNewkirk MR, Nanjan MB, Wang JJ, et al. The prevalence of of age-related maculopathy: the visual impairment project. Ophthalmology 2000; 107:1593–600.
15. Ferris FLI, Fine SL, Hyman LA. Age-related macular degeneration and blindness due to neovascular maculopathy. Arch Ophthalmol 1984; 102:1640–2.
16. AREDS report No 8. A randomized, placebo-controlled clinical trial of high dose supplementation with vitamins C and E, beta carotene, and zinc for age-related macular degeneration and vision loss. Arch Ophthalmol 2001; 119:1417–36.
17. Macular Photocoagulation Study Group. Argon laser photocoagulation for neovascular maculopathy: five-year results from randomised clinical trials. Arch Ophthalmol 1991; 109:1109–14.
18. Fujii GY, De Juan E, Pieramici DJ, et al. Inferior limited macular translocation for subfoveal choroidal neovascularisation: 1-year visual outcome and recurrence report. Am J Ophthalmol 2002; 134(134):69–74.
19. Aisenbrey S, Lafaut B, Szurman P, et al. Macular translocation with 3608 retinotomy for exudative age related macular degeneration. Arch Ophthalmol 2002; 120(451–9).
20. Schmidt-Erfurth U, Hasan T. Mechanisms of action of photodynamic therapy with verteporfin for the treatment of age-related macular degeneration. Surv Ophthalmol 2000; 45:195–214.
21. Algvere PV, Libert C, Lindgarde G, Seregard S. Transpupillary thermotherapy of predominantly occult choroidal neovascularisation in age related macular degeneration with 12 months follow-up. Acta Ophthalmol Scand 2003; 81:110–17.

22. Preclinical and phase 1A clinical evaluation of an anti-VEGF pegylated aptamer (EYE001) for the treatment of exudative age-related macular degeneration. Retina 2002; 22:143–52.
23. Anti-vascular endothelial growth factor therapy for subfoveal choroidal neovascularization secondary to age-related macular degeneration: phase II study results. Ophthalmology 2003; 110:979–86.
24. The Eyetech Study Group. Anti-vascular endothelial growth factor therapy for subfoveal choroidal neovascularisation secondary to age related macular degeneration, phase II study results. Ophthalmology 2003; 110:979–86.
25. The Anecortave Acetate Clinical Study Group. Anecortave acetate as monotherapy for treatment of subfoveal neovascularisation in age related macular degeneration: twelve-month clinical outcomes. Ophthalmology 2003; 110:2372–85.
26. Haim M. Epidemiology of retinitis pigmentosa. Acta Ophthalmol Scand Suppl 2002; 233:1–34.
27. Acland GM, Aguirre GD, Ray J, et al. Gene therapy restores vision in a canine model of childhood blindness. Nat Genet 2001; 28(1):92–5.
28. Kicic A, Shen WY, Wilson AS, et al. Differentiation of marrow stromal cells into photoreceptors in the rat eye. J Neurosci 2003; 23(21):7742–9.
29. Arai S, Thomas BB, Seiler MJ, et al. Restoration of visual responses following transplantation of intact retinal sheets in rd mice. Exp Eye Res 2004; 79(3):331–41.
30. Kim SY, Sadda S, Humayun MS, et al. Morphometric analysis of the macula in eyes with geographic atrophy due to age-related macular degeneration. Retina 2002; 22(4):464–70.
31. Kim SY, Sadda S, Pearlman J, et al. Morphometric analysis of the macula in eyes with disciform age-related macular degeneration. Retina 2002; 22(4):471–7.
32. Humayun MS, Prince M, de Juan EJ, et al. Morphometric analysis of the extramacular retina from postmortem eyes with retinitis pigmentosa. Invest Ophthalmol Vis Sci 1999; 40(1):143–8.
33. Santos A, Humayun MS, de Juan E, Jr., et al. Preservation of the inner retina in retinitis pigmentosa. A morphometric analysis. Arch Ophthalmol 1997; 115(4):511–5.
34. Stone JL, Barlow WE, Humayun MS, et al. Morphometric analysis of macular photoreceptors and ganglion cells in retinas with retinitis pigmentosa. Arch Ophthalmol 1992; 110(11):1634–9.
35. Jones BW, Marc RE. Retinal remodeling during retinal degeneration. Exp Eye Res 2005; 81(2):123–37.
36. Jones BW, Watt CB, Frederick JM, et al. Retinal remodeling triggered by photoreceptor degenerations. J Comp Neurol 2003; 464(1):1–16.
37. Marc RE, Jones BW. Retinal remodeling in inherited photoreceptor degenerations. Mol Neurobiol 2003; 28(2):139–47.
38. Marc RE, Jones BW, Watt CB, Strettoi E. Neural remodeling in retinal degeneration. Prog Retin Eye Res 2003; 22(5):607–55.
39. Potts AM, Inoue J. The electrically evoked response (EER) of the visual system. II. Effect of adaptation and retinitis pigmentosa. Invest Ophthalmol Vis Sci 1969; 8(6):605–12.
40. Potts AM, Inoue J. The electrically evoked response of the visual system (EER). 3. Further contribution to the origin of the EER. Invest Ophthalmol Vis Sci 1970; 9(10):814–9.

41. Potts AM, Inoue J, Buffum D. The electrically evoked response of the visual system (EER). Invest Ophthalmol Vis Sci 1968; 7(3):269–78.
42. Kato S, Saito M, Tanino T. Response of the visual system evoked by an alternating current. Med Biol Eng Comput 1983; 21(1):47–50.
43. Humayun MS, de Juan E, Jr., Dagnelie G, et al. Visual perception elicited by electrical stimulation of retina in blind humans. Arch Ophthalmol 1996; 114(1):40–6.
44. Humayun MS, de Juan E, Jr., Weiland JD, et al. Pattern electrical stimulation of the human retina. Vision Res 1999; 39(15):2569–76.
45. Rizzo J, Wayatt J, Lowenstein J, et al. Acute intraocular retinal stimulation in normal and blind humans. Annual meeting of the Association for Research in Vision and Ophthalmology. Fort Lauderdale, Fla: Abstract 532.S102, 2000.
46. Chen SJ, Humayun MS, Weiland JD. Electrical stimulation of the mouse retina: A study of electricaly elicited visual cortical responses. Annual meeting of the Association for Research in Vision and Ophthalmology. Fort Lauderdale, Fla: Abstract 3886.S735, 1999.
47. Suzuki S, Humayun MS, de Juan E, et al. A comparison of electircal stimulatin threshold in normal mouse retina vs different aged retinal degenerate (rd) mouse rtina. Annual meeting of the Association for Research in Vision and Ophthalmology. Fort Lauderdale: Abstract 3886.S735, 1999.
48. Weiland JD, Humayun MS, Suzuki S, et al. Electrically evoked response (EER) from the visual cortex in normal and retinal degenerate dog. Annual meeting of the Association for Research in Vision and Ophthalmology. Fort Lauderdale, Fla: Abstract 4125.S783, 1999.
49. Rizzo JF, 3rd, Wyatt J, Loewenstein J, et al. Perceptual efficacy of electrical stimulation of human retina with a microelectrode array during short-term surgical trials. Invest Ophthalmol Vis Sci 2003; 44(12):5362–9.
50. Rizzo JF, 3rd, Wyatt J, Loewenstein J, et al. Methods and perceptual thresholds for short-term electrical stimulation of human retina with microelectrode arrays. Invest Ophthalmol Vis Sci 2003; 44(12):5355–61.
51. Richard G, Feucht M, Bornfeld N, et al. Multicenter study on acute electrical stimulation of the human retina with an epiretinal implant: clinical results in 20 patients. ARVO. Fort Lauderdale, USA, 2005.
52. Humayun MS, Weiland JD, Fujii GY, et al. Visual perception in a blind subject with a chronic microelectronic retinal prosthesis. Vision Res 2003; 43(24):2573–81.
53. Hesse L, Schanze T, Wilms M, Eger M. Implantation of retina stimulation electrodes and recording of electrical stimulation responses in the visual cortex of the cat. Graefes Arch Clin Exp Ophthalmol 2000; 238(10):840–5.
54. Kass MA, Gordon MO, Hoff MR, et al. Topical timolol administration reduces the incidence of glaucomatous damage in ocular hypertensive individuals. A randomized, double-masked, long-term clinical trial. Arch Ophthalmol 1989; 107(11):1590–8.
55. Kass MA, Heuer DK, Higginbotham EJ, et al. The Ocular Hypertension Treatment Study: a randomized trial determines that topical ocular hypotensive medication delays or prevents the onset of primary open-angle glaucoma. Arch Ophthalmol 2002; 120(6):701–13; discussion 829–30.
56. Margalit E, Fujii GY, Lai JC, et al. Bioadhesives for intraocular use. Retina 2000; 20(5):469–77.

57. Stieglitz T, Haberer W, Lau C, Goertz M. Development of an Inductively Coupled Epiretinal Vision Prosthesis. 26th Annual International Conference of the IEE EMBS. San Francisco, CA, USA, 2004.
58. Schwahn HN, Gekeler F, Kohler K, et al. Studies on the feasibility of a subretinal visual prosthesis: data from Yucatan micropig and rabbit. Graefes Arch Clin Exp Ophthalmol 2001; 239(12):961–7.
59. Peyman G, Chow AY, Liang C, et al. Subretinal semiconductor microphotodiode array. Ophthalmic Surg Lasers 1998; 29(3):234–41.
60. Chow AY, Pardue MT, Chow VY, et al. Implantation of silicon chip microphotodiode arrays into the cat subretinal space. IEEE Trans Neural Syst Rehabil Eng 2001; 9(1):86–95.
61. Chow AY, Chow VY, Packo KH, et al. The artificial silicon retina microchip for the treatment of vision loss from retinitis pigmentosa. Arch Ophthalmol 2004; 122(4):460–9.
62. Sachs HG, Gabel VP. Retinal replacement–the development of microelectronic retinal prostheses–experience with subretinal implants and new aspects. Graefes Arch Clin Exp Ophthalmol 2004; 242(8):717–23.

2
Artificial Vision: Vision of a Newcomer

Takashi Fujikado[1], Hajime Sawai[2] and Yasuo Tano[3]

[1]*Department of Applied Visual Science, Osaka University Medical School*
[2]*Department of Physiology, Osaka University Medical School*
[3]*Department of Ophthalmology, Osaka University Medical School*

Abstract: The Japanese Consortium for an Artificial Retina has developed a new stimulating method named Suprachoroidal-Transretinal Stimulation (STS). Using STS, electrically evoked potentials (EEPs) were effectively elicited in Royal College of Surgeons (RCS) rats and in rabbits and cats with normal vision, using relatively small stimulus currents, such that the spatial resolution appeared to be adequate for a visual prosthesis. The histological analysis showed no damage to the rabbit retina when electrical currents sufficient to elicit distinct EEPs were applied. It was also shown that transcorneal electrical stimulation (TES) to the retina prevented the death of retinal ganglion cells (RGCs). STS, which is less invasive than other retinal prostheses, could be one choice to achieve artificial vision, and the optimal parameters of electrical stimulation may also be effective for the neuroprotection of residual RGCs.

Introduction

In 2001, the Japanese Consortium for Artificial Retina (Yasuo Tano, Principal Investigator, Osaka University) was organized with supports from the Ministry of Economy, Trade and Industry and the Ministry of Health, Welfare and Labor in Japan.

Because our consortium was a newcomer to the field of artificial retinal and we had to play catch-up with several pioneering groups in USA, Germany, and other countries, we sought our original approach and developed a new stimulation method named STS.

We also investigated the effect of electrical stimulation on the retina and obtained results suggesting the neuroprotective effect of electrical stimulation on RGCs. In this chapter, we summarize the progress during the first 3 years of our consortium with emphasis on the evaluation of STS method and the electrically induced neuroprotection.

Overall Research Goals of Japanese Consortium for Artificial Retina

The goal of our research is to develop within 5 years in animal models a prosthesis for artificial vision that provides visual acuity sufficient for finger counting (the level of visual acuity that allows the patient to identify the number of fingers held at a distance of 30 cm).

The first candidate patients for the artificial retina would be those with advanced retinitis pigmentosa (RP) or autoimmune retinopathy, but in the next step, patients with age-related macular degeneration (AMD) could be included as candidates.

The Concept of Suprachoroidal-Transretinal Stimulation

Many research groups are currently developing electrode arrays that can be attached directly to the retina and used to stimulate the retina in an attempt to restore vision to patients with retinal degeneration [1–3].

In general, there are two types of retinal prostheses: an epi-retinal [4–8] and a subretinal [9–12]. Epiretinal stimulating prostheses are inserted into the vitreous cavity and are attached or placed against the inner retinal surface. The potential risk of this approach is the possible permanent attachment of the device to the retina. Epiretinal implantation can cause damage around the area of the retinal tack [6, 13], or may be unstable without the use of special methods to fix the electrodes [3].

Subretinal prostheses are placed between the pigment epithelial layer and the outer layer of the retina. These arrays are more stable but chronic implantation in this space might lead to a proliferation of glial tissue around the electrode array [12].Although improvements in surgical and electronic technologies may solve some of the problems of implanting retinal stimulating prostheses, there is the potential risk of significant damage to the eye after inserting an electrode intraocularly. To avoid the risks of implanting a stimulating prosthesis into the epiretinal or subretinal space, we have developed a new approach for electrode implantation, namely STS, which avoids direct contact of electrodes with the retina [14].

In STS, the stimulating electrodes were placed on the fenestrated sclera or in the suprachoroidal space and the counter electrode was inserted intravitreally (Figure 2.1). The merits of STS method are that: (1) surgical technique is less complicated because vitreous surgery is not required for this method; (2) electrodes are not in direct contact with the retina because they are placed in the scleral pocket; (3) electrodes are easy to be removed or replaced because they do not contact with the retina; and (4) a wide area of the retina can be stimulated because electrodes are placed in the outer-coat of the eyeball. On the other hand, the potential disadvantages are as follows: (1) the spatial resolution of STS could be limited because the electrodes are situated far away from the

A new approach for the artificial retina
Suprachoroidal-Transretinal Stimulation (STS)

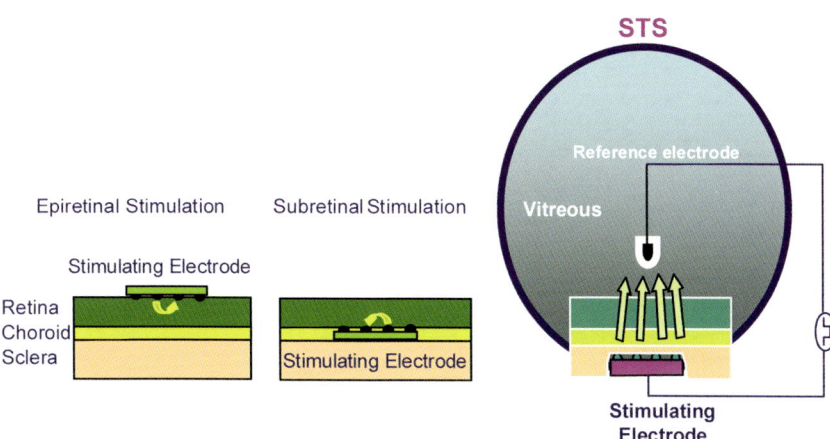

FIGURE 2.1. A new approach for the artificial retina 'suprachoroidal transretinal stimulation (STS)'.

retina; and (2) electrical current needed to attain artificial vision by STS could be higher than that by epi or subretinal electrode because electrodes do not contact the retina. In order to examine the efficacy and safety of STS method, we performed experiments on small animal (rats) and medium-sized animals (rabbits and cats).

The procedures used on all animals conformed to the Institutional Guidelines of Osaka University and the ARVO Resolution on the Use of Animals in Research.

The Effectiveness of STS in Animal Model

Artificial Vision in Royal College of Surgeons Rats

Since it has been reported in patients with RP that phosphenes can be obtained by electrical stimulation with electrodes that are not in direct contact with the retina, we expected that STS would be able to excite RGCs. Thus we tested our expectations regarding the feasibility of STS for an artificial retina, by acute electrophysiological experiments in retinal dystrophic rats (RCS rats). RCS rats are recognized as one of the best animal models for RP. In adult RCS rats (25 ~ 30 weeks of age) that we used, the outer retinal layers, including the photoreceptors, were always completely degenerated, as shown in Figure 2.2a, whereas the laminar organization of the inner retina was well preserved. Owing to such degeneration of the outer retina, flashing light stimuli elicited no detectable responses in the electroretinogram, and no evoked field potentials could be recorded from the superior colliculus (SC) (the upper and the middle traces in Figure 2.2b, respectively).

We first addressed the following issues: Can a focal STS to the retina with degenerated photoreceptors excite the residual neurons, especially RGCs, and can their excitation be transmitted via the optic nerve (ON) to the visual areas of the brain?

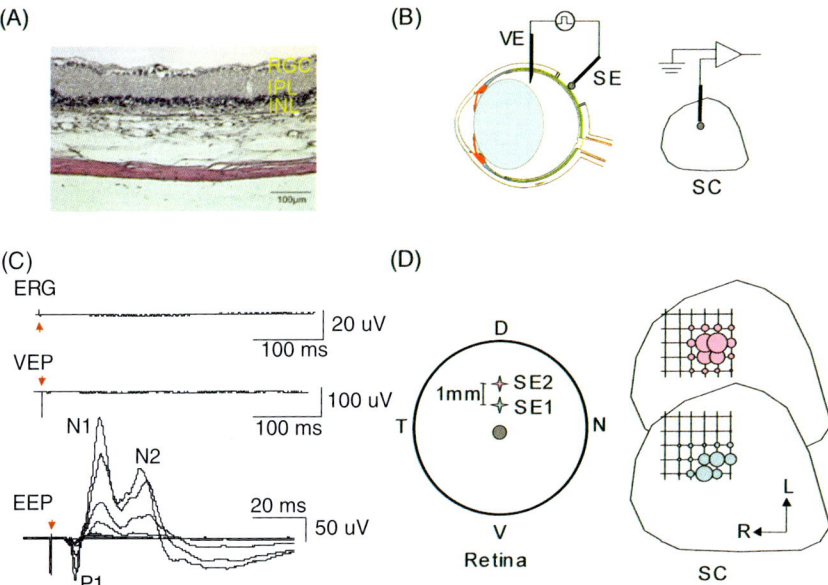

FIGURE 2.2. STS in normal and RCS rats. (a) A photomicrograph of a retinal section of an adult RCS rat stained by hematoxylin-eosin. RGC: retinal ganglion cell, IPL: inner plexiform layer, INL: inner nuclear layer. (b) A design of electrophysiological exper-iment in rats. SE: suprachoroidal electrode (anode), VE: vitreous electrode (cathode). (c) Neuronal responses evoked by flashing light and STS in a RCS rat. Upper trace: electroretinogram (ERG) to flashing light stimulus, showing that the retina is unresponsive to the light stimulus. Middle trace: Averaged visually evoked potentials (VEP) recorded from the SC, also showing no reponse. Bottom trace, Averaged electrically evoked poten-tials (EEP) to STS (10, 30, 60, 80, and 100 μA) recoded from the SC. (d) Differential distribution of STS-evoked responses via two separated electrodes. Left: two crosses indicate retinal positions of STS applied via a suprachoroidal elecrode, separated 1 mm from each other (SE1 and SE2). A filled circle indicates the ON head. D: dorsal, T: temporal, V: ventral, N: nasal, Right: distribution of collicular EPs in response to a single STS of 100 μA to SE1 (the lower), and SE2 (the upper). Recording of EPs was done at every distance of 250 μm. Center and radius of each Circle indicate the recording site of EPs and the relative amplitude of P1–N1 component, respectively. R: rostral, L: lateral.

A single STS, i.e. a brief (0.5 ms) monophasic pulse of electrical current, was applied to the one of the eye balls between an anodic suprachoroidal electrode placed on fenestrated sclera and a cathodic electrode inserted into the vitreous (Figs. 2.1 and 2.2c). The responses to the STS were recorded from the surface of the SC contralateral to the stimulated eye (Figure 2.2c). The threshold intensity of these STS-evoked responses was about $10 \sim 15$ nC of electrical charge ($20 \sim 30 \mu$A, 0.5 msec). Although the electrodes were apart from the retina, the threshold in STS was comparable to the degree that in epiretinal or subretinal stimulation with the electrodes directly attached to the retina.

The evoked potentials (EPs) consisted of a fast positive wave (P1) followed by a slow negative wave (N1), as shown in Figure 2.2c. The peak latencies of the P1 and the N1 were $7.1 +/- 1.5$ ms and $13.0 +/- 2.8$ ms, respectively (mean $+/-$ SD, n = 11). The peak amplitude of these two components was dependent on stimulus intensity. With high-intensity stimulus, N1 was occasionally followed by a negative peak (N2). The mean $+/-$ SD of the N2 was $25.6 +/- 5.0$ ms(n = 7). The same EPs to STS were recorded in normal hooded rats, and no statistical difference was seen between the two strains, in terms of shape, and peak latencies of P0, N1, and P1, as well as threshold.

It was interesting that inverting the polarity of STS greatly reduced the collicular evoked responses. Moreover, the latency of EPs to STS was slightly but consistently longer than that of the EP evoked in response to stimulation at the optic chiasm. The EPs to STS was completely abolished after transection of the ON just behind the eyeball. These results demonstrated that STS induced excitation of RGCs and the excitation was transmitted to the SC via the ON. These results indicate that STS is potentially suitable for retinal prostheses and artificial vision.

Then, we further studied the feasibility of STS, with attention to the spatial resolution that could be achieved with STS-based artificial vision. Because RGC axons are known to project to the SC in a precise topographical order, the N1–P1 amplitude of EPs to supra-threshold intensity of STS was surveyed on the SC at intervals of 0.25 mm along the mediolateral and rostrocaudal axis, in order to evaluate the localization of the retinal excitation induced by STS.

We found that the smallest responsive area was less than a square with sides 200μm, and the mean ($+/-$ SE) of the area was $0.24 +/- 0.12$ mm^2(n = 5). A topographical correspondence was also found between the points beneath the suprachoroidal electrodes and the sites where the maximal responses were evoked in the SC. Finally, the suprachoroidal electrodes were moved 1 mm dorsally. Such relocation of the electrode slightly, but obviously, shifted the localization of collicular EPs to STS from medial to lateral, as shown in Figure 2.2d, suggesting that phosphenes induced by two suprachoroidal electrodes of a STS-based multi-electrode array would be distinguishable from each other if these electrodes are at least 1 mm apart.

These observations in RCS rats strongly suggest that a STS can provide a focal excitation of the RGCs under the suprachoroidal electrodes. Thus, one can expect that patterned STS via an array of suprachoroidal electrodes will provide patterned phosphenes to RP patients.

Artificial Vision in Rabbit by STS

We used rabbits to develop the surgical procedure for implanting the electrode and for the functional assessment of STS method [15].

Surgical Procedure to Insert STS Electrode in Rabbit Eyes

A scleral pocket $(3 \times 5\,mm)$ was created just over the area of visual streak, which was about 12 mm posterior to the limbus (Figure 2.3a). The multichannel electrode array was then implanted into the scleral pocket and sutured with 5-0 Dacron onto the sclera just above the pocket. The counter electrode was a platinum (Pt) wire coated with polyurethane resin and exposed at the tip. It was inserted 4 mm into the vitreous cavity and was fixed 2 mm from the limbus with 5-0 Dacron suture. This surgical procedure was not complicated and can be applicable to human patients.

The Stimulating Electrode and Electrically Evoked Potential

The stimulating electrode consisted of eight dome-shaped electrodes (fabricated by NIDEK CO., Gamagori, Japan). These are mounted in a polyimide

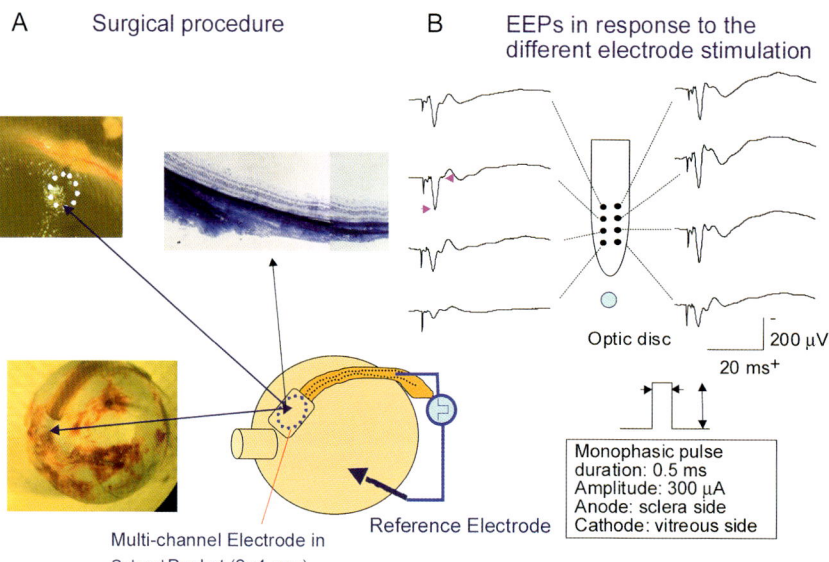

FIGURE 2.3. Surgical procedure and EEPs by STS in normal rabbit. (a) Surgical procedure. (b) EEPs by STS. EEPs with different amplitude were recorded by stimulating different electrode.

strip 3 mm wide, 4 mm long, and 180 μm thick. The dimensions of each electrode were height above the surface = 120–130 μm; diameter = 250 μm; distance between electrodes = 500 μrmm. Their impedance in saline was approximately 10 kΩ at 1 kHz.

The bundle of insulated leads from the microarray was also sutured at the limbus with 5-0 Dacron. A stimulator (SEN-7203, NIHON KOHDEN, Shinjyuku, JAPAN) was then connected through an isolator (A-395R, World Precision Instruments INC., Sarasota FL USA) to the microelectrode array. The recording electrode was a screw in the skull bone above the visual cortex. The EEPs were elicited by monophasic electrical pulses of 0.5 ms duration. The direction of the current was set for inward-flowing currents (electrode array was positive and reference electrode was negative). Fifty responses were averaged. A band-pass filter of 5–1 kHz was used.

Threshold Current by STS in Rabbit

In order to determine the minimum threshold of the EEP, each of the eight electrodes was stimulated with a current of 500 μA, and the amplitudes of EEP waves were compared (Figure 2.3b). The electrode that elicited the largest EEP was selected to determine the threshold current. The electric current was decreased in steps, and the minimum electric current that elicited the first or second positive peaks of the EEP was defined as the threshold current. The relationship between the second and the third peak amplitude of the EEPs and the inward current was examined in six rabbits (Figure 2.4). This regression curve showed that the EEP amplitude increased almost linearly at lower stimulus currents and tended to be saturated with higher stimulus currents. The mean threshold current that elicited a small EEP was 55.0 $\pm \mu$A for the six rabbits (27.5 $\pm$ 5.0 nC or 56.0 $\pm$ 10.2 μC/cm^2).

Humayun et al. reported that a charge density of 8.92–11.9 μC/cm^2 was required to elicit EEPs from rabbits with epiretinal electrodes [5].In experiments on cats, Dawson and Radtke found a threshold charge density of 30.5 μC/cm^2[16]. Chow et al. implanted their electrodes subretinally in rabbits, and reported a threshold charge density of 2.8 nC/cm^2 to elicit EEPs [9]. Although the distance from the electrodes to the retina was farther in our transcleral electrodes than that of the epi or subretinal electrodes, the threshold charge density to elicit EEPs was comparable with the other electrode placements.

McCreery et al. [17] found that the threshold for stimulation-induced neural damages is determined by a synergetic interaction between charge density and charge per phase, and the limit for safe charge density decreased as charge density increased. With our electrodes, 50 NC/Phase would give a charge density of approximately 110 μC/cm^2 and the work of McCreery et al. suggests that this combination would not be injurious.

Our electrodes were not in direct contact with the neural tissue. Therefore, the limit of electrical charge density may be expected to be higher than that with epiretinal stimulation. Indeed, Nakauchi K. et al. [15] reported that for ϕ100μm Pt electrode by STS, retinal tissue was not damaged by a current up

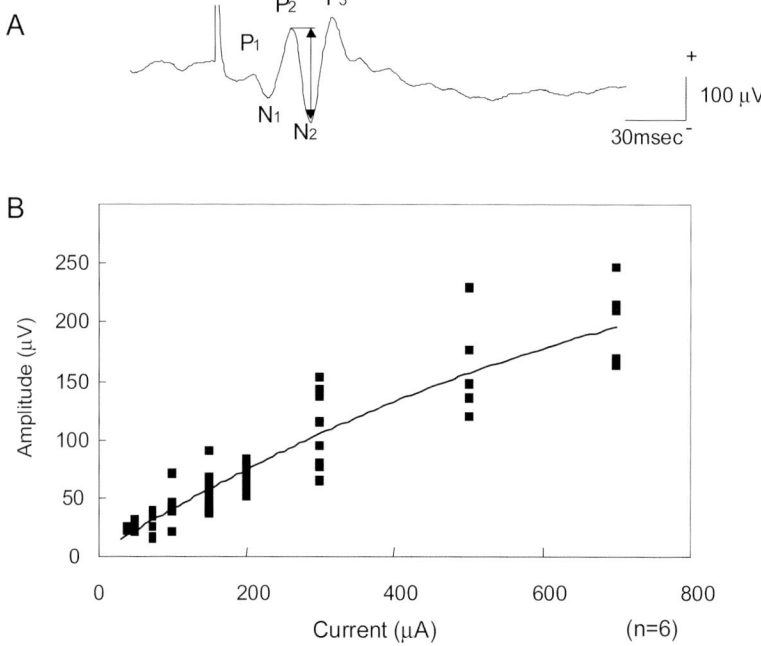

FIGURE 2.4. Amplitude of the EEP induced by STS in rabbit. A typical EEP waveform evoked by 500 uA stimulation. P2–N2 amplitude plotted in relation to the stimulus intensity in rabbit (n = 6). The mean threshold current was $55.0 \pm 10.0\,\mu A$. The increase in response amplitude was approximately linear with low stimulus intensities.

to 1mA $(2100\,\mu C/cm^2)$ using biphasic pulses (anodic first, duration; 0.5 msec, frequency; 20 Hz) continuously for an hour (Chapter 16). These data suggest that our newly developed STS method could safely elicit the percepts necessary for artificial vision.

What We Learned from the Rabbit STS Experiment

As a preclinical study of artificial vision by the STS method, we investigated the surgical procedure, functional analysis, and tissue effects. The surgical procedure to insert an electrode into a scleral pocket was less complicated than the implantation of epi or subretinal electrode and could be applied clinically, although the anatomical difference between rabbit and human should be carefully considered.

The EEP was elicited by electrical stimulation using a combination of charge and charge density that we expect will not be damaging to the retina, suggesting the clinical applicability of the STS method. However, the spatial resolution of STS could not be analyzed by EEP recordings in the rabbit. For this purpose, the cat is more suited as a medium-sized animal, in which the retinotopic mapping onto visual cortex has been more thoroughly investigated.

Artificial Vision in Cats by STS

Although we demonstrate that a single STS induced a localized excitation of RGCs under a suprachoroidal electrode even in retinal dystrophic rats, there still remains the question as to the spatial resolution of STS-based artificial vision. To investigate the question further, a rat eye is too small to give a focal STS to multiple sites on the retina, and the rat visual system, especially the retino-geniculo-cortical system involved in pattern recognition, is not well developed. Therefore, we used adult cats and recorded unitary discharges from a relay cell of the lateral geniculate nucleus (LGN) for analyzing the spatial properties of STS-evoked excitation more precisely (in preparation).

A multichannel electrode array with nine Pt electrodes was set on a fenestrated sclera and a single charge-balanced biphasic pulse of electrical current was delivered as a STS between one of the nine electrodes and a counter electrode inserted into the vitreous. After identification of the receptive field center of a LGN neuron by mapping on a tangent screen in front of the eye, spike responses of the neuron to STS through each stimulating electrode were investigated. A higher stimulus intensity of STS was required to elicit spike bursts from the LGN neurons, when STS was applied through an electrode that was further from their receptive field.

The threshold current for inducing spike response to STS with a probability of 50% was analyzed for every neuron in relation to the distance from a stimulating electrode to the central point of its receptive field. This analysis revealed that the threshold tended to become lower as the distance decreased. For some neurons, the distance was within a visual angle less than $2°$ degrees with a threshold current less than 0.1 mA (50 nC). Therefore, retinal excitation induced by minimal STS seems to be sufficiently localized, ensuring that a STS-based artificial retina can provide spatial resolution that is adequate for finger counting.

Neuroprotection by Electrical Stimulation

Electrical stimulation of the nervous system has profound biological effects on neurons in addition to the induction of neural activities. Recent studies have shown that the electrical activity of neurons can modify neurons themselves, such as neurite extension, dendritic reorganization, and synaptic connectivity in both adulthood and developmental stages [18].

These drastic neuronal modifications by electrical activity are mediated by the changes in extracellular signal networks among the cells in the neural tissues and intracellular signaling systems of the neurons themselves. From the standpoint of treatments for pathologies of the nervous system, it is reasonable to imagine that these cellular effects of electrical activity can also affect the viability of damaged neurons and the functionality of pathological neural systems. In fact, there are several studies that electrical stimulation through the cochlear implant can prevent the secondary degeneration of spiral ganglion neurons after the loss of hair cells [19, 20].

Injury to the ON is one of experimental models for the study of the pathology of RGCs. Axonal injury of RGCs such as ON transection or crush causes their rapid retrograde death. Many models have been developed to investigate protection of ganglion cells after damage to the ON. Most of these studies were extrinsic supplements of the various trophic factors which activate or inhibit intracellular signaling systems involved in cell death and/or survival [21–23]. As another approach, we investigated neuroprotection by electrical activation of the damaged retina.

First, we investigated electrical stimulation of the transected ON stump [24]. After pre-labeling of RGCs with fluorescent dye bilaterally applied to the superior colliculi, the left ON was completely transected at 3 mm behind the eyeball, and the transected stump was stimulated for 2 hours with monophasic pulses ($50\,\mu A$ to $50\,\mu sec$, 20 Hz) via a pair of silver ball electrodes immediately after the transection. One week after the transection, the survival rate of RGCs in the stimulated group was 83% of that of the intact retina, whereas the survival rate of RGCs with transection only was 54%. The intensity of the stimulus current was varied from 20 to $70\,\mu A$. Stimulation at $50\,\mu A$ had the maximum survival-promoting effect. This proved that the electrical stimulation has a neuroprotective effect on the damaged RGCs in vivo.

Although the ON is easy to access surgically and can be stimulated in experimental animals, electrical stimulation of the ON is unlikely to be useful for clinical purposes. Therefore we developed another stimulating method, TES for neuroprotection (Figure 2.5) [25]. In TES, electrical stimulation is applied via a contact lens electrode attached to the cornea. This simulation method has already proved to be able to evoke the sensation of light, phosphene [26]. This makes it possible to reduce the invasiveness and to treat the patient repeatedly only with the surface anesthesia. As shown in Figure 2.5, we examined the neuroprotective effect of TES on the axotomized RGCs. The TES was applied immediately after ON transection, and the survival rate of axotomized RGCs was examined 7 days later. When the TES of 1 ms/phase duration and $100\,\mu A$ intensity was applied for an hour, the survival rate of axotomized RGCs was 85%, which is equivalent to the protective effect of electrical stimulation applied to the transected ON.

We hypothesize that the mechanism of the neuroprotective effect of TES involves neurotrophic factors in the retina, because it was reported that expression of neurotrophic factors are upregulated by electrical stimulation [27]. The changes after TES of four major neurotrophic factors and their receptors, brain-derived neurotrophic factor (BDNF), ciliary neurotrophic factor (CNTF), basic fibroblast growth factor (bFGF), insulin-like growth factor-1 (IGF-1), trkB, CNTF-R alpha, FGF-R1, and IGF-1R were analyzed by reverse transcriptase–polymerase chain reaction (RT-PCR) after TES.

Amongst these, IGF-1 mRNA increased gradually and showed prominent expression on day 7 after TES (Figure 2.6). Northern and Western blotting also confirmed the remarkable upregulation of IGF-1 in the retina treated by TES. Moreover, the intraperitoneal administration of the antagonistic peptide of IGF-1R blocked the effect of TES on survival of axotomized RGCs. These results

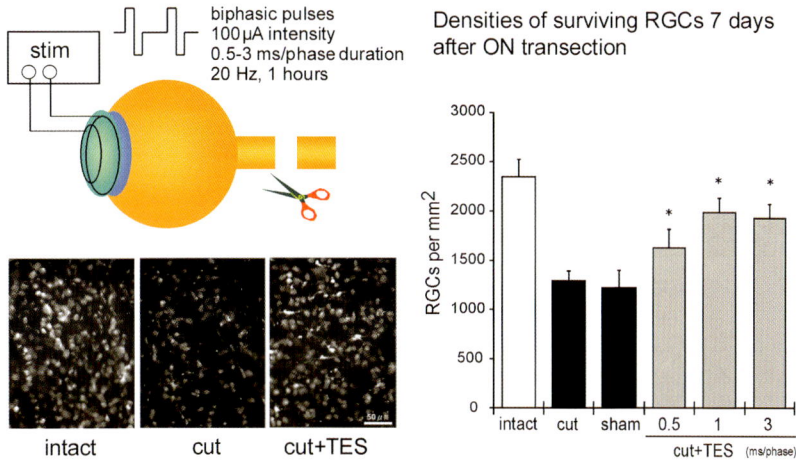

FIGURE 2.5. Protection of retinal ganglion cells (RGCs) by transcorneal electrical stimulation (TES). The RGCs were labeled by fluorescent dye retrogradely from bilateral superior colliculi before optic nerve transection. A biphasic pulse of $100\,\mu A$ was applied through corneal lens-type electrode for an hour after the transection of rat optic nerve. With this current, the EEP was effectively recorded from the SC of the intact rat in other experiment, suggesting that retina was stimulated enough by the current through the contact lens electrode. One week after optic nerve transection, the surviving RGCs were counted. The photomicrographs of the flat-mounted retinas shows more RGCs survived in 'cut + TES' retina than 'cut' retina with transection only. In the sham operated group, 53% of RGCs survived, while 'cut + TES' group showed 70–85% survival. This proved that the electrical stimulation on the retina was effective to promote the survival of RGCs. Intact, the intact retina; cut, the retina with optic nerve transection; sham, the retina with optic nerve transection and attachment of TES electrode but without current through it; cut + TES, the retina with optic nerve transection and TES.

indicate that IGF-1 plays a key role in the TES-induced neuroprotection of the axotomized RGCs.

To investigate in more detail the mechanism of the neural protection, the localization of IGF-1 in the retina after TES was examined immunohistologically (Figure 2.7). In the normal retina, IGF-1 was present only in the inner limiting membrane (ILM) and the nerve fiber layer. After TES, the IGF-1 immunoreactivity expanded from the ILM to the inner nuclear layer (INL) on day 7. An especially intense signal was observed from the radial structure extending from INL to ILM. The double staining of IGF-1 and glutamine synthetase, a marker of Müller cells, showed that the IGF-1 signal was strong on the endfeet and processes of the Müller cells, in addition to the diffuse signals in the inner retina. These immunohistological results showed that Müller cells produced IGF-1 after TES and release it into the extracellular space. Although it remains to be resolved whether activation of RGCs itself is required for neuroprotection by TES and the mechanism of activation of the Müller cells by TES, these series of experiments

RT-PCR

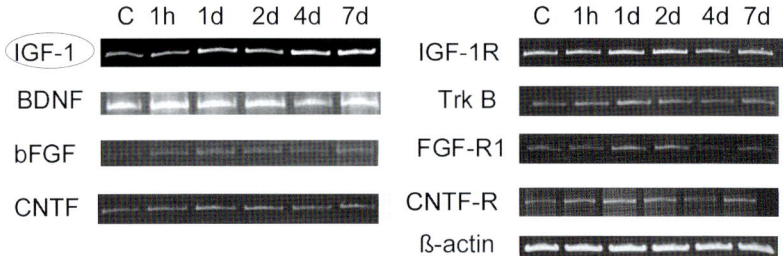

FIGURE 2.6. Expression of mRNA of neurotrophic factors and their receptors after TES. The expression was analyzed by reverse transcriptase–polymerase chain reaction (RT–PCR) at different time points ranging from 1 hour to 7 days after TES without optic nerve transection. The expression of only IGF-1 increased up to day 7. C, control intact retina without optic nerve transection; IGF-1, insulin-like growth factor-1; BDNF, brain-derived neurotrophic factor; bFGF, basic fibroblast growth factor; CNTF, ciliary neurotrophic factor.

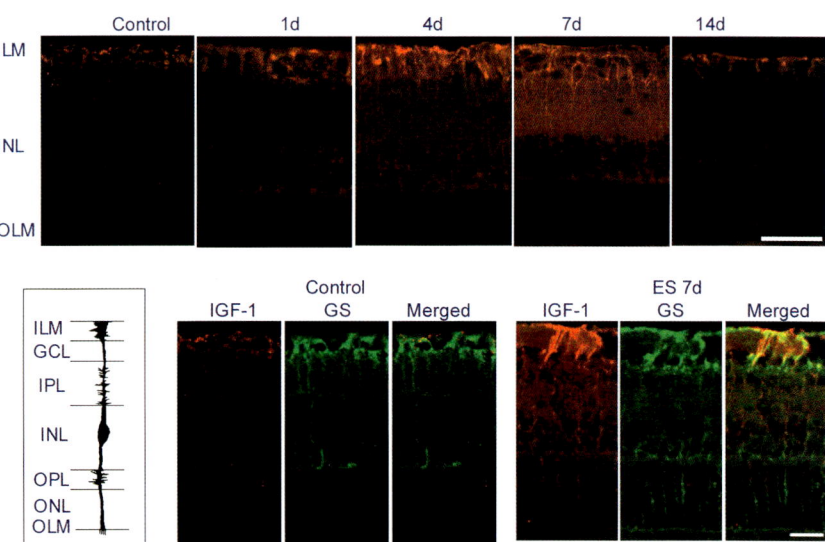

FIGURE 2.7. Immunohistrological analysis of IGF-1 in the retina after TES. Upper panels show the localization of IGF-1 at different time points from day 1 to day 14 after TES. In the control retina (no TES), there is very weak IGF-1 immunoreactivity in the inner limiting membrane (ILM) and in the nerve fiber layer (NFL). After TES, the IGF-1 signal increased and extended to inner plexiform layer diffusely. On day 4 and 7, the radial structure from inner nuclear layer (INL) to ILM was also observed. After the maximum on day 7, the signal decreased on day 14. Lower panels show the double staining of IGF-1 and glutamine synthetase (GS), a marker of Müller cells. In the control retina, weak IGF-1 signals in the ILM and NFL were seen on the endfeet of Müller cells. On day 7 after TES, strong immunoreactivity appeared in the endfeet and processes of Müller cells. Scale; $100\,\mu$m (upper panels), $50\,\mu$m (lower panels).

indicate that the TES activates the intrinsic retinal IGF-1 system and prevents the death of the RGCs.

Our findings are consistent with the concept of electrical stimulation therapy, which activates the intrinsic neuroprotective system. Electrical stimulation can be applied to any nervous system non-invasively, repeatedly, and chronically if an adequate system or devise is developed. TES is simple and non-invasive therapy, and may have a therapeutic potential in other diseases of retinal neurons. In fact, we reported that TES also prolonged the survival of photoreceptors and delayed the loss of retinal function in RCS rats [28].

We also hypothesize that STS-based retinal implants for artificial vision can work as a kind of neuroprotective device for various retinal dystrophic diseases through electrical stimulation. This idea was supported by a recent report that subretinal implants had a neuroprotective effect for photoreceptors of RCS rats, although the neuroprotective contribution of electrical stimulation itself was observed only in the ERG, not in the morphology [29]. In the future, the retinal electrode may not only act as a retinal prosthesis but also may help to prevent the degeneration of retinal cells.

References

1. Humayun MS (2001) Intraocular retinal prosthesis. Trans Am Ophthalmol Soc, 99:271–300.
2. Margalit E, Maia M, Weiland JD, Greenberg RJ, Fujii GY, Torres G, Piyathaisere DV, O'Hearn TM, Liu W, Lazzi G, Dagnelie G, Scribner DA, de Juan E, Jr, Humayun MS (2002) Retinal prosthesis for the blind. Survey of ophthalmology 47:335–356
3. Zrenner E (2002) Will retinal implants restore vision? Science 295:1022–1025.
4. Eckmiller R (1997) Learning retina implants with epiretinal contacts. Opthalmic Res 29:281–289.
5. Humayun MS, Probst R, de Juan E, Jr, McCormick K, Hickingbotham D (1994) Bipolar surface electrical stimulation of the vertebrate retina. Arch Ophthalmol 114:40–46.
6. Majji AB, Humayun MS, Weiland JD, Suzuki S, D'anna SA, de Juan E, Jr (1999) Long-term histological and electrophysiological results of an inactive epiretinal electrode array implantation in dogs. Invest Ophthalmol Vis Sci 40:2073–2081.
7. Nadig MN (1999) Development of a silicon retinal implant: cortical evoked potentials following focal stimulation of the rabbit retina with light and electricity. Clinical Neurophysiology 110:1545–1553.
8. Walter P, Heimann K (2000) Evoked cortical potential after electrical stimulation of the inner retina in rabbits. Graefes Arch Clin Exp ophthalmol 238:315–318.
9. Chow AY, Chow VY (1997) Subretinal electrical Stimulation of the rabbit retina. Neuroscience letters 225:13–16.
10. Chow AY, Pardue MT, Perlman JI, Ball SL, Chow VY, Helting JR, Peyman GA, Liang C, Stubbs EB, Jr, Peachey NS (2002) Subretinal implantation of semiconductor-based photodiodes: Durability of novel implant designs. J Rehabilitation Res Develop 39:313–322.
11. Schwahn HN, Gekeler F, Kohler K, Kobuch K, Sachs HG, Schulmeyer F, Jakob W, Gabel VP, Zrenner E (2001) Studies on the feasibility of a subretinal visual prosthesis:

data from Yucatan micropig and rabbit. Graefes Arch Clin Exp Ophthalmol 239:961–967.

12. Zrenner E, Stett A, Weiss S, Aramant RB, Guenther E, Kohler K, Miliczek KD, Seiler MJ, Haemmerle H (1999) Can subretinal microphotodiodes successfully replace degenerated photoreceptors? Vision Res 39:2555–2567.

13. Walter P, Szurman P, Viobig M, Berk H, Ludtke-Handjery H-C, Deng HR, Mittermayer C, Heimann K, Sellhaus B (1999) Successful long term implantation of electrically inactive epiretinal microelectrode arrays in rabbits. Retina 19:546–552.

14. Kanda H, Morimoto T, Fujikado T, Tano Y, Fukuda Y, Sawai H (2004) Electrophysiological Studies on the Feasibility of Suprachoroidal-Transretinal Stimulation for Artificial Vision in Normal and RCS Rats. Invest Ophthalmol Vis Sci 45:560–566.

15. Nakauchi K, Fujikado T, Kanda H, Morimoto T, Choi JS, Ikuno Y, Sakaguchi H, Kamei M, Ohji M, Yagi T, Nishimura S, Sawai H, Fukuda Y, Tano Y (2005) Transretinal electrical stimulation by an intrascleral multichannel electrode array in rabbit eyes. Graefes Arch Clin Exp Ophthalmol, 243:169–174.

16. Dawson WW, Radtke ND (1977) The electrical stimulation of the retina by indwelling electrodes. Invest Ophthalmol Vis Sci 16:249–252.

17. McCreery DB, Agnew WF, Yuen TG, Bullara L (1990) Charge density and charge per phase as cofacters in neural injury induced by electrical stimulation. IEEE Trans Biomed Eng 37:996–1001.

18. Zhang LI, Poo MM (2001) Electrical activity and development of neural circuits Nat. Neurosci. 4(Suppl): 1207–1214.

19. Leake PA, Hradek GT, Rebscher SJ, Snyder RL (1991) Chronic intracochlear electrical stimulation induces selective survival of spiral ganglion neurons in neonatally deafened cats. Hear Res 54:251–257.

20. Leake PA, Hradek GT, Snyder RL (1999) Chronic electrical stimulation by cochlear inplant promotes survival of spiral ganglion neurons after neonatal deafness. J Comp Neurol 412:543–562.

21. Mansour-Robaey S, Clarke DB, Wang YC, Bray GM, Aguayo AJ (1994) Effects of ocular injury and administration of brain-derived neurotrophic factor on survival and regrowth of axotomized retinal ganglion cells. Proc Natl Acad Sci USA 91:1632–1636.

22. Kermer P, Kloecker N, Labes M, Baehr M (1998) Inhibition of CPP32-like proteases rescues axotomized retinal ganglion cells from secondary cell death *in vivo*. J Neurosci 18:4656–4662.

23. Shen S, Wiemelt AP, McMorris FA, Barres BA (1999) Retinal ganglion cells lose trophic responsiveness after axotomy. Neuron 23:285–295.

24. Morimoto T, Miyoshi T, Fujikado T. Tano Y, Fukuda Y (2002) Electrical stimulation enhanced the survival of axotomized retinal ganglion cells in vivo. Neuroreport 13:227–230.

25. Morimoto T, Miyoshi T, Matsuda S, Tano Y, Fujikado T, Fukuda Y (2005) Transcorneal electrical stimulation rescues axotomized retinal ganglion cells by activating endogeneous retinal IGF-1 system. Inv Ophthal Vis Sci 46:2147–2155.

26. Potts AM, Inoue J, Buffum D (1968) The electrically evoked response of the visual system (EER). Invest Ophthalmol 7:269–278.

27. Al-Majed AA, Brushart TM, Gordon T (2000) Electrical stimulation accelerates and increases expression of BDNF and trkB mRNA in regenerating rat femoral motoneurons. Eur J Neurosci 12:4381–4390.

28. Morimoto T, Choi JS, Miyoshi T, Fujikado T, Fukuda Y Tano Y, Fujikado T (2005) Effects of transcorneal electrical stimulation on the survival of photoreceptors in RCS rats, ARVO 2005 46: E-Abstract.
29. Paedue MT, Phillips MJ, Yin H, Sippy BD, Webb-Wood S, Chow AY, Ball SL (2005) Neuroprotective effect of subretinal implants in the RCS rat. Inv Ophthal Vis Sci 46:674–682.

3
The Effects of Visual Deprivation: Implications for Sensory Prostheses

Ione Fine

Department of Ophthalmology, USC

Introduction

Normally, early stages of sensory processing are organized in a modular fashion with sensory information passing from the relevant sensory organ (whether it be retina or cochlea) to a hierarchy of specialized cortical regions which process increasingly complex attributes of the incoming sensory information. Responses within each pathway are primarily driven by a single sensory modality, especially within early stages of this hierarchy.

In recent years, there has been increasing interest in how these sensory pathways can be modified by experience, both in development and in adulthood. When a sense is lost, especially early in life, significant cortical reorganization is observed. A significant body of research has begun to describe positive aspects of this reorganization, characterizing how processing within the remaining intact senses adapt to compensate for the missing sense. However, though less studied, neural reorganization also has potential negative effects, resulting in a gradual deterioration in the ability to process the missing sense.

Here we discuss the implications of this reorganization for the ability to restore sight, either through standard medical interventions or through implantation of a sensory prosthesis. First, deterioration as a result of sensory deprivation can seriously limit the ability to restore sight. Second, understanding the cortical changes that occur as a result of deprivation can help guide clinicians toward developing rehabilitation strategies that are suited to take advantage of the natural plasticity of the brain.

Sensory Plasticity in Adulthood: Potential Differences between Cortical Areas

In recent years, it has been clearly demonstrated that primary tactile and auditory cortical areas can show remarkable amounts of plasticity even in adulthood.

Classical studies by Recanzone et al. have demonstrated that removal of a finger leads to striking reorganization within primary somatosensory cortex [1, 2]. Similarly, extensive practice on an auditory discrimination task results in an expansion of the region of primary auditory cortex responsive to the trained frequency [3]. The remarkable plasticity of auditory cortex is thought to be a major contributing factor to the success of cochlear implants: language comprehension tends to be quite poor immediately after implantation, and then improves over many months [4].

In the field of visual prostheses the hope is, of course, that visual cortex will show a similar capacity for adult plasticity. However, there is a significant amount of data suggesting that early visual cortex may show much less adult plasticity than early auditory or somatosensory areas. It is surprisingly difficult to find perceptual learning effects that can be definitively attributed to primary visual areas. While some electrophysiology studies in monkeys have found neural changes within primary visual area V1, these tend to be both restricted in magnitude and task dependent [5, 6]. In humans, V1 changes in responsivity have been found as a result of training in an orientation discrimination task using functional magnetic resonance imaging (fMRI), but these changes may be partially mediated by attentional feedback from higher visual areas [7].

One type of cortical plasticity that has been extensively studied is the "filling in" of retinal lesions (a process in which cortical neurons that subserve a retinal scotoma begin to respond to nearby intact retina). While rapid filling in has been found in cat [8, 9], reorganization in response to scotomas seems to occur extremely slowly in both the monkey and the humans. A recent study in macaque found no evidence for remapping of regions of cortex that represented a retinal scotoma over several months [10]. However, in the case of human patients, there is one study examining cortical mapping in patients with well-established (several years) foveal scotomas due to macular degeneration which does demonstrate remapping [11]. Perhaps if adult retinotopic reorganization occurs as a result of restricted visual loss, it happens extremely slowly.

One possible explanation for why early visual areas show so little plasticity is that visual cortex is already performing a demanding host of visual functions. Reorganization of visual cortex in response to auditory deprivation could potentially undermine the ability of visual cortex to perform basic visual functions. As described below, the changes that occur in somatosensory cortex consequent on learning to read Braille with three fingers result in worse performance in discriminating which finger had been stimulated. In the case of early visual cortex, the need to perform normal visual functions may limit the scope for changes consequent on the demands of deafness.

By contrast, one might expect more substantial reorganization in higher-level visual areas that tend to be more experience dependent. Neuronal changes as a function of learning do seem to be larger in higher visual areas [12], though no direct electrophysiological comparisons between low- and high-level plasticity have been made to date. One reason for assuming that higher levels of cortex are dependent on experience is that neurons in these areas show a strong preference for stimuli that have been frequently encountered, or are behaviorally important. For example, neurons in macaque inferotemporal cortex (IT) are tuned for particular shapes, including hands and faces [13], and these representations can be shaped by training [14]. Monkey face selective cells in IT show different responses to different faces, with their responses carrying identity information. The tuning of these cells seems also to be dependent on factors, such as familiarity or social hierarchy [15, 16] that are clearly highly experience dependent.

In support of the notion that plasticity increases across the visual hierarchy, it seems that perceptual learning may be related to the complexity of the task, with simpler tasks demonstrating less learning. Figure 3.1 shows perceptual learning for 16 visual tasks collated as part of a meta-review [17]. The learning index on the y-axis represents an increase in the ability to perform the task on each session compared to performance on the first session (d' on each session/d' on the first session). No improvement in performance would be a flat line of slope $= 0$.

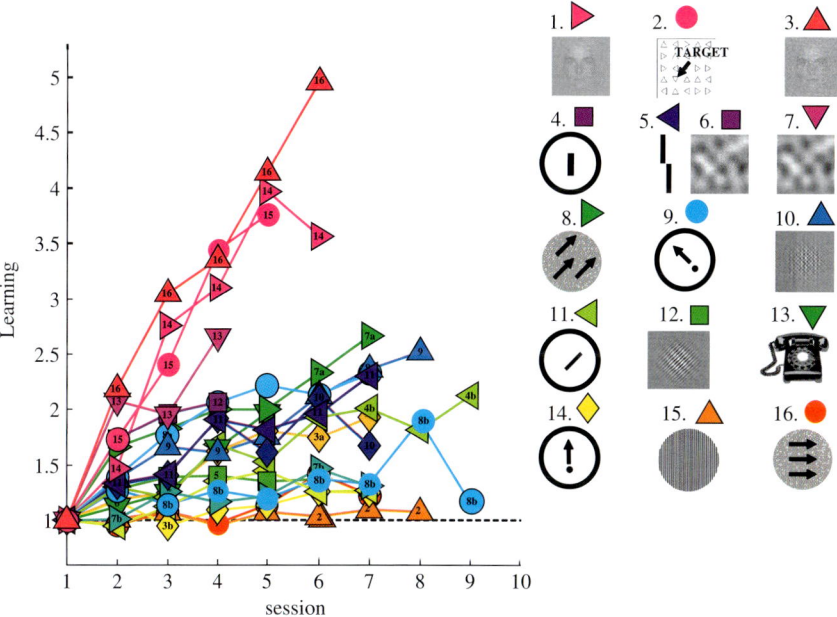

FIGURE 3.1. Learning as a function of session for 16 tasks. These were novel face discrimination in (1) high and (3) low contrast noise, (2) simple shape search, bandpass noise identification with (4) low and (5) high contrast noise, (6) vernier discrimination, (7) cardinal and (13) oblique orientation discrimination, spatial frequency discrimination for a (8) complex and (12) simple plaid, direction of motion for a single dot moving in a (9) oblique or (16) cardinal direction of motion, direction of motion discrimination for (10) oblique and (14) cardinal directions, (11) familiar object recognition, and (15) the resolution limit for high spatial frequency gratings [17]. Copyright 2002 by ARVO. Adapted with permission.

Performance on low-level tasks shows fairly limited improvement with practice; after four sessions, the slope of the learning index averaged across various tasks defined as low level was found to be 0.17. In contrast, complex tasks that required discriminations along more than one perceptual dimension showed more than twice as much learning, with a learning slope of 0.43 (note that many of these stimuli also include external noise). High-level tasks, involving identifying or discriminating real-world natural objects showed three times as much learning as the low-level tasks, with a learning slope of 0.60. The exception was that observers showed little learning for a task where they were asked to identify briefly presented familiar objects, possibly because identification learning had already occurred for these very familiar stimuli.

These data have important implications for visual prostheses. On the positive side, these data suggest that there will be little deterioration in early visual areas as a consequence of blindness in adulthood. This is supported by the observation that no significant visual deficits tend to be observed after the removal of adult cataracts, even when the cataracts are dense and have been present for several years. On the other hand, these data suggest that the ability to adapt to the unnatural input provided by a visual prosthesis may be relatively limited and/or slow within early visual areas.

Compensating for a Missing Sense: After Losing a Sense there are Improvements in the Ability to Use the Remaining Senses

There is a joke in the blind community – when a blind individual forgets a phone number or a name, they may be asked in a concerned tone of voice "Are you quite sure you are really blind?" Because people who are blind cannot write things down, they tend to rely heavily on short- and long-term memory to perform a variety of tasks, such as remembering appointments or phone numbers. Not surprisingly, this seems to result in better memory for tasks such as memorizing verbal lists [18].

Rehabilitation training is based on the understanding that individuals who have a reduced or missing sense can develop alternative strategies and abilities to compensate. However, while there is an extensive animal literature examining both compensatory neural plasticity and the effects of sensory deprivation, within visual and auditory cortex [19–22], the underlying neurophysiological changes that underlie these compensatory improvements in behavior are only now beginning to be studied in humans. This chapter mainly discusses studies examining plasticity as a result of sensory deprivation within humans. While the animal literature has provided great insights into the neural mechanisms underlying plasticity, it is perfectly plausible that the extent and the nature of plasticity differ enormously between humans and animals.

The growing literature examining the neural site of these behavioral modifications in humans is mainly based on using functional magnetic resonance imaging (fMRI). FMRI is based on a modification of standard clinical MRI techniques. However in fMRI the pulse sequence used to acquire images of the brain is designed to differentiate between the different magnetic properties of oxygenated and deoxygenated blood. When a region of cortex is active, the vascular system provides an oversupply of oxygenated blood. By measuring the relative amounts of oxygenated and deoxygenated blood, it is possible to determine which regions of cortex are active for a given task over a time scale of a few seconds.

Blind Individuals are Better at Certain Tactile and Auditory Tasks

People who are blind often have had extensive practice with various tactile and auditory tasks such as reading Braille and navigating using a cane. There is now growing evidence that blind individuals can be better than sighted individuals in certain tasks involving both touch and hearing [21]. As far as auditory tasks are concerned, blind subjects seem to be particularly good at auditory spatial tasks (determining where in space a tone came from) in the periphery [23, 24]. This is true even for tones localized in distant space beyond the reach of the subject [25]. Curiously, this superior ability for the blind to localize sounds is particularly pronounced with monaural presentation, possibly because blind subjects are better able to make use of the subtle spectral cues produced from "shadowing" by the head and ear pinna [26]. Blind subjects are also better than the sighted at identifying changes in pitch [27].

Sighted teachers of Braille read using the shadows of the dots on the paper, not by touch, suggesting that the tactile discriminations required in Braille are difficult for the sighted (Pascual-Leone, personal communication). Consistent with this observation, blind subjects tend to be better than sighted subjects at some (though not all) tactile discrimination tasks [28]. For example, in a tactile grating discrimination task, blind subjects had thresholds that were, on average, 50% lower than sighted subjects [29], with differences between blind and sighted subjects being largest for the dominant "reading" finger.

In many, though not all, of the tactile and auditory tasks that show differences between blind and sighted individuals, there is a difference in performance between early and late blind subjects. One possibility is that there is a greater potential for the cross-modal plasticity that is thought to underlie fluent use of sensory substitution strategies when deprivation occurs at an early age. However, it should be noted that early blind subjects tend to both use a cane more frequently and competently, and read Braille more fluently than late blind subjects (though there are, of course, many exceptions), so the level of practice presumably differs between the two groups. How plasticity and practice may interact is discussed later in this chapter.

Deaf Individuals are Better at Certain Visual Tasks

Sign language contains large amounts of visual form and motion information. There is now extensive evidence for differences in motion processing between deaf signers, fluent hearing signers (hearing children of deaf parents), and hearing subjects that seem to correlate with the unique demands of under-standing sign language. Both deaf and fluent-hearing signers are better at detecting a motion target or discriminating direction of motion in the right visual field than non-signers. The right visual field advantage is thought to be due to the right visual field projecting to the left hemisphere which is responsible for language processing in the majority of individuals. Enhanced motion processing in the right visual field may provide a route by which high-quality visual motion information is provided to language areas of the brain [30–33].

Subjects who are deaf also tend to be better at distributing attention to the periphery and at performing peripheral visual tasks [34–40]. One possibility is that an enhanced sensitivity to peripheral visual stimuli may help compensate for the loss of the auditory cues that often direct our attention to objects outside the central focus of attention.

Sign language also relies heavily on facial expression, which is used to provide both emphasis and punctuation. Facial expressions signal lexical and syntactic structures, such as questions and adverbs. Linguistic facial expressions differ from emotional expressions in their scope and timing and in the facial muscles that are used. To take an example from McCullough et al.[43], the lips pressed together and protruded indicate an effortless action, whereas the tongue protuded slightly means carelessly. These expressions accompanying the same verb (e.g. "DRIVE") convey very different meanings (to drive effortlessly vs. drive carelessly). Obviously being sensitive to these facial expressions is crucial for learning sign language. It has been found that fluent-hearing and deaf signers perform significantly better than non-signers in distinguishing among similar faces (e.g. the Benton Faces Test), in identifying emotional facial expressions, and in discriminating local facial features [41–43].

Compensating for a Missing Sense: What is the Neural Basis?

So, what are the neural substrates of these improvements in performance after loss of a sense? Take for example, the improved visual processing in deaf subjects. One possibility is that these improvements are due to modifications in the way that *visual* cortex processes visual information (compensatory hyper-trophy). The alternative is that these improvements are due to cross-modal plasticity whereby *auditory* cortex, deprived of its usual input, begins to respond to visual information.

Similarly within blind subjects, improved auditory and tactile performance might be due to either changes within auditory and somatosensory cortex (compensatory hypertrophy) or changes within visual cortex (cross-modal plasticity).

Compensatory Hypertrophy

Blind Individuals Show Compensatory Hypertrophy within Both Somatosensory and Auditory Cortices

There is a significant amount of evidence for compensatory hypertrophy in blind subjects. For example, stimulation of the reading finger of blind Braille readers results in expanded areas of scalp-recorded somatosensory-evoked potentials [44] and in changes within the representation of reading fingers within somatosensory cortex, measured using magnetic source imaging [45, 46]. Analogous topographic changes have been noted in animals trained with tactile discrimination tasks [1, 2]. These changes include a doubling of the size of the hand representation in blind subjects who read with three fingers, as well as a topographic disorganization that seems to result in perceptual "smearing" or uncertainty about which finger was stimulated. This "smearing" of information across fingers may be due to the need to fuse input transmitted over different fingers, so that the incoming information can be processed as a whole [46]. This result provides the first evidence to demonstrate that dramatic reorganization of an area has the potential to undermine the ability of the reorganized region of cortex to perform its primary functions. Presumably, the improved ability to integrate information across the three fingers while reading Braille outweighs any functional disadvantages resulting from deterioration in the ability to determine which finger was stimulated.

Analogous changes have also been found in the auditory cortex of blind subjects using magnetic source imaging to measure responses within primary and secondary auditory areas for tone bursts. Regions responding to auditory stimulation, measured using tonotopic mapping [47], were expanded by almost a factor of 2. N1m latencies were also considerably shorter in blind individuals.

Compensatory Hypertrophy in Deaf Individuals

As described above, deaf subjects perform better at a variety of visual tasks. Might this better performance be mediated by changes within visual cortex? Two independent studies that measured the size of early retinotopic areas using fMRI have failed to find any difference in either the overall size of these visual areas or the relative amounts of cortex allocated to the fovea and the periphery (Dougherty and Wandell, personal communication, [48]). It seems that there are no major differences in the size or responsivity of primary visual areas (V1–V4) between deaf and hearing subjects.

There is some evidence that attentional effects on fMRI responses to motion within MT+ (the cortical region associated with motion processing) may be slightly different in deaf subjects. Bavelier et al. found larger attentional modulation in the visual cortex of deaf than hearing subjects, and also reported an asymmetry in the extent of activity in area MT+ for deaf and hearing subjects [34, 49]. In these experiments the stimulus consisted of a full field of moving dots, and subjects were asked to attend to an annulus within the center or the periphery of the dot field. While these effects were significant, they were not particularly large, and a recent study of ours that examined lateralization of motion processing in MT+ using a similar stimulus and task [48] failed to replicate that finding. One possible explanation for the difference between these two studies may be that deaf subjects are different in their ability to allocate attention to the periphery in the presence of competing stimuli, as there were no distracting dots in the Fine et al. study (see [48] for a discussion). However, in any case, both studies suggest that changes in early areas of visual cortex as a consequence of auditory deprivation may be fairly limited in scope and magnitude.

The effects of deafness seem to be greater within higher-level visual areas. For example, a recent fMRI study has demonstrated that the higher-level areas of the brain involved in recognizing the emotional and linguistic content of faces (in the superior temporal sulcus and the fusiform gyrus) are reorganized in deaf signers as compared to hearing non-signers [50], consistent with the behavioral observation that sign language results in an enriched processing of faces [43]. Differences between deaf and hearing subjects have also been observed within posterior parietal cortex for attentional tasks with moving dot stimuli [49].

A second reason why higher visual areas might show larger amounts of cortical reorganization than primary sensory areas is that higher visual areas seem to normally show more cross-modal responsivity than lower visual areas. For example, attentional modulation of responses to visual stimuli by auditory information has been shown to increase across the visual hierarchy (Ciaramitaro and Boynton, personal communication). It is plausible that areas which normally show significant cross-modal modulation can more easily be biased toward the remaining senses [51, 52]. If neurons in higher-level visual areas already have access to auditory information, becoming purely visual would involve biasing their selectivity rather than responding to an entirely novel input modality.

There are therefore two reasons to expect greater reorganization as a function of deafness in higher-level than in lower-level visual areas. First, neuronal responses in higher visual areas seem to be more adaptable by experience than in lower sensory areas and second, neuronal responses in high-level visual areas are influenced by auditory information, even in sighted individuals.

Cross-Modal Plasticity

Blind Individuals Show Cross-Modal Plasticity within Visual Cortex

Both auditory and tactile tasks seem to activate visual cortex in blind individuals. For example, Sadato et al. [53–55] have found that early visual areas are activated in early blind subjects for Braille reading and tactile discrimination tasks (though not passive stimulation). Disruption of occipital activation due to a stroke [56] or transcranial magnetic stimulation (TMS, a coil on the surface of the head emits a short electromagnetic pulse that temporarily disrupts processing in cortical areas local to the coil) disrupts Braille reading, suggesting that these occipital responses play a functional role [57].

It is not clear whether these cross-modal responses are dependent on loss of visual input early in development. There are data showing that cross-modal responses are larger in early and congenitally blind subjects. For example, responses within occipital cortex to Braille are larger in congenital and early blind than in late blind subjects [53, 57]. Moreover, in a study examining the disruption of Braille reading TMS, only congenital and early blind subjects were affected, suggesting that the development of functionally useful cross-modal responses has a critical period [57].

On the other hand, a recent study found that learning Braille in sighted subjects was easier when subjects were temporarily blindfolded for 5 days, and that the blindfolding resulted in occipital responses to Braille measured using fMRI [58]. This last study suggests that some potential for cross-modal responses within visual cortex may exist even in sighted individuals with a normal visual history, and that these responses can be "unmasked" with a relatively short period of visual deprivation.

The advantages of cross-modal plasticity may go beyond simply colonizing unused regions of cortex. For example, cross-modal plasticity in Braille may be important in allowing tactile Braille to take advantage of areas such as the "visual word" area – a region within the midfusiform gyrus that has been shown to be selectively responsive to visual graphemes and is thought to play an important role in reading and lexical processing [59].

The anatomical substrates of these cross-modal influences on visual cortex are still to be determined. One possible substrate is direct monosynaptic projections from primary auditory to primary visual cortex. Direct monosynaptic projections from primary auditory cortex to both V1 and V2 have been found in adult monkeys. These auditory projections are denser in V2 than V1, and are found in portions of V1 and V2 representing the more peripheral and lower visual field [60, 61]. Alternatively responses could be due to feedback projections from higher-level visual or multimodal areas within parietal and prefrontal cortex [61]. These cross-modal connections exist in adult animals with normal sensory experience, and it is plausible to assume that these connections are likely to be more pronounced in dark reared animals and humans deprived from vision at an early age.

This evidence for cross-modal plasticity has serious potential implications for prosthetic vision. For example, it is possible that implanting a prosthetic device might interfere with cross-modal processing. If so, patients implanted with a prosthetic might possibly find that their ability to navigate with a cane or read Braille was disrupted by the implant. This is discussed further below.

Cross-Modal Plasticity in Deaf Individuals

The majority of studies of cross-modal plasticity in deaf subjects have used visual stimuli. This is because sign languages are visual in nature, making it plausible that cross-modal effects will be more powerful for visual than for tactile stimuli.

However, tactile cross-modal responses may exist: one magnetoencephalography study demonstrated tactile responses within auditory cortex of a congenitally deaf subject [62] and these cross-modal tactile responses do seem to have a behavioral correlate [63].

Cross-modal plasticity for visual stimuli has been found by our group within the primary auditory cortex of deaf individuals [64, 65] using non-linguistic moving dot stimuli. Hearing children of deaf adults did not show cross-modal plasticity, suggesting that lack of normal auditory input (rather than extensive experience with a visual sign language) was necessary for cross-modal plasticity to occur [48]. Significant cross-modal responses were only found within right auditory cortex. This region appeared to include primary, secondary, and association auditory areas (Brodmann's areas 41, 42 and 22; [66–68]). The amount of activity within the region that showed cross-modal plasticity was approximately a third of the activity produced in this same region for music in hearing subjects. This recruitment of auditory cortex for visual function may explain why despite significant changes in white matter, deaf subjects do not show a reduction in gray matter volume within primary auditory cortex [69]. One obvious advantage of this cross-modal plasticity may be to allow visual sign language to take advantage of connections between auditory cortex and neighboring semantic and language processing areas.

Somewhat surprisingly, cross-modal plasticity seems to play a much more important role in compensating for deafness than compensatory hypertrophy. As described above, there is little evidence for changes within visual cortex as a consequence of deafness, whereas there are significant changes within auditory cortex [48]. One possible explanation, as discussed above, is that visual cortex only has a limited potential to reorganize; since drastic reorganization would have the potential to seriously undermine the ability of visual cortex to perform its primary functions. In the case of visual cortex, any disruption of the ability to perform everyday visual tasks could easily be catastrophic, consequently the scope for plasticity within visual cortex as a consequence of deafness may be limited.

The anatomical route for cross-modal connections to auditory cortex is still a matter of speculation. As in the case of cross-modal visual responses, there are several possible anatomical routes, and it is of course possible that more than one anatomical pathway may be involved. There are exuberant projections during infancy from the visual thalamus to the primary auditory cortex [70, 71]. Normally, these visual projections to the auditory cortex get pruned away over the course of development. In the case of auditory deprivation, however, the lack of functional auditory input to the auditory cortex may result in the stabilization of visual input to this area [72]. Recently, Pallas et al. [20] have demonstrated direct projections from the visual thalamus to the primary auditory cortex in early deafened ferrets, leaving open the possibility of such projections in deaf individuals (see Figure 3.1).

A second possible source of visual responses in the auditory cortex is from visual cortical areas such as V1 or V2. Projections from visual areas V1 and V2 to the auditory cortex have recently been reported in adult monkeys [73]. It is plausible that these projections might be more pervasive in infancy, and would fail to be pruned in the course of development in deaf children.

Finally, these cross-modal responses may be the result of feedback from cortical areas responsible for processing language. At this point a large number of studies have demonstrated responses to visual stimuli that are linguistically meaningful within auditory regions of deaf subjects [74–76] For example, Brodmann's areas 42 and 22 are activated in deaf subjects in response to visual images of sign language. Some of the areas showing visual activation in deaf subjects may normally play an analogous role in visual language processing in hearing subjects. For example, similar regions of auditory cortex are activated by a silent lip-reading task in hearing subjects as are activated by sign language in deaf subjects [52].

Molyneaux's Question: The Role of Experience in Maintaining Sensory Function

Visual Processing Deteriorates in the Early Blind

"Suppose a man born blind, and now adult, and taught by his touch to distinguish between a cube and a sphere … Suppose then … the blind man made to see … Query: whether by his sight, before he touched them, he could distinguish and tell which is the globe, which is the cube?"

(Locke, Essay Concerning Human Understanding, part of a correspondence between Locke and a lawyer, Molyneaux).

What is the effect of cross-modal plasticity on the ability to make use of a sense if it is ever restored? One possibility is that cross-modal plasticity maintains neuronal function and connectivity within regions of cortex that might otherwise be susceptible to deterioration and cell death. According to this model, cross-modal plasticity would facilitate later recovery of vision. Alternatively, the rewiring involved in cross-modal plasticity, especially when it occurs early in development, may permanently limit the ability to regain useful sensory function.

Despite the philosophical and psychological interest of Molyneaux's question, cases of adult sight restoration are so rare that even now little is known about perceptual experience after long-term visual deprivation. Although the first report of recovery from blindness was in AD 1020 [77] and the first clinical study of sight recovery after long-term blindness was carried out in 1728 [78], until recently only sporadic studies have been carried out [79–85]. While these studies provide an interesting body of observations, most studies only had access to small numbers of observers (often only one), and generally only had crude measures of preoperative visual abilities. Studies were often anecdotal and qualitative, focusing on patients' ability to use their sight functionally and on their visual acuity, rather than on the neural changes consequent on visual deprivation. In addition, many patients only suffered partial deprivation. For example, Gregory and Wallace's famous patient, SB, could, in fact, count fingers preoperatively [79, 82], suggesting significant form vision.

A recent study of ours used a combination of psychophysical and neuroimaging techniques to characterize the effects of long-term visual deprivation on human cortex [86]. At three-and-a-half years old, MM lost one eye and was blinded in the other after chemical and thermal damage to his cornea. Limbal epithelium damage prevented successful replacement of MM's cornea for 40 years. As a result he had some light perception, but no experience of contrast or form. One unsuccessful corneal replacement was attempted in childhood, but he reported no visual memories or imagery. At the age of 43, MM received a corneal and limbal stem-cell transplant in his right eye.

MM's postoperative performance was tested on a wide array of tasks designed to crudely characterize form, motion, depth, object, and face processing. Stimuli presented to control observers were blurred using a filter designed to match MM's spatial resolution. MM had no discernable deficits in simple form tasks postoperatively (Figure 3.2). When first tested, 5 months after surgery, he perceived slight changes in the orientation of a bar and easily recognized simple shapes. He reported perceiving simple 2D shapes, even immediately after surgery.

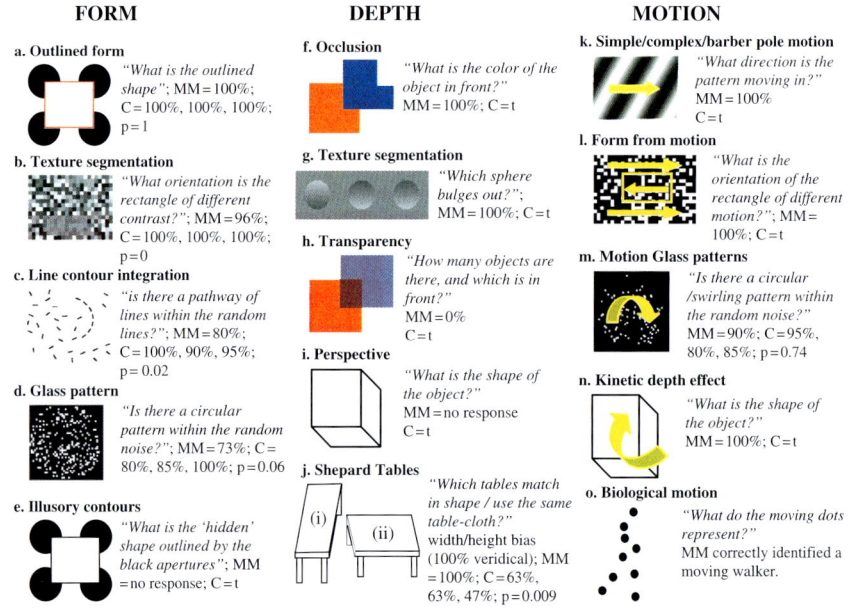

FIGURE 3.2. Stimuli, tasks, and performance for tests of MM's form, depth, and motion processing. Stimuli presented to control observers were always blurred using a low pass filter (cutoff of 1 c.p.d.) designed to match MM's spatial resolution losses. Some tasks were trivial for control observers (t) and were not formally tested. C – control observers. P values, 1-tailed t-tests [86]. Copyright 2003 by Nature Publishing Group. Adapted with permission.

Like most [82, 83], though not all [79], sight-recovery patients, MM identified colors easily (though he occasionally had difficulty remembering the correct color name), and his equiluminance settings were normal. The only patient who was thought to have had difficulty discriminating colors was Ackroyd et al.'s patient HB [79]. However, this may have been due to uncertainty about what property of the stimulus (brightness, hue, saturation) was referred to by the color name, since HB spoke English as a second language. Although HB was also blinded at the age of three, unlike MM, she had no experience of English color names before the onset of blindness. Differences in color processing that were observed between MM and visually normal observers were limited to color illusions based on seeing images as being 3D.

Similarly motion processing also appears to be relatively normal in sight recovery patients. "His [SB] only signs of appreciation were to moving objects, particularly the pigeons in Trafalgar square"[82]. Similarly, Ackroyd et al. (1974) [79] reported of their patient, "She could see the pigeons as they alighted in Trafalgar Square but she said that they appeared to vanish as they came to rest." MM's motion processing also appears normal: he easily identified the direction of motion of simple and complex plaids and was susceptible to the barber pole illusion. He could segregate textured fields based on motion, and could distinguish rotational Glass motion patterns (in which two successive frames differ by rotation) from random noise patterns. Surprisingly MM could use motion cues to compute 3D shape, a computation thought to involve MT+ [87–89]. This sensitivity to form from motion was in striking contrast to his insensitivity to most pictorial depth cues. For example, a stationary Necker cube immediately "popped-out" as a cube when motion in depth was simulated. MM was also sensitive to biological motion, recognizing a point-light "Johansson" biological motion figure and even being able to determine whether the figure walked with a male or female gait. Consistent with MM's ability to perform motion tasks, when we measured fMRI responses in MM, we found that responses in MT+ were normal in area and amplitude (surface area: left MT+, MM $= 9.23\,\text{cm}^2$, Controls $= 7.06\,\text{cm}^2$, $9.14\,\text{cm}^2$, $7.78\,\text{cm}^2$; right MT+, MM $= 7.6\,\text{cm}^2$, Controls $= 6.46\,\text{cm}^2$, $6.47\,\text{cm}^2$, $9.61\,\text{cm}^2$). There was also evidence of retinotopic phase map encoding in MT/MST [90].

MM could segment texture patterns based on luminance contrast but was slightly worse than control observers at other form tasks requiring integration of texture elements, such as identifying whether a field of line contours contained a sequence of nearly collinear line segments [91] and discriminating Glass patterns from random noise [92–94]. MM also had severe difficulty with "subjective contours"; though he recognized outlined 2D shapes, he could not identify the same shapes in "Kanisza figures".

The literature on sight recovery seems to be in fairly good agreement that patients have difficulties projecting their 2D retinal image into a 3D world. Cheselden reported of his patient that he "thought no Objects so agreeable as those which were smooth and regular, though he could form no Judgment of their Shape". It was reported of Sacks' patient Virgil, "Sometimes surfaces of objects would seem to loom ... when they were still quite a distance away; sometimes he would get confused by his own shadow ... [Steps] posed a particular hazard. All he could see was a confusion, a flat surface of parallel and crisscrossing lines". Though Gregory and Wallace's patient, SB, seems to have been somewhat better at object recognition than many earlier patients (possibly due to his better preoperative acuity), he nonetheless was immune to many illusions based on perspective. One of Valvo's [84] patients, HS, described his initial experiences after sight recovery: "I had no appreciation of depth or distance; street lights were luminous stains stuck to window panes, and the corridors of the hospital were black holes" (It is worth noting that this patient, HS, was not deprived of sight until the age of 15.) Similarly, MM had difficulty constructing a 3D interpretation from his retinal image. He could exploit occlusion cues but not shading, transparency, or perspective. For example, he could not identify wire drawings of stationary Necker cubes or pyramids (regardless of their 3D orientation), describing them respectively as "a square with lines" and "a triangle with lines". MM was immune to illusions based on perspective cues such as Roger Shepard's "tables"[95], correctly choosing quadrilaterals of equal aspect ratios as being rotated versions of each other. Normal observers tend to (mistakenly) choose very different aspect ratios of the tables even when asked to match the 2D image shapes.

Possibly as a consequence of these difficulties with constructing a 3D percept, sight recovery patients also consistently show difficulties recognizing even familiar objects and faces [77]. Sacks observed of Virgil, "[his] cat and dog bounded in ... and Virgil, we noted, had some difficulty telling which was which". One of the earliest recorded cases [78] describes the same confusion, "Having often forgot which was the cat and which the dog, he was ashamed to ask, but catching the cat which he knew from feeling, he was observed to look at her steadfastly and then ... have said, So puss, I shall know you another time". Similar difficulties have been reported in the case of sight recovery that we studied [86]. MM identified only 25% of common objects [96], and he had difficulty judging gender or expression in faces. Like observers with prosopagnosia [97, 98], MM seems to rely entirely on individual features, such as the hair length or eyebrow shape. Other sight-recovery cases seem to have suffered from similar difficulties [78, 82], and short-term binocular deprivation in infancy seems to result in analogous (though less striking) behavioral deficits [98, 99]. Face and object images evoked little fMRI activation within MM's lingual and fusiform gyri, whereas controls showed strong responses in these areas. These data were the first to show deficits in lingual and fusiform areas as a result of abnormal visual experience rather than as a result of neurophysiological damage.

MM showed no improvement in basic visual functions over more than 2 years. Admittedly, no rehabilitation was carried out. However, other studies have made efforts to improve vision after sight restoration, with very limited success [80, 81, 84, 100]. Both the motivation and intelligence of the patient seems to be critical in making cognitive use of the confusing visual world of the sight restoration patient. However, to date there is little or no evidence suggesting that the effects of early deprivation on early visual cortex are anything other than permanent after adolescence [86, 100].

These results clearly have important implications for sight restoration in the early or congenitally blind. Even if it were possible to implant a "perfect" prosthesis – i.e. one that would result in perfect vision within a normal visual system – it is still the case that within early blind patients this would result in only limited useful vision.

Deterioration within Auditory Cortex in Deaf Individuals

This same process of deterioration also seems to occur after early deafness. The success of cochlear implants depends greatly upon the age of implantation. A variety of studies find that intervention during the first 6 months of life (whether with a hearing aid or a cochlear implant) significantly increases the level of language development, speech intelligibility, and emotional stability compared with children with later diagnosis and intervention. Even though the critical period for language development is thought to last until 7 years of age, it has generally been found that infants implanted at a very early age (younger than 18 months) have better outcomes than children implanted later in life [101–103]. Neurophysiological markers show a similar pattern: in children implanted after the age of 7 years, the cortical auditory evoked potential is aberrant and shows slower latencies as compared to children implanted before the age of 3 years [104].

There is some evidence that this deterioration in the ability to process auditory input may be due to cross-modal plasticity. Lee et al. [105] found that the resting metabolic rate within auditory areas was inversely related to years of deafness. Children who had been deaf for a relatively short period of time showed a large drop in the resting metabolic rate within auditory cortex. In contrast, young deaf adults who had been deaf for many years had a resting metabolic rate within auditory cortex that was close to normal levels. Performance on speech perception after cochlear implantation was inversely related to the metabolic rate pre-implantation. Speech processing outcomes were worst in those patients who showed near-normal resting metabolic activity in auditory cortex. This is consistent with the hypothesis that the lack of a reduction in metabolic rate may be due to cross-modal plasticity, which in turn interferes with the ability to recover auditory processing abilities post-implantation.

Interestingly, the "critical period" for auditory development seems to depend on the task used to assess post-implantation performance. Better performance on simple sound identification within a closed set of alternatives seems to depend on whether implantation occurred before the age of 4 1/2 years. More difficult tasks involving identifying a sound from an open set of alternatives depends on whether implantation occurred before the ages of 5 1/2–8 1/2 years [106]. This is consistent with work examining the effects of visual deprivation which suggests that different regions of visual cortex vary in their developmental time courses, with higher-level visual areas continuing to retain plasticity at later ages [17, 86].

In postlingually deaf adults, it has been found that performance recovering speech perception is inversely related to how well the cochlear implant emulates the cortical input that had been provided by the lost cochlear mechanisms, and to the duration of deafness pre-implantation [4]. This impact of deafness duration in adulthood demonstrates that in the case of auditory cortex deterioration in function as a result of deprivation is not limited to the effects of early deafness [106].

Implications for Sensory Prostheses and Rehabilitation

As described above, plasticity as a result of sensory deprivation depends on a variety of factors. One important factor is age of deprivation. Early blind and early deaf subjects tend to be much more fluent at skills such as Braille, use of a cane, or sign language. One reason for this may be that these skills are easier to learn at an early age, when cortex is more plastic. The greater skills of the early deprived may also be partially due to behavioral/cultural factors – early blind and deaf subjects tend to rely on Braille and sign language much more heavily than those deprived at a later age. What is most likely is that these two factors may reinforce each other: early blind and deaf subjects learn sensory substitution skills more easily; this makes them rely on these skills more heavily, giving them practice and thereby further improving those skills. In those that are deprived later in life this "circle of competency" may be much more difficult to initiate.

It has already been shown that patients implanted with cochlear implants can be very successful at learning to make use of limited and abnormal auditory input, but for many patients this process takes months or years. There are also large individual differences in development of auditory proficiency over time. In the case of visual prosthetics, it will be interesting to see whether a similar learning process will result in improvements in visual performance over the first months and years of implantation. It will also be important to develop rehabilitation paradigms that maximize the speed and extent of learning.

We described above how plasticity increases across the visual hierarchy. Our growing understanding of these differences in plasticity across sensory areas has important implications for choosing effective strategies for training prosthetic patients. For many years, the sensory input provided by prostheses will differ substantially from normal sensory input. The advantage of targeting low-level stages of processing is that any rehabilitative learning will presumably mediate a general improvement in performance, since neurons in primary areas feed to all stages of sensory processing. Such low-level stages of processing would presumably be modified by simple tasks such as tactile or auditory discrimination tasks. However, much of the evidence described above suggests that the effects of perceptual learning may be very limited for these simple tasks, especially in the case of visual cortex.

Alternatively, practicing higher-level tasks may result in faster and larger plasticity effects, but these learning effects would be expected to show much greater specificity for the task and stimulus that subjects were trained on. One possibility is that training may benefit from a mixed strategy whereby time and attention is divided between key aspects of low-level processing (which would be expected to show limited improvements that would apply to a wide range of visual tasks) and higher-level tasks that are functionally important (which might show greater learning effects that are more specific in their scope). The nature of rehabilitation training might also depend on patient age. In the case of early deprived subjects especially those in whom sight is also restored early in life, a low-level strategy is more likely to be beneficial than for those deprived later in life or those in whom sight is restored later in life.

The possible advantages of developing coherent strategies of mixed training are not limited to rehabilitation after implantation with a prosthetic device. There are a wide variety of rehabilitation paradigms that might take advantage of such an approach, including learning sensory substitution skills, strategies for dealing with low vision (such as the development of a preferred retinal location in macular degeneration or training to reduce the effects of amblyopia), or rehabilitative training after neural damage (e.g. stroke).

Normally the cortical changes that occur as a result of sensory deprivation are of benefit to the individual, playing an important role in allowing him or her to make better use of his or her remaining senses. However, as described above, if there is an opportunity for the missing sense to be restored by a prosthesis (or some other type of restorative surgery), then these adaptations become maladaptive, and can seriously limit an individual's ability to make use of the restored sense.

This is particularly troubling since restoring the missing sense can also have the unintended consequence of interfering with cross-modal abilities. The decision to implant a prosthetic device or carry out a restorative surgery needs to consider not only the potential loss to any residual vision or hearing but also the potential of the restorative intervention to interfere with cross-modal processing.

This leaves the field of prosthetics and "cutting edge" restorative surgeries with a dilemma. Many of the devices (or restoration procedures) currently available or potentially available in the near future (cortical auditory implants, visual retinal implants, epithelial stem cell replacements, retinal transplants) offer either limited sensory restoration or have a high probability of tissue rejection or some other type of failure at some point in the individual's lifetime. As described earlier in this chapter, there are clear neurophysiological advantages to providing a prosthetic as early as possible. However, if implanted early in life, a prosthetic device may interfere with the ability to learn cross-modal skills, such as being able to navigate with a cane or read Braille. There are three possible ways in which a prosthetic device (or some other sensory restoration procedure) can interfere with the development of cross-modal plasticity.

First, implantation may limit the degree to which sensory substitution skills are practiced and encouraged, and thereby impede establishment of a "circle of competency" as described above. Hearing children of deaf signers often speak sign language extremely fluently, suggesting that there is no neurophysiological reason for it to be difficult to be fluent in both languages. Nonetheless, a child implanted with a cochlear implant, especially one born to hearing parents, will tend to associate with hearing rather than deaf children in school and play, and will often be encouraged to rely on spoken English rather than sign language. As a consequence, children with cochlear implants are less likely to learn fluent sign language. For those deaf children that learn fluent spoken English, this is no bad thing. But for the minority of children that never become fluent in spoken English, this strategy can result in a failure to become fluent in any language. When cochlear implants first came on the market, the question of whether the improvement in auditory function was worth the potential loss of cross-modal abilities became a hotly debated issue. We can expect to see analogous discussions in the blind community in the future as visual prosthetics begin to reach the marketplace.

Second, prosthetics may directly compete for cortical resources. For example, Kauffman et al. [107] have demonstrated that TMS to early visual cortex disrupts Braille reading in blind, and that sighted subjects who are (over 5 days) temporarily deprived of vision and taught Braille begin to show responses to Braille within visual cortex. Subjects who were taught Braille but were not deprived of vision were less fluent with Braille at the end of the 5-day period, and showed no responses in visual cortex. Interestingly, sighted teachers of Braille tend not to read by touch, but instead read Braille "visually" by looking at the shadows cast by the paper indentations. These data, while not conclusive, do suggest that lack of visual input is necessary for visual cortex to respond to Braille, and that this cross-modal plasticity plays an important functional role.

As well as considering the possible impact of a prosthetic on the ability to develop cross-modal skills, it is also important, when evaluating outcomes in adult patients, to consider possible deleterious effects on existing cross-modal skills. Given that the ability to make use of a restored sense may be limited in adulthood, any deterioration in a patients' ability to navigate with a cane, read Braille, or understand sign language can have serious consequences. Take for example, the description of SB's ability to deal with crossing the road after a sight recovery operation [82], "He found the traffic frightening, and would not attempt to cross even a comparatively small street by himself. This was in marked contrast to his former behavior, as described by his wife, when he would cross any street in his own town by himself. In London, and later in his home town, he would show evident fear, even when led by a companion whom he trusted, and it was many months before he would venture alone." Clearly new visual information interfered with SB's ability to use cross-modal skills that he had relied on very effectively pre-operatively.

Currently the default assumption is that the decision to implant a prosthetic device (or carry out some other type of restorative surgery) should be based on the potential to restore useful vision or audition, as compared to the risk of losing any residual sensory function and the discomfort and risks of the operation. The data described above shows that cross-modal plasticity is likely to play a very significant role in compensating for sensory loss. This suggests that the possibility of a detrimental effect on skills mediated by cross-modal activity should also be borne in mind when considering sensory restoration procedures.

Acknowledgments. Research supported by the National Institute of Health, EY014645 and a Career Development award from Research to Prevent Blindness.

References

1. Recanzone, G.H., Jenkins, W.M., Hradek, G.T., and Merzenich, M.M. (1992a). Progressive improvement in discriminative abilities in adult owl monkeys performing a tactile frequency discrimination task. *J Neurophysiol, 67* (5), 1015–1030.
2. Recanzone, G.H., Merzenich, M.M., Jenkins, W.M., Grajski, K.A., and Dinse, H.R. (1992b). Topographic reorganization of the hand representation in cortical area 3b owl monkeys trained in a frequency-discrimination task. *J Neurophysiol, 67* (5), 1031–1056.
3. Recanzone, G.H., Schreiner, C.E., and Merzenich, M.M. (1993). Plasticity in the frequency representation of primary auditory cortex following discrimination training in adult owl monkeys. *J Neurosci 1993 Jan;13(1 87-103), 13* (1), 87–103.
4. Oh, S.H., Kim, C.S., Kang, E.J., Lee, D.S., Lee, H.J., Chang, S.O., Ahn, S.H., Hwang, C.H., Park, H.J., and Koo, J.W. (2003). Speech perception after cochlear implantation over a 4-year time period. *Acta Otolaryngol, 123* (2), 148–153.
5. Ghose, G.M., Yang, T., and Maunsell, J.H. (2002). Physiological correlates of perceptual learning in monkey V1 and V2. *J Neurophysiol, 87* (4), 1867–1888.
6. Schoups, A., Vogels, R., Qian, N., and Orban, G. (2001). Practising orientation identification improves orientation coding in V1 neurons. *Nature, 412* (6846), 549–553.
7. Furmanski, C.S., Schluppeck, D., and Engel, S.A. (2004). Learning strengthens the response of primary visual cortex to simple patterns. *Curr Biol, 14* (7), 573–578.
8. Gilbert, C.D., and Wiesel, T.N. (1992). Receptive field dynamics in adult primary visual cortex. *Nature, 356* (6365), 150–152.
9. Kaas, J.H., Krubitzer, L.A., Chino, Y.M., Langston, A.L., Polley, E.H., and Blair, N. (1990). Reorganization of retinotopic cortical maps in adult mammals after lesions of the retina. *Science, 248* (4952), 229–231.
10. Smirnakis, S.M., Brewer, A.A., Schmid, M.C., Tolias, A.S., Schuz, A., Augath, M., Inhoffen, W., Wandell, B.A., and Logothetis, N.K. (2005). Lack of long-term cortical reorganization after macaque retinal lesions. *Nature, 435* (7040), 300–307.
11. Baker, C.I., Peli, E., Knouf, N., & Kanwisher, N.G. (2005). Reorganization of visual processing in macular degeneration. *J Neurosci, 25* (3), 614–618.
12. Yang, T., and Maunsell, J.H. (2004). The effect of perceptual learning on neuronal responses in monkey visual area V4. *J Neurosci, 24* (7), 1617–1626.
13. Desimone, R., Albright, T.D., Gross, C.G., and Bruce, C. (1984). Stimulus-selective properties of inferior temporal neurons in the macaque. *J Neurosci, 4* (8), 2051–2062.
14. Logothetis, N.K., Pauls, J., and Poggio, T. (1995). Shape representation in the inferior temporal cortex of monkeys. *Curr Biol, 5* (5), 552–563.
15. Rolls, E.T., and Tovee, M.J. (1995). Sparseness of the neuronal representation of stimuli in the primate temporal visual cortex. *J Neurophysiol, 73* (2), 713–726.

16. Young, M.P., and Yamane, S. (1992). Sparse population coding of faces in the inferotemporal cortex. *Science, 256* (5061), 1327–1331.

17. Fine, I., and Jacobs, R.A. (2002). Comparing perceptual learning across tasks: A review. *Journal of Vision, 2* (5), 190–203.

18. Amedi, A., Raz, N., Pianka, P., Malach, R., and Zohary, E. (2003). Early 'visual' cortex activation correlates with superior verbal memory performance in the blind. *Nat Neurosci, 6* (7), 758–766.

19. Hunt, D.L., King, B., Kahn, D.M., Yamoah, E.N., Shull, G.E., and Krubitzer, L. (2005). Aberrant retinal projections in congenitally deaf mice: How are phenotypic characteristics specified in development and evolution? *Anat Rec A Discov Mol Cell Evol Biol, 287* (1), 1051–1066.

20. Pallas, S.L., Razak, K.A., and Moore, D.R. (2002). Cross-modal projections from LGN to primary auditory cortex following perinatal cochlear ablation in ferrets. *Society for Neuroscience* (Orlando, Florida).

21. Rauschecker, J.P. (1995). Compensatory plasticity and sensory substitution in the cerebral cortex. *Trends Neurosci, 18*, 36–43.

22. Sur, M., Pallas, S.L., and Roe, A.W. (1990). Cross-modal plasticity in cortical development: differentiation and specification of sensory neocortex. *Trends Neurosci, 13* (13), 227–233.

23. Lessard, N., Pare, M., Lepore, F., and Lassonde, M. (1998). Early-blind human subjects localize sound sources better than sighted subjects. *Nature, 395* (6699), 278–280.

24. Roder, B. (1999). Improved auditory spatial tuning in blind humans. *Nature 400, 400*, 162–166.

25. Voss, P., Lassonde, M., Gougoux, F., Fortin, M., Guillemot, J.P., and Lepore, F. (2004). Early- and late-onset blind individuals show supra-normal auditory abilities in far-space. *Curr Biol, 14* (19), 1734–1738.

26. Doucet, M.E., Guillemot, J.P., Lassonde, M., Gagne, J.P., Leclerc, C., and Lepore, F. (2005). Blind subjects process auditory spectral cues more efficiently than sighted individuals. *Exp Brain Res, 160* (2), 194–202.

27. Gougoux, F., Lepore, F., Lassonde, M., Voss, P., Zatorre, R.J., and Belin, P. (2004). Neuropsychology: pitch discrimination in the early blind. *Nature, 430* (6997), 309.

28. Grant, A.C., Thiagarajah, M.C., and Sathian, K. (2000). Tactile perception in blind Braille readers: a psychophysical study of acuity and hyperacuity using gratings and dot patterns. *Percept Psychophys, 62* (2), 301–312.

29. Van Boven, R.W., Hamilton, R.H., Kauffman, T., Keenan, J.P., and Pascual-Leone, A. (2000). Tactile spatial resolution in blind Braille readers. *Neurology, 54*, 2230–2236.

30. Bosworth, R.G., and Dobkins, K.R. (1997). Visual field asymmetries for motion processing in hearing vs. deaf subjects. *Investigative Ophthalmology and Visual Science*

31. Bosworth, R.G., and Dobkins, K.R. (1999). Left hemisphere dominance for motion processing in congenitally deaf individuals. *Psychological Science, 10*, 256–262.

32. Bosworth, R.G., and Dobkins, K.R. (1999a). Left-hemisphere dominance for motion processing in deaf signers. *Psychological Science, 10* (3), 256–262.

33. Bosworth, R.G., and Dobkins, K.R. (2002b). Visual field asymmetries for motion processing in deaf and hearing signers. *Brain Cogn, 49* (1), 170–181.

34. Bavelier, D., Tomann, A., Hutton, C., Mitchell, T., Corina, D., Liu, G., and Neville, H. (2000). Visual attention to the periphery is enhanced in congenitally deaf individuals. *J Neurosci, 20* (17), RC93.

35. Bosworth, R.G., and Dobkins, K.R. (2002a). The effects of spatial attention on motion processing in deaf signers, hearing signers, and hearing nonsigners. *Brain Cogn, 49* (1), 152–169.

36. Loke, W.H., and Song, S. (1991). Central and peripheral visual processing in hearing and nonhearing individuals. *Bulletin of the Psychonomic Society, 29*, 437–440.

37. Neville, H.J., and Lawson, D. (1987). Attention to central and peripheral visual space in a movement detection task: an event-related potential and behavioral study. II. Congenitally deaf adults. *Brain Res, 405* (2), 268–283.

38. Parasnis, I., and Samar, V.J. (1985). Parafoveal attention in congenitally deaf and hearing young adults. *Brain Cogn, 4* (3), 313–327.

39. Proksch, J., and Bavelier, D. (2002). Changes in the spatial distribution of visual attention after early deafness. *Journal of Cognitive Neuroscience, 14* (5), 687–701.

40. Rettenbach, R., Diller, G., and Sireteanu, R. (1999). Do deaf people see better? Texture segmentation and visual search compensate in adult but not in juvenile subjects. *J Cogn Neurosci, 11* (5), 560–583.

41. Arnold, P., and Murray, C. (1998). Memory for faces and objects by deaf and hearing signers and hearing nonsigners. *Journal of Psycholinguistic Research, 27*, 481–497.

42. Goldstein, N.S., Sexton, J., and Feldman, R.S. (2000). Encoding of facial expressions of emotion and knowledge of American Sign Language. *Journal of Applied Social Psychology, 30*, 67–76.

43. McCullough, S., and Emmorey, K. (1997). Face processing by deaf ASL signers: evidence for expertise in distinguishing local features. *Journal of Deaf Studies and Deaf Education, 2*, 212–222.

44. Pascual-Leone, A., and Torres, F. (1993). Plasticity of the sensorimotor cortex representation of the reading finger in Braille readers. *Brain, 116 (Pt 1)*, 39–52.

45. Sterr, A., Muller, M., Elbert, T., Rockstroh, B., and Taub, E. (1999). Development of cortical reorganization in the somatosensory cortex of adult Braille students. *Electroencephalogr Clin Neurophysiol Suppl, 49*, 292–298.

46. Sterr, A., Muller, M.M., Elbert, T., Rockstroh, B., Pantev, C., and Taub, E. (1998). Changed perceptions in Braille readers. *Nature, 391* (6663), 134–135.

47. Elbert, T., Sterr, A., Rockstroh, B., Pantev, C., Muller, M.M., and Taub, E. (2002). Expansion of the tonotopic area in the auditory cortex of the blind. *J Neurosci, 22* (22), 9941–9944.

48. Fine, I., Finney, E.M., Boynton, G.M., and Dobkins, K.R. (2005). Comparing the effects of auditory deprivation and sign language within the auditory and visual cortex. *J Cogn Neurosci, 17* (10), 1621–1637.

49. Bavelier, D., Brozinsky, C., Tomann, A., Mitchell, T., Neville, H., and Liu, G. (2001). Impact of early deafness and early exposure to sign language on the cerebral organization for motion processing. *J Neurosci, 21* (22), 8931–8942.

50. McCullough, S., Emmorey, K., and Sereno, M. (2005). Neural organization for recognition of grammatical and emotional facial expressions in deaf ASL signers and hearing nonsigners. *Brain Res Cogn Brain Res, 22* (2), 193–203.

51. Bavelier, D., and Neville, H.J. (2002). Cross-modal plasticity: where and how? *Nat Rev Neurosci, 3* (6), 443–452.

52. Calvert, G.A., Bullmore, E.T., Brammer, M.J., Campbell, R., Williams, S.C., McGuire, P.K., Woodruff, P.W., Iversen, S.D., and David, A.S. (1997). Activation of auditory cortex during silent lipreading. *Science, 276* (5312), 593–596.

53. Sadato, N., Okada, T., Honda, M., and Yonekura, Y. (2002). Critical period for cross-modal plasticity in blind humans: a functional MRI study. *Neuroimage, 16* (2), 389–400.

54. Sadato, N., Pascual-Leone, A., Grafman, J., Deiber, M.P., Ibanez, V., and Hallett, M. (1998). Neural networks for Braille reading by the blind. *Brain, 121 (Pt 7)*, 1213–1229.

55. Sadato, N., Pascual-Leone, A., Grafman, J., Ibanez, V., Deiber, M.P., Dold, G., and Hallett, M. (1996). Activation of the primary visual cortex by Braille reading in blind subjects. *Nature, 380* (6574), 526–528.

56. Hamilton, R., Keenan, J.P., Catala, M., and Pascual-Leone, A. (2000). Alexia for Braille following bilateral occipital stroke in an early blind woman. *Neuroreport, 11*, 237–240.

57. Cohen, L.G., Weeks, R.A., Sadato, N., Celnik, P., Ishii, K., and Hallett, M. (1999). Period of susceptibility for cross-modal plasticity in the blind. *Ann Neurol, 45* (4), 451–460.

58. Pascual-Leone, A., and Hamilton, R. (2001). The metamodal organization of the brain. In: C. Casanova, and M. Ptito (Eds.), *Progress in Brain Research*, 134.

59. Hillis, A.E., Newhart, M., Heidler, J., Barker, P., Herskovits, E., and Degaonkar, M. (2005). The roles of the "visual word form area" in reading. *Neuroimage, 24* (2), 548–559.

60. Falchier, A., Clavagnier, S., Barone, P., and Kennedy, H. (2002). Anatomical evidence of multimodal integration in primate striate cortex. *J Neurosci, 22* (13), 5749–5759.

61. Rockland, K.S., and Ojima, H. (2003). Multisensory convergence in calcarine visual areas in macaque monkey. *Int J Psychophysiol, 50* (1–2), 19–26.

62. Levanen, S., Jousmaki, V., and Hari, R. (1998). Vibration-induced auditory-cortex activation in a congenitally deaf adult. *Curr Biol, 8* (15), 869–872.

63. Levanen, S., Uutela, K., Salenius, S., and Hari, R. (2001). Cortical representation of sign language: comparison of deaf signers and hearing non-signers. *Cereb Cortex, 11* (6), 506–512.

64. Finney, E.M., Clementz, B.A., Hickok, G., and Dobkins, K.R. (2003). Visual stimuli activate auditory cortex in deaf subjects: evidence from MEG. *Neuroreport, 14* (11), 1425–1427.

65. Finney, E.M., Fine, I., and Dobkins, K.R. (2001). Visual stimuli activate auditory cortex in the deaf. *Nat Neurosci, 4* (12), 1171–1173.

66. Penhune, V.B., Zatorre, R.J., MacDonald, J.D., and Evans, A.C. (1996). Inter-hemispheric anatomical differences in human primary auditory cortex: probabilistic mapping and volume measurement from magnetic resonance scans. *Cereb Cortex, 6* (5), 661–672.

67. Rademacher, J., Morosan, P., Schormann, T., Schleicher, A., Werner, C., Freund, H.J., and Zilles, K. (2001). Probabilistic mapping and volume measurement of human primary auditory cortex. *Neuroimage, 13* (4), 669–683.

68. Westbury, C.F., Zatorre, R.J., and Evans, A.C. (1999). Quantifying variability in the planum temporale: a probability map. *Cereb Cortex, 9* (4), 392–405.

69. Emmorey, K., Allen, J.S., Bruss, J., Schenker, N., and Damasio, H. (2003). Morpho-metric analysis of auditory brain regions in congenitally deaf adults. *Proc Natl Acad Sci U S A, 100* (17), 10049–10054.

70. Catalano, S.M., and Shatz, C.J. (1998). Activity-dependent cortical target selection by thalamic axons. *Science, 281* (5376), 559–562.

71. Ghosh, A., and Shatz, C.J. (1992). Pathfinding and target selection by developing geniculocortical axons. *J Neurosci, 12* (1), 39–55.
72. Sur, M., Angelucci, A., and Sharma, J. (1999). Rewiring cortex: the role of patterned activity in development and plasticity of neocortical circuits. *J Neurobiol, 41* (1), 33–43.
73. Schroeder, C.E. (2004). Cooperative Processing of Multisensory Cues in Auditory Cortex and Classic Multisensory Regions. *International Multisensory Research Forum* (Barcelona, Spain).
74. MacSweeney, M., Woll, B., Campbell, R., McGuire, P.K., David, A.S., Williams, S.C., Suckling, J., Calvert, G.A., and Brammer, M.J. (2002). Neural systems underlying British Sign Language and audio-visual English processing in native users. *Brain, 125* (Pt 7), 1583–1593.
75. Nishimura, H., Hashikawa, K., Doi, K., Iwaki, T., Watanabe, Y., Kusuoka, H., Nishimura, T., and Kubo, T. (1999). Sign language 'heard' in the auditory cortex. *Nature, 397* (6715), 116.
76. Petitto, L.A., Zatorre, R.J., Gauna, K., Nikelski, E.J., Dostie, D., and Evans, A.C. (2000). Speech-like cerebral activity in profoundly deaf people processing signed languages: implications for the neural basis of human language. *Proc Natl Acad Sci USA, 97* (25), 13961–13966.
77. von Senden, M. (1960). Space and Sight. (Great Britain: Butler and Tanner).
78. Cheselden, W. (1728). An account of some observations made by a young gentleman, who was born blind, or who lost his sight so early, that he had no remembrance of ever having seen, and was couch'd between 13 and 14 years of age. *Phil. Trans., 402*, 447–450.
79. Ackroyd, C., Humphrey, N.K., and Warrington, E.K. (1974). Lasting effects of early blindness. A case study. *Q J Exp Psychol, 26* (1), 114–124.
80. Carlson, S., and Hyvarinen, L. (1983). Visual rehabilitation after long lasting early blindness. *Acta Ophthalmol (Copenh), 61* (4), 701–713.
81. Carlson, S., Hyvarinen, L., and Raninen, A. (1986). Persistent behavioural blindness after early visual deprivation and active visual rehabilitation: a case report. *Br J Ophthalmol, 70* (8), 607–611.
82. Gregory, R.L., and Wallace, J.G. (1963). Recovery from early blindness: a case study. In: *Exp. Psychological Soc. Monograph 2* (Cambridge: Heffer and Sons).
83. Sacks, O. (1995). To see and not to see. In: *An Anthropologist on Mars* (pp. 108–152). New York: Vintage Books, Random House.
84. Valvo, A. (1971). Sight restoration after long-term blindness: the problems and behavior patterns of visual rehabilitation. (New York: American Foundation for the blind).
85. Wright, M.J., Geffen, G.M., and Geffen, L.B. (1995). Event related potentials during covert orientation of visual attention: effects of cue validity and directionality. *Biol Psychol, 41* (2), 183–202.
86. Fine, I., Wade, A., Boynton, G.M.B., Brewer, A., May, M., Wandell, B., and MacLeod, D.I.A. (2003). The neural and functional effects of long-term visual deprivation on human cortex. *Nature Neuroscience, 6* (9)
87. Bradley, D.C., Chang, G.C., and Andersen, R.A. (1998). Encoding of three-dimensional structure-from-motion by primate area MT neurons. *Nature, 392* (6677), 714–717.

88. Dodd, J.V., Krug, K., Cumming, B.G., and Parker, A.J. (2001). Perceptually bistable three-dimensional figures evoke high choice probabilities in cortical area MT. *J Neurosci, 21* (13), 4809–4821.

89. Orban, G.A., Sunaert, S., Todd, J.T., Van Hecke, P., and Marchal, G. (1999). Human cortical regions involved in extracting depth from motion. *Neuron, 24* (4), 929–940.

90. Huk, A.C., Dougherty, R.F., and Heeger, D.J. (2002). Retinotopy and functional subdivision of human areas MT and MST. *J Neurosci, 22* (16), 7195–7205.

91. Field, D.J., Hayes, A., and Hess, R.F. (1993). Contour integration by the human visual system: evidence for a local "association field". *Vision Res, 33* (2), 173–193.

92. Glass, L. (1969). Moire effect from random dots. *Nature, 223* (206), 578–580.

93. Lewis, T.L., Ellemberg, D., Maurer, D., Wilkinson, F., Wilson, H.R., Dirks, M., and Brent, H.P. (2002). Sensitivity to global form in glass patterns after early visual deprivation in humans. *Vision Res, 42* (8), 939–948.

94. Ross, J., Badcock, D.R., and Hayes, A. (2000). Coherent global motion in the absence of coherent velocity signals. *Curr Biol, 10* (11), 679–682.

95. Shepard, R.N. (1990). Mind Sights. Original Visual Illusions, Ambiguities, and Other Anomalies With a Commentary on the Play of Mind in Perception and Art. (New York, NY: W.H. Freeman and Co).

96. Kanwisher, N., McDermott, J., and Chun, M.M. (1997). The fusiform face area: a module in human extrastriate cortex specialized for face perception. *J Neurosci, 17* (11), 4302–4311.

97. Farah, M.J., Wilson, K.D., Drain, M., and Tanaka, J.N. (1998). What is "special" about face perception? *Psychol Rev, 105* (3), 482–498.

98. Le Grand, R., Mondloch, C.J., Maurer, D., and Brent, H.P. (2001). Neuroperception. Early visual experience and face processing. *Nature, 410* (6831), 890.

99. Le Grand, R., Mondloch, C.J., Maurer, D., and Brent, H.P. (2003). Expert face processing requires visual inuput to the right hemisphere during infancy. *Nature Neuroscience, 6* (10), 1108–1112.

100. Fine, I., Smallman, H.S., Doyle, P.G., and MacLeod, D.I.A. (2002). Visual function before and after the removal of congenital bilateral cataracts in adulthood. *Vision Research, 42*, 191–210.

101. Geers, A.E., Nicholas, J.G., and Sedey, A.L. (2003). Language skills of children with early cochlear implantation. *Ear Hear, 24* (1 Suppl), 46S–58S.

102. Svirsky, M.A., Teoh, S.W., and Neuburger, H. (2004). Development of language and speech perception in congenitally, profoundly deaf children as a function of age at cochlear implantation. *Audiol Neurootol, 9* (4), 224–233.

103. Waltzman, S.B., and Cohen, N.L. (1998). Cochlear implantation in children younger than 2 years old. *Am J Otol, 19* (2), 158–162.

104. Sharma, A., Dorman, M.F., and Kral, A. (2005). The influence of a sensitive period on central auditory development in children with unilateral and bilateral cochlear implants. *Hear Res, 203* (1–2), 134–143.

105. Lee, D.S., Sung Lee, J., Ha Oh, S., Kim, S., Kim, J., Chung, J., Lee, M.C., and Kim, C.S. (2001). Cross-modal plasticity and cochlear implants. *Nature, 409*

106. Harrison, R.V., Gordon, K.A., and Mount, R.J. (2005). Is there a critical period for cochlear implantation in congenitally deaf children? Analyses of hearing and speech perception performance after implantation. *Dev Psychobiol, 46* (3), 252–261.

107. Kauffman, T., Theoret, H., and Pascual-Leone, A. (2002). Braille character discrimination in blindfolded human subjects. *Neuroreport, 13* (5), 571–574.

4
Prosthetic Vision Simulation in Fully and Partially Sighted Individuals

Matthias Walter[1], Liancheng Yang[2] and Gislin Dagnelie[2]

[1]*Biomedical Optics and Ultrafast Lasers, Kirchhoff Institute für Physik, Ruprecht Karls Universität*
[2]*Department of Ophthalmology, Johns Hopkins University School of Medicine*

Abstract: This article examines performance and learning under conditions of simulated prosthetic vision, using the tasks of counting white squares on modified checker boards, and placing black checkers on these squares. Aim of our study was to determine how much environmental information people would be able to obtain from crude pictures without tactile feedback. For this, we used video camera images which were convolved by a dedicated computer program in real time to represent phosphene vision as future chip implants might be able to do. The resolution was equivalent to a vision of 20/2400. The program allowed us to vary a broad set of variables such as contrast, number of gray levels, number of phosphenes per column and row, etc. In our tests we used a matrix of 6×10 dots with Gaussian intensity profile. Test subjects varied in age, gender, and educational background.

Introduction

Human vision is mediated by one of the most highly developed sensory systems found in nature. The capacity to combine high spatial resolution near the center of fixation with a wide peripheral field of view, accurate depth perception, color discrimination, and light–dark adaptation over 12 orders of magnitude is unparalleled [1]. The information from over 100 million photoreceptor cells is pre-processed in subsequent layers of retinal neurons, so that 1 million fibres in the optic nerve are sufficient to transport the information to the visual cortex, where further information processing takes place.

The most common causes of adult visual impairment in North America and Western Europe include age-related macula degeneration (AMD), retinitis pigmentosa (RP), and diabetic retinopathy. AMD is caused by a combination of genetics and environmental and other individual factors, and is characterized by progressive loss of central vision. Six million Americans are affected, and it is the leading cause of blindness among individuals over age 50. About 15 million more Americans are pre-symptomatic. For RP there is currently no effective treatment or cure. Early symptoms of this retinal degeneration include night blindness and increasing loss of peripheral vision. These symptoms most often begin in the first three decades of life and are followed by a progressive loss of vision over several decades, often leading to complete blindness. [2]. The disorder is inherited and about 100,000 people in the US are affected. Worldwide one in every 4000 persons is afflicted with RP.

Blindness caused by these disorders is due to failure of the outer retina. i.e. the photoreceptors and retinal pigment epithelium (RPE) layers. Secondary nerve cell layers throughout the retina appear to remain viable for a period of years or even decades [3]. Recent studies by Marc et al. show that deprivation of input from the photoreceptor layer leads to rewiring in the nerve cell layer [4]. This plasticity results in new connections of the secondary neurons which may extend up to several hundred micrometers, corresponding to approximately 1° in visual space. This has an important implication for prosthetic vision: Either the rewiring of the secondary neurons has to be stopped as early as possible or the wearer of a visual prosthesis may at best see only a phosphene pattern with size and spacing on the order of 1°.

A prosthesis could replace the function of the degenerated photoreceptors by stimulating the remaining secondary retinal neurons, the optic nerve fibers, or the visual cortex. Over the past few years, several groups have shown that it is possible to stimulate the inner retina with small electrodes, producing a sensation of points of light called phosphenes. The location of the stimulated area on the retina and the visual field location where the elicited phosphene is perceived match closely [5, 6].

In the experiments reported here, we assumed that the wearer of a retinal prosthesis will be able to see a regular field of fuzzy dots. With this assumption in mind, we modeled prosthetic vision by convolving camera images with a very low-resolution spatial grid filter, in real time, and presenting the results to sighted volunteers. The experiments included tasks in which vision had to guide motor activities (hand–eye coordination). Subjects' performance under these conditions of extremely poor vision was recorded.

Methods

The experiments were performed during an internship of Matthias Walter at the Lions Vision Center, Department of Ophthalmology, Johns Hopkins University, Baltimore, MD, USA.

Equipment

The headset in this study is a modified Low Vision Enhancement System (LVES; no longer in production) head-mounted display (HMD). It has a 36° × 48° field of view (8 mm exit pupil) and is capable of monochrome VGA resolution (640×480 pixels). The headset has built-in refractive correction, IR pupil illumination, and a charge coupled device (CCD) camera allowing recording of the test subject's eye movements. The scene in front of the HMD wearer was captured by a CCD camera, Watec model WAT-660D (Genwac, Orangeburg, NY), mounted centrally on the front cover of the headset, as shown in Figure 4.1. The optical axis of the camera is at a 90° downward angle to the subject's line of sight. The test subject was allowed to use just the left eye (the right eye monitor of the headset was switched off), hence there was an offset of about 3.2 cm between the axis of the camera view and the subject's eye position, the camera position did correspond to the subject's cyclopean eye and body axis. The video feed from the downward-looking camera was processed by a Dell Inspiron 8200 laptop computer with a dedicated filter implemented under the DirectShow operating environment.

Stimulus

The DirectShow filter program allowed many different settings such as size of the phosphene grid and of each phosphene, gray levels of phosphenes and background, etc. In our experiments we used a grid of 6×10 dots with 1° effective diameter and 2° center-to-center spacing, representative of prosthesis resolutions that may be achieved in the near future. The dots had a Gaussian intensity profile on a black background. The dot raster was shown to subjects only while they were allowed to perform the experimental task.

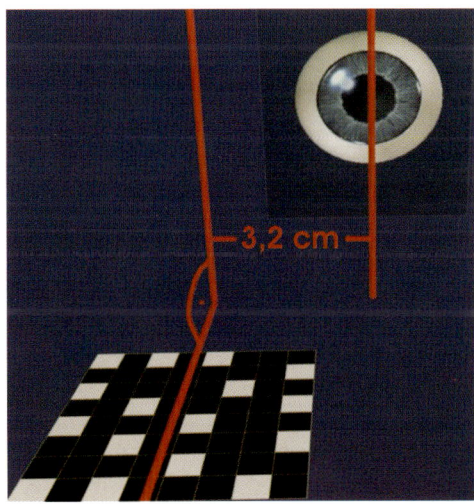

FIGURE 4.1. Angle between the view planes of the camera and the test subject and the offset between the middle of the camera view and the middle of the test subjects view.

Modified checkerboards (8×8 squares) were printed on white paper with a square size of 2×2 cm. Only a small number of squares (1 to 16) were white; the rest were black. The white squares were positioned in quasi-random positions limited only by the fact that they were not allowed to share a common border. The boards were framed by a 2 cm white border to define the area of interest for the test subjects. The circular checker pieces used in the experiments were made of black plastic and had a diameter of 2.2 cm. This size was chosen so that coverage of a white square was accomplished without too much interference from checkers in diagonally adjacent positions.

Experimental Procedures

Test subjects participated in the experiments on a weekly basis, in one hour-long test sessions. The one-hour limit was chosen to limit errors caused by fatigue, while the weekly schedule was kept to maintain a sufficient level of practice. We noticed in the first few test sessions that subjects had more problems to get back to their former performance level if this follow-up interval was stretched beyond a weekly schedule.

The test subjects were told only the following facts about the experiment:

- The checker boards consist of 8×8 squares; most of them black, some white; there is also a white rim around the whole board to define the area of the 8×8 squares.
- Your first task is to count the number of white squares; later you will be given a number of black checkers that either matches or exceeds the number of white squares on the board and you will have to cover the white squares with these for at least 50%. If the coverage is less than 50%, the square will be counted as not covered.
- We will measure the time you take to complete the task, but the main focus is for you to correctly count the number of white squares, and to achieve sufficient coverage of the white squares by the checkers.
- The white squares may be placed directly adjacent to the white rim or even in the corners of the board; two or more white squares may be placed diagonally so that they share one corner, but they will never share a border.
- The size of the squares perceived in the headset depends on the distance between the camera and the board; you will be looking straight ahead, but the camera is looking down at the board on the table.
- Tilting your head may help you in finding squares that are incompletely covered but it will also distort the image you see.
- Your hand will show as dark in the pixelized image; this means that it will obstruct your view if it is between the camera and the board, which may lead you to believe that you covered a square with a checker while you actually did not do so.

During the first few sessions, these facts were repeated every time before testing was started. This was done not only to remind subjects of the test conditions, but also to assure them that there had been no changes compared to the previous time they did the experiment. During the entire series of test visits, the subjects never saw the boards other than through the headset, in filtered form, which forced them to create their own mental image. They also were never given feedback regarding their performance, to simulate the condition of total self-learning faced by a blind prosthesis wearer, and to see what problem-solving strategies each subject would develop. To decrease the probability of pattern recognition for the counting task, we designed three different arrangements of white squares for the boards with 4 or fewer white squares. Board designs can be found in the Appendix. To further decrease possible pattern recognition, we randomly rotated the boards by 180°. Checkerboards were named according to their orientation and the number of white squares. Thus "Top 3A" refers to a board with 3 white squares in standard orientation; while "Bottom 3B" would refer to a different board with 3 white squares, turned 180°.

Time to completion of the counting and checker placing tasks was measured with a stopwatch. Only during this time was the video feed from the camera switched on, so subjects never had an opportunity to observe the removal, placement, or rotation of the board. The board was always removed between trials even to just be turned around, so no sound cues were given to the subject as to what action was taken. Figure 4.2 shows a pixelized view of a checker board.

Tests were split into three parts:

- Experiment 1: Subjects familiarized themselves with the test equipment, the boards, and checkers and the two different tasks of counting white squares and placing the checkers.
- Experiment 2: Subjects practiced their skills in counting white squares while they were given the boards in random order.

FIGURE 4.2. Picture of the view the test subject has. It shows the upper left edge and parts of the rim (seen from the position of the test subject) of the board and some white squares.

- Experiment 3: Subjects practiced their skills in placing checkers while they were given the boards in random order.

Subjects

We are reporting results from five test subjects (3 male, 2 female; aged 25–51 years) who were all practiced in the use of simulated phosphene vision through their participation in other tests (reading tests and face/object recognition) that did not involve object manipulation. Four subjects were normally sighted; the remaining subject was severely visually impaired (legally blind) due to a chemical burn of both corneas in early childhood.

All testing was done according to the tenets of the Helsinki convention. The subjects were informed of the purpose and procedures of the experiment, and read and signed a consent form approved by the IRB of the Johns Hopkins University School of Medicine. In accordance with HIPAA rules the identities of the test subjects were coded with numbers to mask their identities.

Results

To evaluate the results of the tests the following rules were applied:

1. Errors in counting tasks were "punished" by using the formula

$$FinalTime = UsedTime +$$
$$|(Real\ number\ of\ squares - Counted\ number\ of\ squares)|$$
$$\times AverageTime \qquad (1)$$

where *UsedTime* is the time measured with the stop watch, *Real number of squares* is the number of white squares on the board, *Counted number of squares* is the number that was called out by the test subject, and *AverageTime* is the calculated time it takes the test subject to count one white square on a board. This *AverageTime* is calculated for every test subject respectively using the formula:

$$AverageTime = \frac{Sum\ of\ time}{Sum\ of\ squares} \qquad (2)$$

with *Sum of time* as the sum of the measured time for all the boards in the session and *Sum of squares* as the total number of squares summed over all boards used in the session.

2. To evaluate the "placing checker" parts of the experiments the following approach was chosen:

Errors were divided into groups with their respective error values:

(a) Checkers which did not fulfill the 50%-coverage rule but were still close to a white square resulted in an error value of −0.5
(b) Checkers which were placed nowhere near a white square resulted in an error value of −1
(c) White squares that were not covered resulted in an error value of −1

As before, Eq. 2 was used to calculate an average time to place one checker on one white square. This average time was then multiplied by the error value of each board and added to the measured time to obtain the Final Time.

The number of correctly placed checkers is denoted as percent correct for each board and test subject. Percent correct is calculated using the following formula:

$$Percent\ correct = \frac{Number\ of\ correctly\ placed\ checkers}{Number\ of\ white\ squares} \times 100 \quad (3)$$

The error scoring method under 2b above was introduced in experiment 3, when we noticed that some subjects would cover all white squares but placed one or more additional checkers on the board. For experiment 1, this rule had not been necessary, as the test subjects did not place any additional checkers during the tests, and no or few additional checkers were given to the subjects.

Experiment 1

In the first experiment, we introduced the test subjects to the experimental setup, starting with boards on which the number of white squares was less than or equal to three. The number of squares was then increased in +1 steps for the checker placement task. Initially, the counting and checker placing tasks were combined in such a way that the subject had to count the same number of white squares on three different boards, and then would place the checkers, also three times. The orientation of the boards was varied randomly to avoid recognition, and subjects frequently did not realize that the number remained constant due to the difficulty of the task. Subsequently, the difficulty level was increased by interjecting boards with slightly different numbers of white squares during the counting task.

Checkers to be placed on the board were stacked on the side of the board corresponding to the subject's handedness. The number of checkers available to the subject depended on the number of white squares on the board: For boards with a number of white squares less or equal to four, six checkers were placed next to the board, and for boards with more white squares the number of checkers was the number of white squares plus two. This approach was chosen to limit the possibility of false placement. Only if the test subject was searching for more checkers after placing all of them (and thus misplacing some of them) were more checkers provided.

At the end of the first experiment, we asked the test subjects if they found any connection between the number of white squares which they had counted and the number of checkers they subsequently placed. Subjects unanimously reported that they did not count the number of checkers they placed, and they could not tell if there was a connection between the numbers.

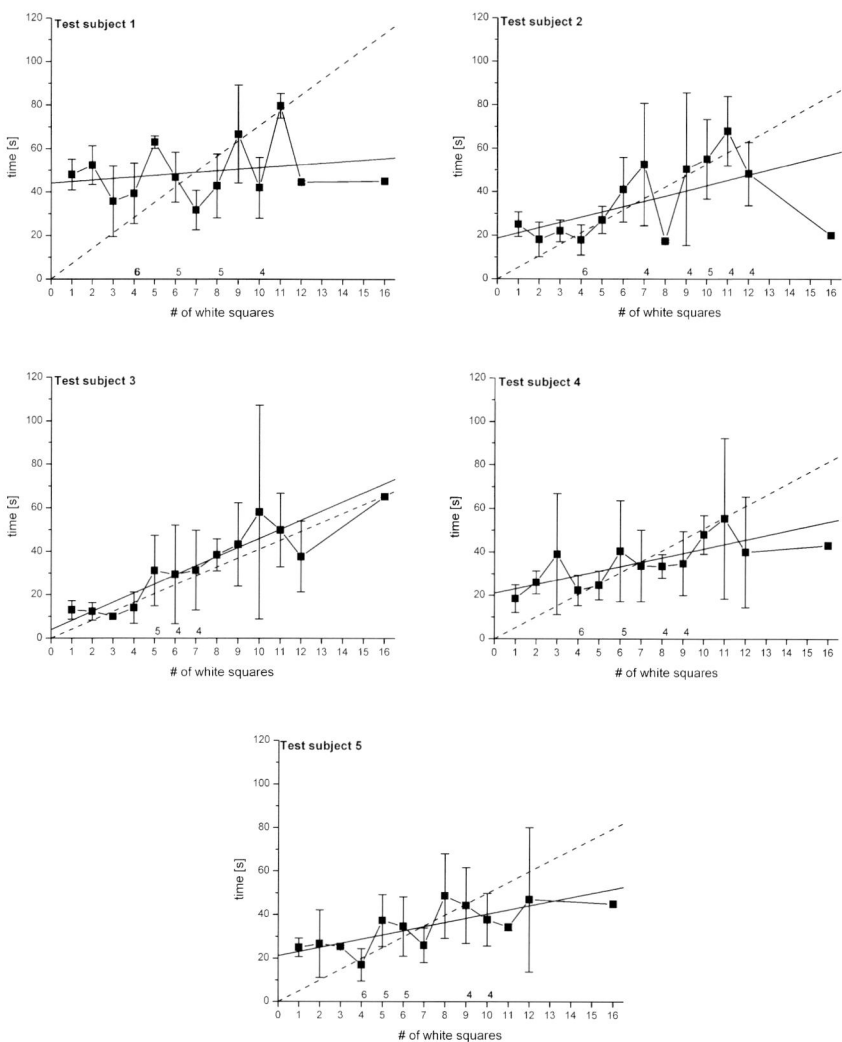

FIGURE 4.3. These graphs show the results of the counting task for all the test subjects and the linear regression of the data (solid straight line). Board Top 16 was used only once at this stage of the experiment, so there is no standard deviation data available and thus no error bar. Top 1 was one time used once without noting the time to familiarize the test subjects with the task. The number of counts for the other boards varied from test subject to test subject but was at least 3 times for every board and at the maximum 6 times (small numbers at the x-axis denote if there were more than three times and give the number of times). The dashed line shows the regression of the calculated average time value (see Eq. 2).

Counting Task

Figure 4.3 shows the results for the counting part of this experiment. The plots present the results for each subject averaged across multiple trials. Error bars represent the standard deviation of the mean. A board with a single white square was used once without keeping time, to familiarize the subjects with the task, and twice after that with timing. For boards with all other white square counts three or more trials were administered, varying between subjects; small numbers along the abscissa denote higher repetition numbers. Board Top 16 was only used once, so the data point has no error bars; this point was not included in regression line calculations. Mean counting time by number of squares, regression line (solid; computed from individual trials), and dashed line indicating the expected counting times if the time per counted square was constant throughout the experiment, all rise with the number of white squares for all subjects. This is to be expected: Task difficulty increases, albeit with the density rather than the number of white squares, since double counts may be a bigger risk than skips. For most subjects the regression line rises less steeply than the line of constant counting speed, suggesting that the time per counted square decreases for increasing practice and/or white square number. The graph for subject 3 most closely approximates the use of a constant time interval per counted square: The linear regression and calculated average time lines are parallel to each other, with a small offset. This can be interpreted by saying that this subject required only minimal learning, adapted quickly to the test environment, and utilized a constant amount of time per white square. The graph of test subject 1, on the other hand, shows the strongest difference between the linear regression and the line of constant counting speed: The slope of the linear regression is almost zero. This can be attributed to the fact that test subject 1 is visually impaired and needs to perform the same scan pattern regardless of the number of white squares. This is plausible because somebody who cannot trust his or her vision has to scan the whole board meticulously to be sure to not overlook a white square. For subjects 2, 4, and 5, the regression line slopes are very similar: 1.9 to 2.4 s/square, and the Y-intercepts fall between 18.6 and 21.2 s. This can be interpreted as subjects needing a near constant amount of time initially, and becoming more agile with increasing white square numbers, possibly trading off intercept and slope.

Checker Placing Task

Figure 4.4 depicts the results for the checker-placing part of the experiment as a function of the number of white squares on the board, with bars indicating the percentage of white squares covered correctly (left vertical axis), and connected symbols indicating mean time to completion (right vertical axis; note different time scales by subject). Every white square number was repeated at least three times; small numbers along the abscissa denote higher repetition numbers. The board with 16 white squares was used only once. As was the case for the counting task, completion times and linear regression rise with increasing numbers of white squares for all subjects. Subjects were quite accurate in placing the checkers

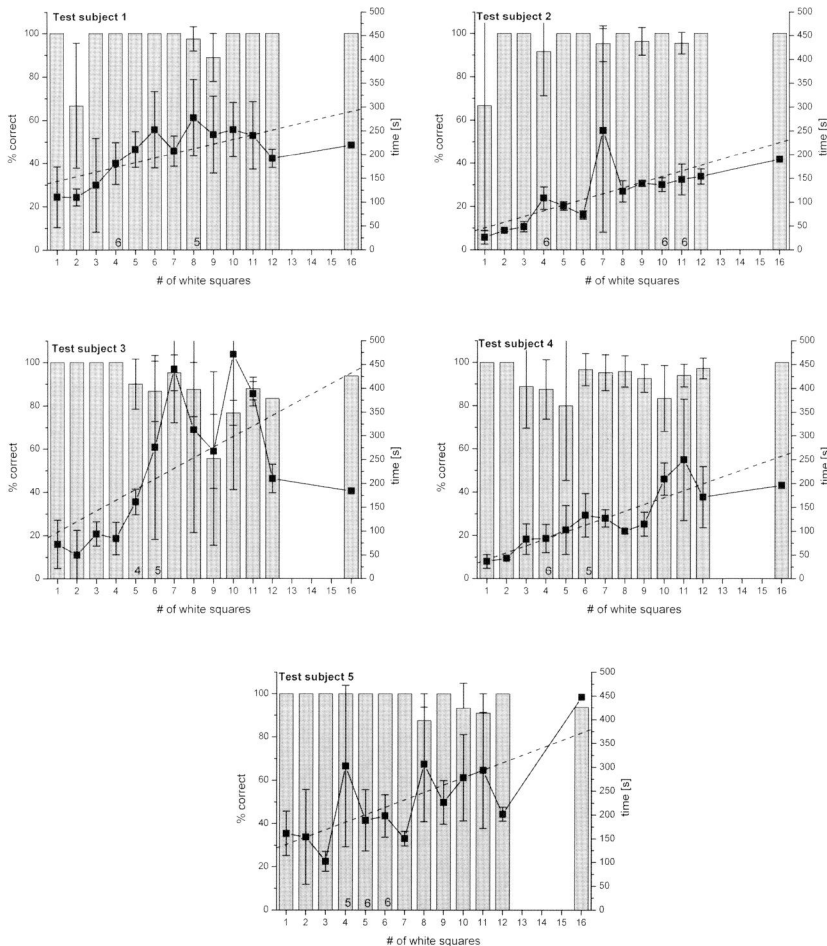

FIGURE 4.4. These graphs show the results of the checker-placing task for all the test subjects. The bars show the percentage of the correctly placed checkers calculated with Eq. 3 and the dashed line represents the linear regression of the time. Board Top 16 was used only once at this stage of the experiment, so there is no standard deviation data available and thus no error bar. The number of repetitions for the other boards varied from test subject to test subject but was at least 3 times for every board and at the maximum 6 times (small numbers at the x-axis denote if there were more than three times and give the number of times).

at the start, but a slight drop in correct percentage is seen for boards with more white squares. This drop may be attributed to the increase in the number of white squares per board as well as to their positions on the boards: With more white squares on a board the squares were located more often at borders or in corners of the board as well as diagonally adjacent to another (see Figure 4.10, board designs). The regression time slopes show similar narrow intervals as in

Figure 4.3 for subjects 2, 4, and 5 (time/checker: 12.0–15.6 s), but the intercept for subject 5 (123 s) is much higher than that for the other two (26 and 34 s); this same subject also has much larger within session variability (error bars), although the same can be seen for a few board types (i.e. sessions) in subjects 2 and 4 as well. In most subjects, both mean time and variability drop below the regression line for the largest numbers of white squares, indicating the effect of practice reached by the last session of this experiment. This practice effect is also suggested by the high intercept values in subjects 1, 3, and 5.

Experiment 2

Experiment 2 was used to monitor an increase in subjects' accuracy on the counting task. Boards were given to the test subjects in random order, for four consecutive test sessions. In addition to the prior instructions, subjects were allowed to start over if they felt that they had lost count; trials in which this happened were disregarded. The results of this task are depicted in Figure 4.5. The plots show the counting time as a function of the number of white squares. Similar to Experiment 1, a regression line was calculated from individual data points. In addition, 95% confidence interval bands for the regression line were added using Origin (Ver. 7.0, OriginLab Corp Northampton, MA). As can be seen from the time scale, all subjects have become significantly faster than in Experiment 1: Regression line slopes range from 0.5 to 1.2 s/square. Also the y-axis intercept is lower, at 5–17 s. Note that all five subjects now fit into these intervals, and that subject 1 (visually impaired) performs nearly as well as the other subjects. The regression lines provide a better fit to the mean values than those in Figure 4.3, and the SEM values are lower, suggesting that the subjects have a more even, i.e. practiced, performance.

Experiment 3

The purpose of Experiment 3 was to achieve proficiency in the placing task. The subjects placed the checkers on the boards without first counting the white fields, and boards were given to them in random order. The difficulty here was to coordinate hand and eye only by proprioceptive information because the hand was not directly visible in the pixelized view: Only the obstruction of the white border or white squares by the hand was noticeable. The possibility to re-start a trial was again provided. In this case the video image was switched off, all checkers were removed, and the orientation of the board was changed. Unlike in Experiment 1, subjects were always given 16 checkers. This was done to eliminate any possible cues about the number of white squares on the board. Figure 4.6 shows a test subject placing a checker and the corresponding pixelized camera image as it presented itself to the test subject at that moment. Figure 4.7 indicates the approach one of the test subjects developed for placing a checker. Other test subjects used different techniques. This experiment was more time consuming, and the number of trials per session varied from subject to subject,

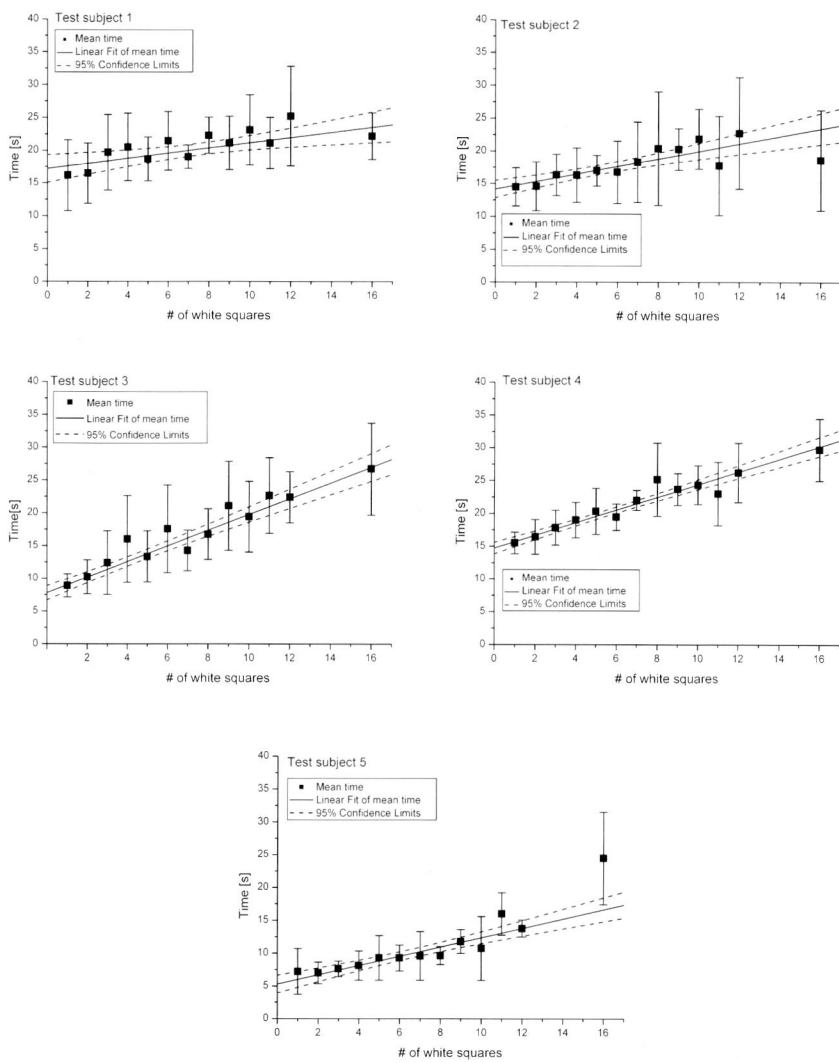

FIGURE 4.5. Results of the counting test. The boards were given randomly and so the number of times one board was used is different for every test subject.

ranging from 4 to 9 depending on the initial proficiency and further progress of the subject. The results of this experiment are depicted in Figure 4.8 (time vs. number of white squares) and 9 (points earned vs. number of white squares).

The regression line slopes for four subjects in Figure 4.8 are nearly identical: 9.58–9.73 s/checker. Only subject 3, with 7.8 s/checker, performs somewhat faster. The Y-intercepts range from 7.6 to 24.3 s, lower and more narrowly distributed than in Experiment 1. These data strongly suggest that during the 4 months it took to reach this point, all subjects reached very similar performance

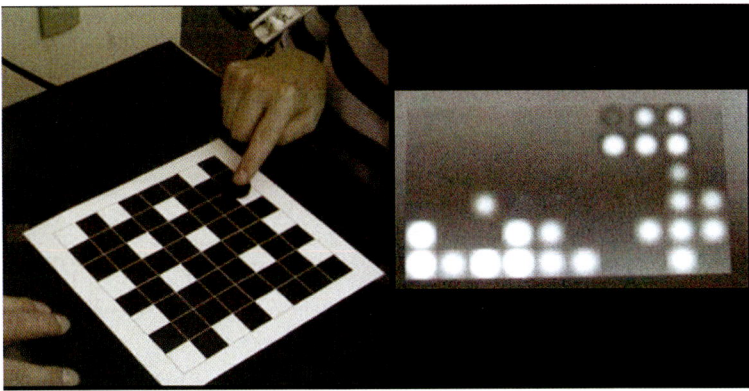

FIGURE 4.6. Correspondent images of the board and the pixelized view presented to the test subject while one checker is placed on a white square.

levels, and times that depend primarily on the number of squares. Figure 4.9 shows that error rates for most subjects was consistently low even with a rigorous penalty system. That also indicates the rise in proficiency.

Questionnaire/Self-Assessment

Results of a brief self-assessment questionnaire are discussed in the Appendix.

Discussion

The three experiments described here were designed to study subjects' ability to perform visual inspection (finding and counting white squares) and eye-hand coordination (placing checkers accurately on white squares) tasks. While the visual properties of these high-contrast, low-resolution tasks may seem to lend themselves particularly well for detection through a retinal prosthesis, our results show that it may take considerable practice to reach proficiency, especially for more realistic daily activities. On the other hand, a real prosthesis wearer would live with phosphene vision continuously and reduce his or her adaptation time through intensive practice. Given adequate practice, our subjects attained a level of proficiency that allowed them to complete both tasks quickly and with few errors. The initially widespread in performance levels became narrower with practice, irrespective of age and gender differences in our small subject group. Even the visually impaired subject, though not quite as proficient as the others, readily performed both tasks.

It is obvious that the simulations are idealizations of what a real prosthesis wearer will perceive. Most notably, our grids contained high and uniform contrast, a regular phosphene distribution, and low noise levels. The real phosphene image may not show ordered rows and columns or

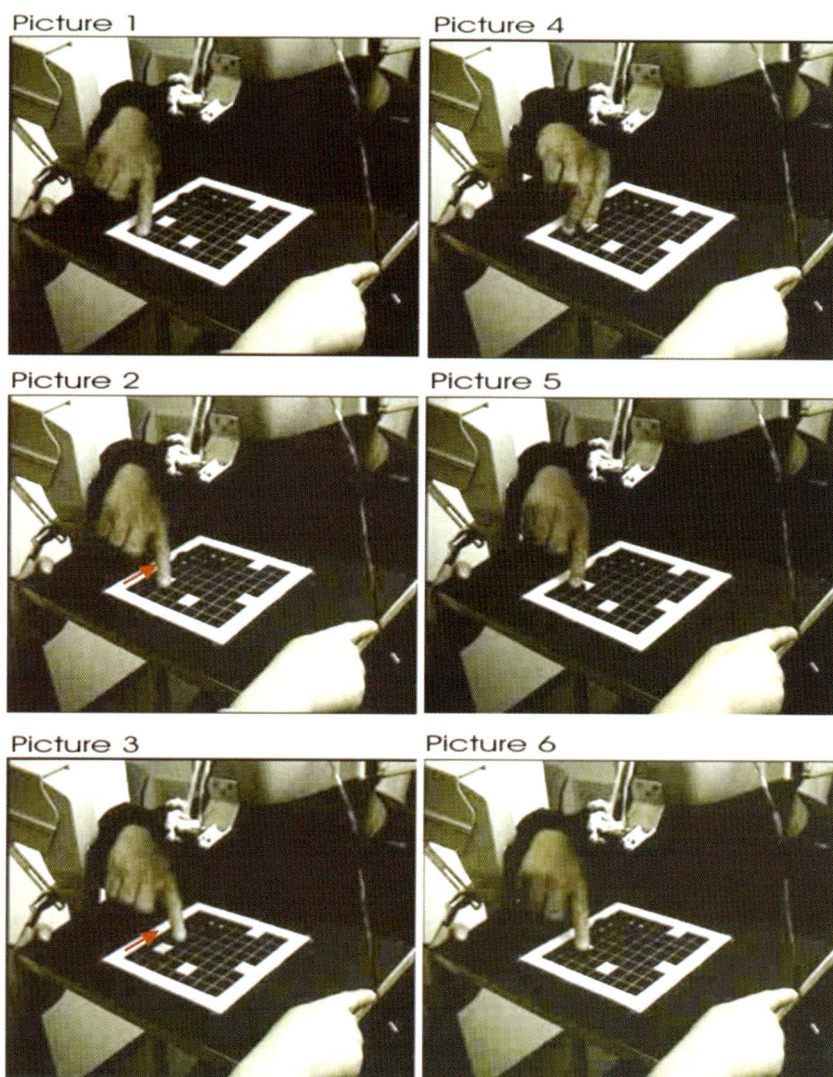

FIGURE 4.7. Image series to show one of the techniques developed by the test subjects: First a white square is chosen; then the subject uses the index finger to cover the white square (depicted in images 1 to 3; movement is indicated by the red arrow); after finding the position of the chosen square the checker is placed (image 4) and its position is gradually corrected as necessary to cover the square (images 5 and 6).

have such a consistent contrast distribution as was the case in the experiments. In reality, phosphenes may be fused together, appear inhomogeneous and distorted, and there may be ongoing background noise which will reduce the visibility of the dots. None of these complexities were taken

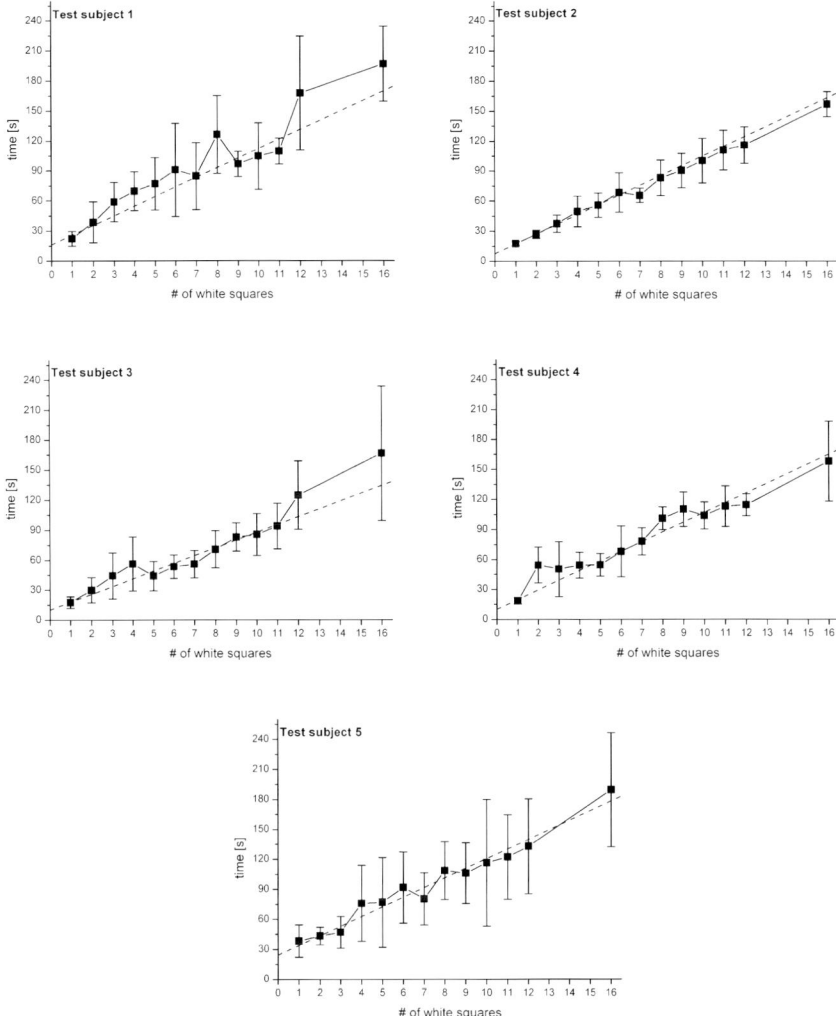

FIGURE 4.8. The graphs show the times for this part of the experiment. The dashed line is the linear regression of the data.

into account during these experiments. In part, this was done because there is little knowledge about the appearance of prosthetic vision at this time. But we also kept our simulations "clean" because, contrary to a chronic implant wearer, our subjects did not continuously experience this unaccustomed visual perception, and so their visual system would not learn to interpret the incoming stimuli into meaningful data to the extent that real prosthesis wearers will.

Our results show that simple visual inspection and hand–eye coordination tasks can be accomplished with pixelized vision corresponding to a visual acuity

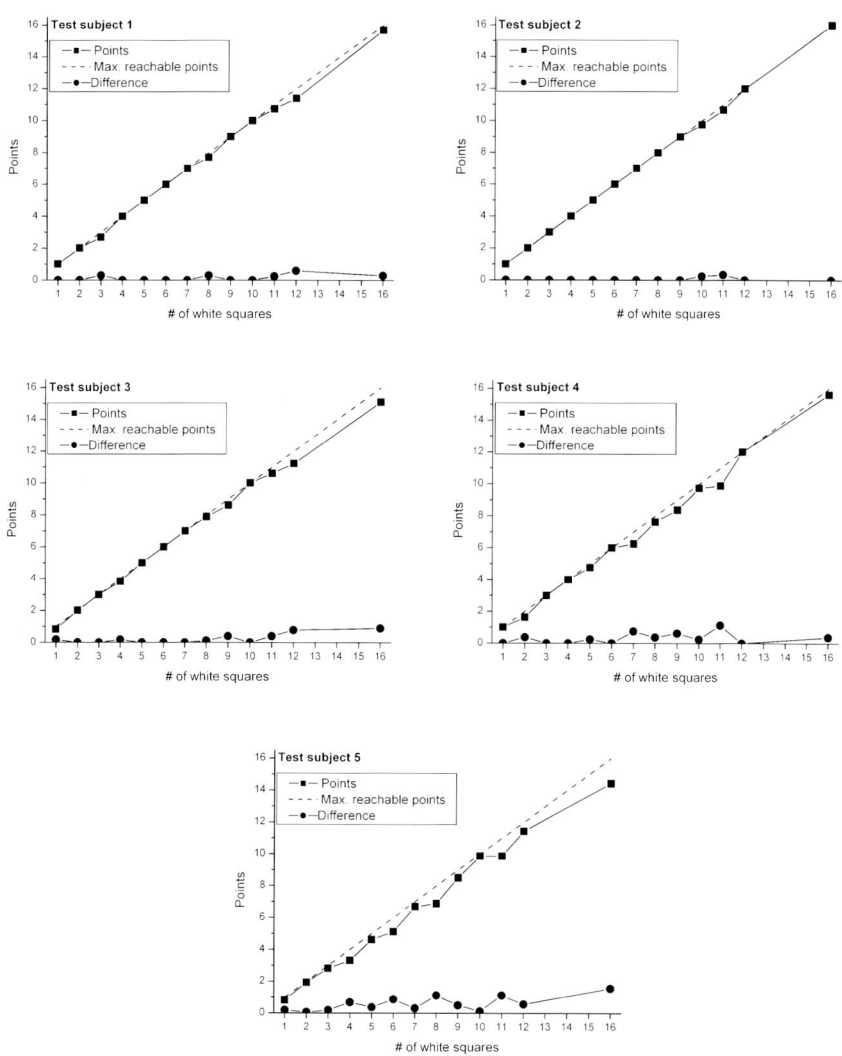

FIGURE 4.9. These graphs show the points for each test subject.

of 20/2400. A brief questionnaire about strategies for the square counting and checker placing tasks (Table 4.1) showed that each subject developed personal ones. This suggests that there is no single best approach to solve the given tasks. Another important benefit to be gained from simulation experiments is that a skilled rehabilitation instructor can use the results and reported strategies to develop training methods for future prosthesis wearers.

The data obtained in these experiments strongly suggests that even with extremely poor vision it is possible to obtain reliable information about the environment and to react accordingly.

TABLE 4.1. Answers to questions of the questionnaire as given by the respective test subject (no changes to spelling or grammar were done).

	Counting	Playing
test subject 1	I start at the top left and scan to the right over the top 1/3 or so, then scan right to left over the middle portion and then scan left to right over the bottom portion. If there are more than 9 or 10 to count, sometimes I will double check by recounting by scanning from the top left to the bottom left over about halfway and then scanning the second half from bottom right to top right.	I start at the top left and cover any squares in this area first. I move the checker off the board far enough for it to partially appear in the white. I try to move straight towards the square and usually will cover enough to know that I'm near it. This process works on squares on or near the any edge. It becomes more difficult to cover squares in the middle of the board with this method because of not knowing if the checker is being moved straight over that distance. Covering the interior squares sometimes comes down to guessing where to move the checker until a small part of the square gets blocked out indicating that the checker is in the right area.
test subject 2	Scan left to right, bottom to top. If I think I may have double counted/missed one, I'll try left to right, top to bottom.	Scan left to right, bottom to top and cover white squares I see. Then I move away and make sure there are no more white squares left.
test subject 4	I move from the bottom right around the edge to the top left and down and then count what is in the middle.	cover the outside squares first and then move to covering the inner squares. I try to save the right edge until last so that I don't move any checkers.
test subject 5	start at bottom left corner of grid. Keep left border in sight scan up count squares. Shift to top right corner. Scan down while keeping right border in sight;count squares. Problem in any overlapping centre area were squares counted or not?	start bottom left keep left border in view. Use border to identify vertical location of my finger (left hand). Then when finger aligned with square slide in to place checker. Same goes for the right side (use right hand). Similar top and bottom of board. If not possible to have square and border in view at same time position headset camera over square and sweep right hand in spirular motion in to cover the square.

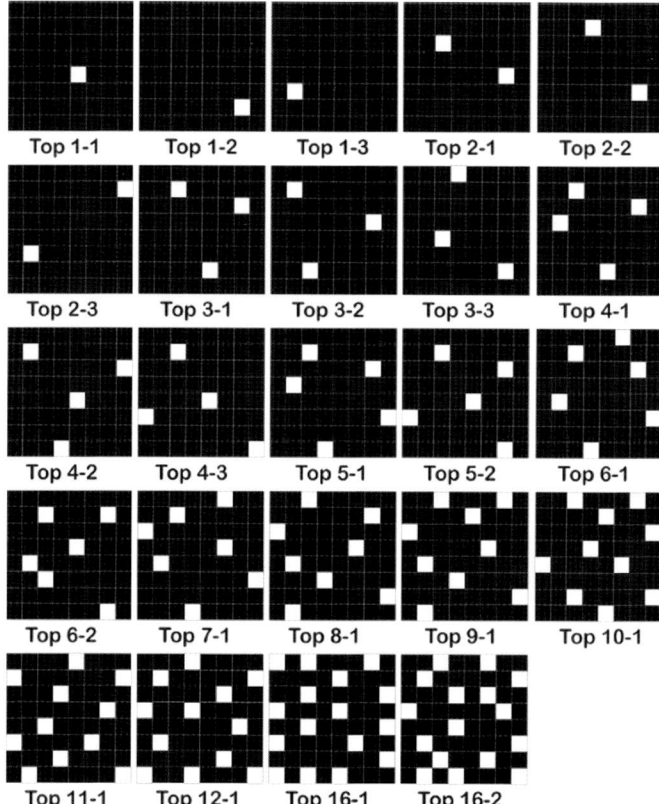

FIGURE 4.10. Board designs used in the experiments. The white rim is not shown; it is equal in width to one square (2 cm). The black checkers used in the placing task have a diameter of 2.2 cm.

Conclusion

It is possible to perform simple tasks under pixelized conditions with very low resolution (equivalent to 20/2400 visual acuity). The results of these experiments give us confidence that a visual prosthesis with the characteristics simulated here (a 6×10 electrode grid with electrode-diameters of approximately $300\,\mu m$ and a $600\,\mu m$ center-to-center spacing) is likely to enable the wearer to perform similar tasks, provided such a device mediates any vision at all. A retinal prosthesis with these characteristics, while far from able to restore normal vision, may provide a solution for people who once were sighted, but lost their sight due to destruction of the outer retina. They may at least gain crude images of their environment and interpret those on the basis of their former knowledge of the visual world. With these crude images, they can attain a higher level of autonomy than they have in their present state.

Acknowledgments. This research was supported by the Landesstiftung Baden-Württemberg (MW) and R01 EY12843 (GD). The assistance of our test subjects is gratefully acknowledged.

References

1. G. Dagnelie, M. S. Humayun, R. W. Massof. Principles of Tissue Engineering, Chapter 54, pp. 761–72. Academic Press, 2nd edition, 2000.
2. Foundation Fighting Blindness. Treatments and cures; breakthroughs and beyond. http://www.blindness.org/pdfs/FFBPresentation.pdf.
3. A. Santos, M. S. Humayun, E. de Juan Jr, R. J. Greenberg, M. J. Marsh, I. B. Klock, A. H. Milam. Preservation of the inner retina in retinitis pigmentosa. a morphometric analysis. Archives of Ophthalmology, 115:511–515, 1997.
4. R. E. Marc, B. W. Jones, C. B. Watt, E. Strettoi. Neural remodeling in retinal degeneration. Progress in Retinal and Eye Research, 22:607–55, 2003.
5. M. S. Humayun, E. de Juan Jr, G. Dagnelie, R. J. Greenberg, R. H. Propst, D. H. Phillips. Visual perception elicited by electrical stimulation of retina in blind humans. Archives of Ophthalmology, 114:40–46, 1996.
6. J. F. Rizzo III, J. Wyatt, J. Loewenstein, S. Kelly, D. Shire. Methods and perceptual thresholds for short-term electrical stimulation of human retina with microelectrode arrays. Investigative Ophthalmology and Vision Science, 44:5355–61, 2003.

Appendix

To collect voluntary responses from subjects who were not previously familiar with the task, study participants were asked to provide feedback regarding the tests and describe their approach through a brief questionnaire administered immediately after the last session of Experiment 3. A brief summary follows.

Question 1: Eye dominance. This was important because subjects were limited to using only the left eye during the experiments, so persons with a right eye dominance (subjects 2 and 3) might have been at a disadvantage compared to those with a left eye dominance (subjects 4 and 5) and subject 1, who has no useful vision in the right eye. However, an ANOVA (SAS Institute, Research Triangle Park, NC) of test results by eye dominance shows a significant difference in errors (0.17 vs. 0.34; $P < .005$) and in time (97.3 vs. 82.2 s; $P < .002$) per trial, suggesting that the two subjects claiming left eye dominance required more time per trial, yet made fewer errors, and thus that using the dominant eye to perform this task at very poor acuity may be a mixed blessing. Note, however, that in both cases the difference is less than the standard deviation across all subjects (0.20 errors, 18.4 s), so the difference is most likely unrelated to eye dominance.

Question 2: Hand dominance. All subjects chose to use their right hand, and in fact indicated that this was their dominant hand. Therefore no conclusions regarding a possible effect of handedness can be drawn.

Question 3: "Are you playing video games in your spare time?" targeted the possible superior hand-eye coordination of such a person. Subjects 2 and 3 responded affirmatively, so the difference under question 1 is equally (un)likely to be related to this activity.

Questions 4 and 5 asked for descriptions of the subjects' techniques for the counting and checker-placing tasks. Table 4.1 provides the answers as written by 4 subjects; subject 3 chose not to answer these two questions. The answers show that every subject developed his/her own method.

Question 6: When asked what would make a very difficult board, all subjects indicated that the difficulty level rises with the increasing number of white squares, and is attributable to the increasing difficulty of placing the checkers without moving previously placed ones. They also indicated that squares in the corners are hard to detect and that a diagonally adjacent squares increase the difficulty level.

Figure 4.10 shows the board-designs used in the experiments.

5
Testing Visual Functions in Patients with Visual Prostheses

Robert Wilke[1], Michael Bach[2], Barbara Wilhelm[3], Wilhelm Durst[1],
Susanne Trauzettel-Klosinski[1] and Eberhart Zrenner[1]

[1]*University Eye Hospital Tübingen*
[2]*University Eye Hospital Freiburg*
[3]*Steinbeis Transfer Center for Biomedical Optics Ofterdingen*

Abstract: A number of different technical devices for restoring vision in blind patients have been proposed to date. They employ different strategies for the acquisition of optical information, image processing, and electrical stimulation. Devices with external cameras or with integrated components for light detection have been developed and are designed to stimulate such different sites as the retina, optic nerve, and cortex. First clinical trials for these devices are being planned or already underway. As vision with these artificial vision devices (AVDs) may differ considerably from natural vision and as it may not be possible to predict visual functions provided by such devices on the basis of technical specifications alone, novel test strategies are needed to comprehensively describe visual performance. We propose a battery of tests for standardized well-controlled investigations in these patients that allow for objective assessment of efficacy of these devices.

Introduction

Natural vision depends on the integrity of a complex biological system, containing a sensory system (optical media retina and early visual cortex) and a perceptual and cognitive system (higher cortical areas). As a whole, a structure forms that shows a high degree of interaction and constrained functions. Therefore, testing visual functions means exploring a system of considerable complexity and intricacy.

In contrast to this, multiple system architecture of natural vision, AVDs are currently far less complex. Several different types of AVDs have been proposed and developed so far. They are designed to just restitute single subsystems of the visual system to some extent. As biological and technical systems join together to build a human–machine interface, a new functional entity will be

91

established featuring new visual properties. We cannot expect this new system to behave in a manner fully analogous to the natural system, nor can we assume that a technical device in a biological environment will perform as predicted on technical requirement specification sheets. For this reason it is indispensable to develop a specialized set of tests for evaluating visual functions in patients with an AVD.

In this chapter, we will briefly review standard methods to assess visual functions and their limitations for testing patients with AVDs. Next we will describe the theoretical concept of a new test design based on three main pillars. These pillars will then be discussed in detail and actual implementations for each subtest will be presented.

Testing Visual Functions in Normal Individuals and its Limitations in AVD

Tests used today for assessing natural vision are designed to measure psychophysical thresholds for specific subfunctions like light sensitivity and resolution, or they qualitatively assess functions like stereopsis or certain color vision tests. The most widely used method to describe visual function is visual acuity testing. This test gives a fast first impression about the integrity of the entire system, as it depends on the proper function of optic, sensory, and neuronal systems. Visual acuity is the measurement of the ability to discriminate two stimuli separated in space at high contrast relative to the background. Different charts and characters can be used, and the test result can be expressed with various types of measurements like Snellen notation, logMAR, decimal acuity, or log (decimal acuity) $= -$logMAR [1, 2]. Some specialized tests for the assessment of very low vision have also been developed (e.g. LoVE [3]).

The visual field can be examined using either static or kinetic perimetry. The first of these presents small light stimuli on a hemisphere with various degrees of eccentricity from the center. By changing light intensity the threshold of perception for each individual spot can be determined. Kinetic perimetry uses a moveable light point that is gradually shifted from the far periphery into the center. The eccentricity is recorded at which the patient can first see this target. This procedure is repeated for several directions, light intensities, and target sizes. Thus rings of equal light sensitivities ("isopters") are determined. Prerequisite for useful test results are good compliance of the patient and stable fixation of gaze on a central target.

Electrophysiological examinations are capable of detecting fluctuations in locally measured electrical potentials generated by neural activity. Depending on the positioning of electrodes and test procedures, different functional elements can be tested. The electroretinogram (ERG) detects light-evoked potentials generated in the outer retina (a-wave) and inner retina (b-wave) [4]. Inner retina function can be assessed by the pattern ERG [5].

Visually evoked cortical potentials (VEP) can be detected as responses of the visual cortex to a visual stimulus [6]. Area V1 of the brain is responsible for primary image processing. VEPs recorded over this site can help to detect whether the information of a visual stimulus is perceived and processed in the brain. In animal models, variations on this technique employing multielectrode recordings are widely used to confirm that electrical stimulation by an AVD elicits a response in the visual cortex.

A variety of additional tests exist, e.g. to evaluate contrast or glare, to measure the range of accommodation, to test binocular functions and stereopsis, to test the light reflexes of the pupil, and to measure the ability of the sensory system to adapt to darkness (Table 5.1).

Additional experimental setups exist to assess qualities such as temporal resolution or perception of motion. For severely visually impaired subjects, it is of particular relevance to be able to evaluate their performance in routine activities encountered in daily life, but no widely accepted standard routine exists for this task.

All of the tests mentioned above have certain limitations when applied to the assessment of visual functions in patients with visual prostheses. The way stimulation is performed with electronic devices differs from physiological stimulation, e.g. via neurotransmitters. In the natural system information about color, contrast, fluctuation of brightness over time, motion, and additional characteristics are encoded at the retinal level. This coded information is relayed to the brain and analyzed there [7, 8]. AVDs, however, will encode this information in a different way. Consequently there will be input to the visual cortex that is somewhat different from that experienced earlier in life. In addition to this

TABLE 5.1. A sample of tests used for assessing visual functions.

Tested function	Measured value/quality	Test name (examples)
Minimal angle of resolution	Visual acuity	Snellen chart
		EDTRS-chart
		Landolt-C chart
Contrast sensitivity	Contrast sensitivity function	Vistec chart
		Pelli-Robson chart
		Ginsburg chart
Visual field	Kinetic or static light sensitivity	Goldmann perimetry
		Humphrey, Octopus perimetry, Tübingen Perimetry
Conversion of light stimuli into electric information	Potentials generated at retinal level	ERG
Transmission of information into the brain	Potentials generated at the visual cortex	VEP
Color perception	Color discrimination	Ishihara plates
		CCT
		Nagel anomaloscope
Dark adaptation	Light sensitivity	Adaptometer
Pupillary light reflex	Changes in pupil diameter and their dynamics	Pupillography

different modulation of information, there are limits to the information that can be provided by AVDs. Most of the AVDs have a limited number of stimulation electrodes. For retinal prostheses they range from about 4 to 1500 electrodes. Even with this limited number of stimulation sites, however, one should be able to generate useful vision, as some studies have demonstrated [9, 10].

For these reasons, a test designed to evaluate visual functions in patients with visual prosthesis must meet special requirements that may differ from those that apply to natural vision.

Designing a Test for Visual Functions with Visual Prostheses

When defining specifications for a new test design, some universally valid or general specifications must be considered first (Table 5.2).

The test has to be reliable and meaningful with a high test–retest reproducibility. Furthermore, the test should be as objective as possible, minimizing investigator- or patient-driven bias. It has to be sensitive enough to detect even small changes in visual performance. Finally, it must be valid, i.e. it should really measure what it claims to measure, which is a basic demand any test has to meet and to prove.

There are additional, more specific requirements due to the nature of the human–machine interface (see Table 5.2). Up to now visual prostheses are not able to restore natural vision. The impression a person gets from an artificial

TABLE 5.2. Requirement specifications for a new test.

1. General specifications	
Specification	*Description*
Reliable	Providing the same results in test and retest
Objective	Independence of bias due to expectations of investigator or patient
Sensitive	Capability of determining even slight changes
Valid	Measurement of the intended value: visual function, and not, for example, intellectual performance
2. Particular specifications	
Specification	*Description*
Relevant	Delivery of substantial information about the patient's benefit of artificial vision
Wide scope	Ability to measure very basic visual function as well as advanced visual performance with reasonably high acuity
Impact for further development	Ability to provide better understanding about the performance of an artificial vision device and the human-machine interface
Generic use	Suitable, irrespective of the technical specifications of the artificial vision device

vision system will still be somewhat different from what that person experienced with natural sight. With this in mind, a new test must test functions that are relevant to the patient. In other words, it should not simply measure technical specifications, but rather reflect the patient's benefit in daily living.

As the development of technical devices for artificial vision is still at the beginning, the first devices under clinical investigation may provide quite basic visual impressions like perception of light or movement of light. For a blind person even the simple ability to gain orientation by being able to localize light sources like windows or doors can mean a substantial improvement in mobility and self-determined life. As the development of these devices proceeds, there is good reason to expect further improvements. Thus, a new test has to cover a wide scope of visual functions, including very basic visual perceptions (see Table 5.2).

With a view to ongoing developments, new tests must be designed to provide a better understanding of the properties of the human–machine interface and functions of the device. They must provide substantial data about the properties of the human–machine interface and the technical data needed to allow the enhancement of the next generation of devices.

A test aimed at characterizing visual abilities provided by AVD should be suitable for generic use. Different prostheses that are under development so far use a variety of coupling methods to connect to the biological system. Devices for epiretinal and subretinal coupling as well as for stimulation of the optic nerve and cortex are being developed [11–15]. All these approaches have special benefits and technical or biological and surgical limitations. A new test should not be tailored to the specific features of one individual prosthesis design, but rather be suitable for generic use. This will facilitate comparability of the functional results of different technical approaches in humans.

Implementation of a New Test Battery

The specifications listed above provide the basis for developing a new test design. However, the demands for a new test may be conflicting to some extent. For example, a test designed for evaluating visual performance may be highly objective and reliable, but may not be suitable for assessing how well a device facilitates everyday activities. Conversely a setup for testing performance in daily activities may show deficiencies in objectivity and validity, because such tests by their very nature are subjective and influenced by habituation to the test.

To approach this problem of conflicting requirements a battery of sequential tests may be advantageous. Individual tests are designed to provide partial information focused on one main topic. The synopsis of this battery of tests will provide sufficient information about the efficacy of an AVD in severely visually impaired patients as well as information needed for further development.

Figure 5.1 illustrates schematically the concept of a set of tests based on three different methodological approaches.

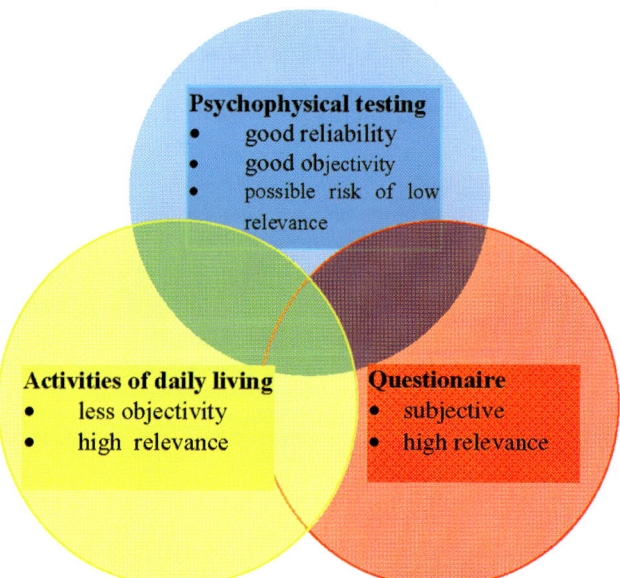

FIGURE 5.1. Test battery design consisting of three main pillars to address conflicting requirements. These pillars are 1st: a set of psychophysical tests, 2nd: a set of tests to estimate performance in tasks of daily living, 3rd: a set of questionnaire to address subjective perception.

The first pillar is based on classical psychophysical testing utilizing and extending regular clinical ophthalmologic tests. These tests can be highly standardized, and examiner and patient bias can be minimized. Thus this pillar of the test design mainly contributes to good reliability, validity, and objectiveness of the whole set of tests. However, this kind of tests could turn out to be the least relevant ones in terms of the needs of a patient in coping with everyday life.

The second pillar therefore consists of a setup for testing performance in everyday tasks. This should consist of tests that are as close to real conditions as possible to be relevant, but as "artificial" as needed to be well standardized. As AVDs may provide visual impressions that are different from natural vision, this test is highly important to gain an estimate about the efficacy of such devices in supporting blind patients in leading a more independent life. Testing performance in activities of daily life is not a well-standardized task and there are some methodological obstacles. As there is no direct physical correlation that can be measured, the test has to rely on indirect measuring techniques. The patient himself, or better in cooperation with an experienced, independent mobility trainer, has to give an assessment of performance in this test. As this proceeding is prone to reasonable subjectivity by nature, strategies to control this handicap gain great importance. These strategies include a double-blind or placebo-controlled test design as well as the use of standardized subjective scales rather than free descriptions or even narrations of the patient.

The third pillar of the test battery is meant to get direct information about the impressions and constitution of the patient while carrying an AVD. While testing specifications and performances of visual functions with visual prostheses, it is important to evaluate the subject's well-being and its implications for test performances. Strong expectation as well as anticlimax or even dysthymic conditions may influence the outcome of the tests and must be controlled. Furthermore information must be recorded about the particular impression using an AVD, about special sensations or discomfort. This information can be provided only by the patient himself and must be addressed in specialized questionnaires. Undoubtedly this pillar is the most subjective one and, when designed to detect short-term fluctuations in psychological conditions, it may not be of high reproducibility, though it is of high reliability.

Test Battery Pillar One: Psychophysical Testing

Since a wide range of visual functions has to be covered, we propose a battery of psychophysical tests with increasing demands on visual performance. At least for the first series of clinical trials presently underway, in most cases, only blind patients can be included, and special consideration must be taken in testing the remaining with very low vision. Even in most blind people some residual visual functions can be noted. If routine methods for assessing visual functions fail, there is a fallback to less standardized methods to categorize remaining functions. Patients are then asked in clinical practice to count fingers held in front of them, or to detect hand movement at a defined distance in front of the patient's face, and it is noted whether the patient can recognize movement or the direction of movement. Finally a bright light beam is directed into the eye and light perception from different directions is evaluated. These commonly used clinical examinations lack standardization, as ambient light, distance, and dimensions of the hand and fingers, as well as the brightness of the light source may be subject to considerable variations.

There is a need for more sensitive and more standardized methods for evaluating these very low visual functions. In analogy to clinical testing we propose a computerized test that is capable of assessing four basic visual qualities. These are *light perception, temporal resolution, location of light*, and *detection of motion*. We developed a software and a technical setup, called the "Basic Light and Motion (BaLM) test," that will be described later on in detail.

With increasing visual performance, additional tests assessing resolution of the visual system are needed. As can be seen in Table 5.3, we propose a three-level procedure. The very basic routine is provided by the BaLM test (level 1) and higher performances can be tested with standardized techniques for determining visual acuity (level 3). However, there might be a gap between these two levels, leaving an area of uncertainty in the case where a patient easily passes the BaLM test, but cannot recognize correctly even large optotypes. This scenario is not unlikely, as various situations can be imagined when resolution gained with artificial vision is sufficient for determining a large optotype, but the patient is

TABLE 5.3. Elements of psychophysical testing.

Basic Light and Motion (BaLM) Test	Evaluation of four basic qualities of vision: light, temporal resolution, light localization, and motion
Grating acuity (BaGA)	Definition of the theoretical capacity of resolution of the device
Freiburg Visual Acuity and Contrast Test (FrACT)	Computerized measurement of visual acuity

still not able to recognize it. Distortion due to various reasons might be one obstacle. Another possible situation that must be considered in which resolution and recognition do not match is the generation of a patchy visual impression. AVDs do not necessarily create a visual image that covers one contiguous area. Rather a kind of patchwork of seeing and not-seeing areas may be present within a certain area with restituted functions. Nevertheless, it is important even in these situations to gain valid information about the resolution and visual acuity an AVD could provide under optimal conditions.

For this reason we suggest a grating acuity test as a second level of psychophysical testing. These tests are well known for determining optical transfer functions of optical devices and use patterns with black and white lines with different spatial frequencies. Certain limitations of this test method in the human visual system are well known and will be discussed later on in detail.

Psychophysical testing can be completed by a standardized test for measuring visual acuity, as in the Freiburg visual acuity and contrast test (FrACT) (see <http://www.michaelbach.de/fract.html>). This is a fully computerized test to estimate the psychometric threshold.

All three of these tests can be performed sequentially, using one common platform, i.e. a computer with a customized display.

Electrophysiological tests in clinical trials of AVDs are of limited value. Their principal application is in experimental animal testing. Measuring the ERG may demonstrate the ability of an AVD to generate electrical potentials at a retinal level or prove function and integrity of the device by recording electrical pulses delivered by an AVD under biological conditions. This is suitable only for retinal prostheses, and even in this case the conclusions from these results may be of limited value.

The VEP can measure activation of the visual cortex after stimulation by an AVD. However, this examination is not necessarily correlated with a useful visual impression. For this reason we expect electrophysiological tests to be of low validity and relevance in clinical trials and would not suggest using them as a tool to determine the efficacy of such devices.

Technical Setup for Psychophysical Testing

All three psychophysical tests mentioned above are computerized tests. Herein we describe a technical implementation as it has been used for evaluating Basic

Light and Motion (BaLM) test and Basic Grating Acuity (BaGA) test in first patients with very low vision, and as it is intended to be used in first clinical trials with retinal prostheses.

Visual prostheses may place particular demands on the technical implementation of these tests. As most of them contain electronics for image transformation of any kind, they are paced with a certain refresh rate. Hardware design of a test has to incorporate this factor to avoid interference with other frequency-driven devices. It appears reasonable to choose a display technology that is capable of generating a quasi flicker-free image, accepting slight losses in contrast, as significant flicker could interfere unpredictably with an AVD. For this reason we propose an LCD projector rather than a DLP projector with color wheel, or CRT.

To be able to cover a wide range of luminances, we designed a setup for maximal brightness. For this reason we propose a projector with 2000 ANSI lumen light flux or higher in conjunction with a short projection distance and small image size. In our setup the projector is mounted upside-down over the patient's head and projects onto a screen at a distance of 60 cm. We use a commercially available screen that is optimized for brightness and contrast. An image of 34×27 cm is created, covering a visual field of $32° \times 25°$ when viewed by the patient (Figure 5.2). This setup generates a luminance at the center of screen of about $5100 \, cd/m^2$ and a Michelson contrast of 99.5%. Using a model eye with a pupil of 6 mm, we measured 140 lux at the retinal level. Lower light intensities can be obtained by using neutral density filters.

FIGURE 5.2. Photograph of the technical setup for testing visual functions. An LCD projector is mounted upside-down over the patient's head and projects onto a screen 60 cm ahead of the patient. The patient uses a keypad for entry.

The software developed for this test is based on Macromedia Flash 7.0, allowing for usage on Macintosh, Linux, and Windows systems. It is designed to run self-paced, meaning that a sequence of tests runs automatically, advanced by the reaction of the patient. This minimizes bias due to interactions between investigator and patient.

Software Solutions: 1. BaLM Test

This test consists of a battery of four modules for perception of light, time, location, and motion. In a preferences screen, the variables of the four tests can be determined and the test can be calibrated to projecting distance and image size.

Upon starting the first subtest the program runs self-paced, requiring the patient to enter an alternative choice response via a numeric keypad. Most visually impaired subjects are familiar with a numeric keypad, and the central key is marked with a tactile prominence, so that this task can be handled easily. At the beginning of each test there is a short sound to prompt a patient's response. After the patient has pressed the corresponding key, two different sounds are played to indicate either a right or a wrong response.

In the first test module the ability of detecting a single flash of maximal brightness is tested. After the initial sound a full-field flash appears, or the display stays black in a random fashion (maximal screen luminance in our setup is $5100\,\mathrm{cd/m^2}$, Michelson contrast of 99.5%). The patient has to indicate whether he has seen a flash or not. The duration of the flashes can be adjusted in the preferences screen. As some AVDs may perform better when stimulated with a high contrast pattern instead of a homogenous Ganzfeld-flash, there is an option to create a black-and-white checkerboard (3° check size). Luminance is varied using neutral density filters to estimate the individual threshold. Hit rate at different brightness as well as the stimulation parameters (Ganzfeld or checkerboard, number of trials) are recorded.

The second module is designated to test temporal resolution. After an auditory prime, one or two flashes are presented in a random fashion with constant time-integrated energy. The patient indicates the number of flashes perceived. The duration of the flashes and the interval between them can be varied. The duration of the single flash is automatically adapted in accordance to the duration of the two flashes. This is necessary to avoid patients solving the test by determining the duration of the total light–dark–light phase rather than by actually detecting two single flashes. Hit rates and stimulation parameters (flash duration, inter-stimulation interval, Ganzfeld or checkerboard, brightness, and number of trials) are recorded.

Recognition of the localization of a light can be tested in the third module. A fixation target appears in the middle of the display, followed after a few seconds by a wedge-shaped bright field with its tip on the central fixation dot. The base is oriented in one of eight directions (up, down, left, right, and four obliques). Figure 5.3 shows a screenshot. The patient is asked to enter the direction, then a randomly chosen new direction is presented. The number of possible directions can also be limited to four (up, down, left, and right). Hit rates and stimulation

FIGURE 5.3. Example image of the localization test: First the fixation spot appears in the middle and will stay for the entire test. After a few seconds a wedge-shaped bright stimulus appears at one of eight directions, in this example on the upper left side.

parameters (size of fixation target, eccentricity of the wedge, brightness, number of trials, and number of choices) are recorded.

The fourth module tests the ability to detect motion (Figure 5.4). A random pattern of white and black hexagons is moving in one of eight directions (up, down, left, right, and oblique); again, the patient has to indicate the direction of movement. The size of the hexagons, the speed of movement, and contrast of the checkerboard can be adjusted. The black and white patterns are designed as hexagons to create a smooth appearance of motion even for oblique directions.

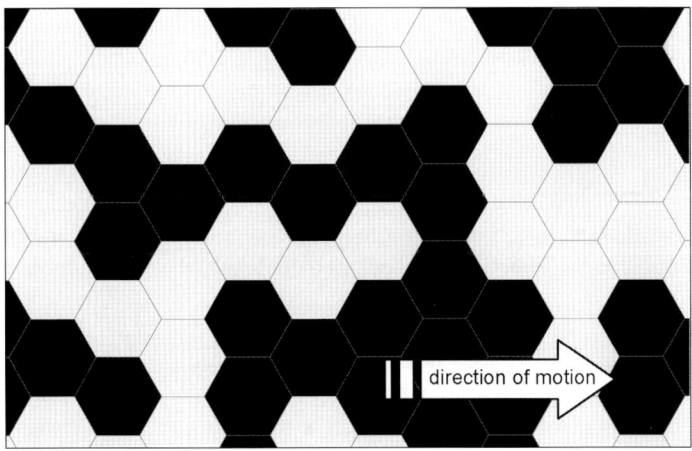

FIGURE 5.4. Example image of the motion test: random pattern of white and black hexagon moving to the right.

Hit rates and stimulation parameters (hex size, contrast, brightness, speed of movement, number of trials, and number of choices) are recorded.

The four tests can be adjusted toward increasing demands for visual functions. The software is not designed to automatically find an individual threshold, it rather tests for an implicit threshold. That is, thresholds are predefined by choosing certain stimulation parameters, the test results are then simply "above threshold," or "below threshold". Changes during a therapeutical trial can be monitored by hit rates and reaction times.

Individual thresholds can be found by repeating these tests with different settings (e.g. different brightness), finding thresholds is described later in the Section "Analysis of test results".

Software Solutions: 2. BaGA Test to Asses Grating Acuity

Gratings are patterns of lines whose brightness changes in a sinusoidal or rectangular pattern. The density of these lines is described by the spatial frequency. Figure 5.5 shows a sinusoidal grating. Gratings have some limitations for testing visual performance in natural systems due to the specific image-processing properties of the visual system [16, 17]. It was demonstrated that the visibility of orientation of the lines does not correlate under all circumstances with visual acuity as commonly measured.

In amblyopic patients (who suffered deprivation of adequate visual stimulation during a sensitive period in childhood) grating acuity can be considerably higher than visual acuity measured with common optotypes. Hess et al. [18] gave a good explanation for this phenomenon. They asked amblyopic patients to draw exactly the pattern they can see when watching a grating. These drawings are shown in Figure 5.6.

As can be seen, the gratings as seen by these patients show a high degree of distortion. This is thought to be due to inadequate neuronal wiring at the visual

FIGURE 5.5. Example image of a high-contrast sinusoidal grating as used in the BaGA test.

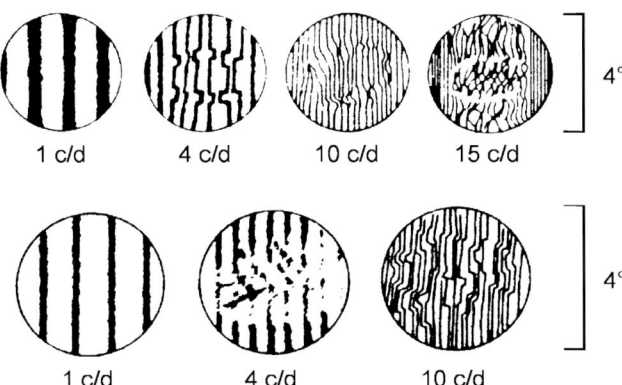

FIGURE 5.6. Gratings as seen by amblyopic patients. With increasing spatial frequency there is considerable distortion, but the main orientation can still be determined [18].

cortex responsible for detection of lines. Despite this distortion, these patients are able to detect the gross direction of the pattern, but they may not be able to detect single optotypes of a similar spatial frequency. Single optotypes bear only one discriminating characteristic that is related to a certain spatial frequency, e.g. the gap in a Landolt C, whereas in gratings this characteristic is repeated several times.

In testing visual functions in patients with retinal prostheses precisely, this feature of grating acuity can be valuable. Grating acuity can provide an estimate of the theoretical resolution that an AVD can provide, even if the individual black and white bars might appear to be distorted or interrupted. As mentioned before, these devices have a limited amount of stimulation electrodes and may create a field of view that is relatively small and possibly non-coherent or even patchy. In this case, single optotypes may not be well identified as they are distorted or might even not be seen as a whole entity.

We have developed a standardized computerized grating acuity test. The test is called "Basic Grating Acuity (BaGA) Test" and presented with the same setup described for the BaLM Test. A round pattern of variable size is presented on the display with a sine-wave grating with a spatial frequency of 0.1, 0.33, 1.0, or 3.3 cycles per degree (Figure 5.7). By using adequate gamma correction an equal mean luminosity of the test pattern and the background is assured. The grating is presented with random orientation in one of four directions and the patient is asked to indicate the orientation via the keypad.

Software Solutions: 3. Freiburg Visual Acuity and Contrast Test (FrACT)

The "Freiburg Visual Acuity & Contrast Test" (FrACT) assesses visual acuity and contrast sensitivity. It is a free computer program that uses dithering, anti-aliasing, and psychometric methods to provide automated, self-paced measurement of visual acuity [19, 20]. It is used by many vision

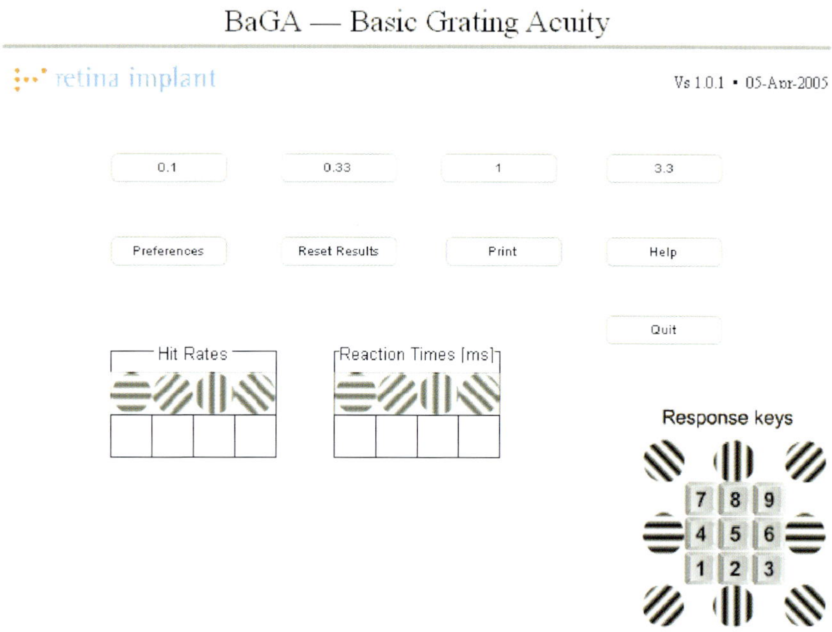

BaGA — Basic Grating Acuity

FIGURE 5.7. Main screen of the BaGA Test. Four different spatial frequencies (0.1; 0.33; 1.0, and 3.3 cycles/degree) can be chosen. In a preferences screen (not shown) stimulus distance, image size and gamma value can be adjusted. After each test hit rates and reaction times are displayed.

labs, optometrists, and ophthalmologists worldwide. FrACT complies with the European Norm for acuity testing EN ISO 8596. It is well established and has recently been demonstrated to provide accurate data even in patients with very low vision [21].

Analysis of Test Results

The BaLM and BaGA tests automatically calculate percentages of correct responses of each test, as well as mean reaction times for each test.

When analyzing individual performance in a psychophysical test, the psychometric function has to be considered [22, 23]. This function describes the correlation between rate of perception of a certain stimulus and strength of that stimulus. For the BaGA test this could be the number of correct responses as it correlates with the spatial frequency of the grating (compare Figure 5.8).

With increasing spatial frequency the number of correct responses will decrease until a basic plateau is reached. This plateau equals the guessing probability.

The psychometric function shows a turning point at which the slope of the curve is steepest. This means that at the turning point changes in the number of correct responses correlate to very small changes in spatial frequency. This is

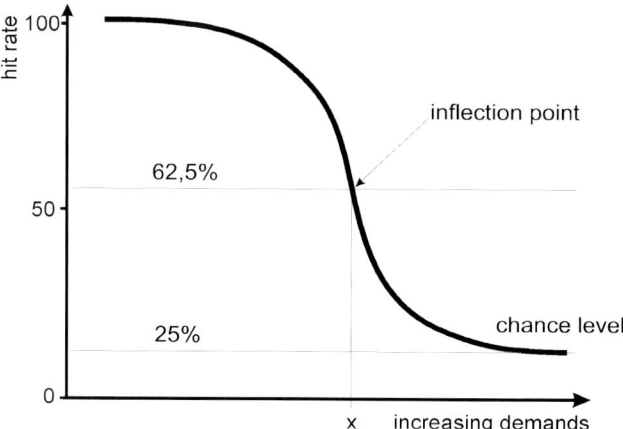

FIGURE 5.8. Psychometric function. The rate of correct responses to a visual test in response to increasing demands like higher spatial frequency is shown. The curve demonstrates that there is no sharp transition between seeing and not seeing but rather an area of uncertainty. The inflection point of the curve where the slope is highest should be defined as the psychometric threshold.

the most sensitive point of the test. For this reason this point should be used to determine a psychometric threshold, as it ensures a good reliability of this test.

The psychometric threshold can be estimated as the mean between guessing probability and 100% correct responses. It can be calculated as follows:

$$\text{Psychometric Threshold } [\%] = (100/n_c) + (100 - (100/n_c))/2$$

where n_c = Number of choices

Accordingly the psychometric threshold for an eight-alternative forced choice, like in the grating test, is 56.25%; for a four-alternative forced choice it would be 62.5%.

As the psychometric threshold is approached, there is increasing uncertainty and the patient has to guess. This is rather uncomfortable for the tested individual and he will tend to give up before actually reaching the threshold. Canceling the test at this point would lead to an underestimation of the psychometric threshold. To avoid this, the BaLM test applies the forced choice principle, in which the patient is required to make a decision even if uncertain. In this way the influence of different criteria ("personalities") can be minimized and the interindividual reproducability of the test can be increased.

Test Battery Pillar Two: Testing Activities of Daily Living and Orientation

Visual prostheses will be available in first clinical trials only to severely visually impaired or completely blind subjects. These people have adopted specialized coping strategies to manage the various tasks of daily living and to maintain autonomy. Restoring basic visual functions by visual prostheses will not necessarily improve these patients' ability to master everyday tasks. However, improvement in the ability to pursue such activities would be a valuable endpoint in determining the efficacy of a visual prosthesis and could turn out to be more relevant than mere visual resolution values.

Therefore, a standardized method of testing performance in everyday activities is needed. In order to avoid as many uncontrollable biases as possible, the test must be carried out under well-defined, slightly artificial conditions. Nevertheless, actual situations encountered in daily life must be represented. In cooperation with an experienced mobility trainer, we have developed a setup for testing activities connected with daily life and orientation.

The test consists of two basic parts. The first one represents tasks in familiar surroundings, like sitting at a dining table and finding the plate, the glass, the cutlery, and other typical objects. The second part tests orientation and the ability to detect prominent markers or obstacles in a new surrounding when moving in public areas.

All tests are performed under the guidance of a mobility trainer. The time needed to fulfill a task is recorded, as is the patient's subjective impression about the difficulty of the tasks. In addition, the trainer records his or her subjective assessment of the performance of the patient.

This test is highly subjective by nature. To achieve the greatest possible objectivity and reliability, a well-planned test design is obligatory. Thus, these tests should be carried out strictly under blinded and controlled conditions to be able to detect even slight improvements in activities of daily living and orientation (ADL/O) as an endpoint of a clinical trial.

The two parts of testing ADL/O are described in detail below.

Testing ADL: the Dining Table Scenery

The patient is seated at a table next to the mobility trainer. The setup is standardized with respect to ambient light and contrast as bright items are used on a black cloth on the surface of the table and ambient light is measured before each session. The patient is informed about the presence of several typical items like a plate, a knife, a spoon, a fork, or a cup in front of him on the table. He is asked to identify these items and to indicate the location of these items according to the hours on a clock face without touching or moving them. It is noted whether the patient can manage this task, and the time needed is recorded. In a second step, he is asked to change the position of certain items (eye–hand coordination). For example, he could be asked to move the cup to 2 o'clock position and the knife to the 3 o'clock position. Again, the time needed is noted. Applying real objects rather than standardized videotapes allows to account for tactile abilities which these patients are trained in very well and to test for eye–hand coordination which could be a demanding task for these patients. This might give a more realistic assessment of the benefit of an AVD in daily living.

The entire test is videotaped for documentation. After each test the patient himself rates how challenging he or she found the task on a seven-grade scale ranging from very easy to very hard. The mobility trainer indicates his estimation of the performance of the patient on a visual analog scale.

Testing Orientation: Simulation of Outside Scenarios

This test evaluates the ability of the patient to orientate himself or herself in unknown surroundings. Typical markers and obstacles encountered when moving on the street are presented, like a pedestrian crossing, a traffic light, or a pillar. In order to avoid hazards and to create standardized conditions, we do not suggest performing these tests in actual outside scenarios at least in first clinical trials. We recommend creating a well-controlled situation, where images can be created under standardized conditions. For this reason, we developed a series of standardized images that can be displayed in a big scale using a video projector. These images are presented as they would be viewed from different distances. For example, a picture of a pedestrian crossing is presented as it can be seen from 10 m, 5 m and 1 m distance (compare Figure 5.9).

It is noted if these objects can be determined correctly, and from what distance they are first recognized. For each picture only a limited period of time is given to solve the task.

Test Battery Pillar Three: Subjective Scales

The last pillar of the test battery takes advantage of standardized questionnaires to acquire information about the patient's personality and psychological condition. Using questionnaires that are well established and evaluated can provide more accurate information than free questioning and narration of the patient.

Subjective scales are valuable tools for assessing the psychological condition of patients in clinical trials. They are gaining significance as when considering quality of life and value-based medicine in clinical trials. A number of well-described tests exist, some of which are specially designed for

FIGURE 5.9. Example images for testing object recognition in outside scenarios. Different typical scenes can be displayed like pedestrian crossings, or common obstacles like pillars. All photos were taken from three different distances.

evaluating quality of life in patients suffering from eye diseases, like the National Eye Institute – Visual Function Questionnaire-25 (NEI-VFQ25, <http://www.proqolid.org/public/nei-vfq25.html>). However, most of these tests are of limited value in severely visually impaired or even blind patients. To be suitable for evaluating the psychological condition in blind patients during first clinical trials, possibly facing tremendous expectations and dissapointments, such tests must meet special requirements. We suggest using the "Brief Symptom Inventory (BSI)" test by L.R. Derogatis (Pearson Assessments) because it is capable of detecting even short-term fluctuations [24–26].

The BSI is a 53-item self-report scale used to measure 9 primary symptom dimensions (somatization, obsessive–compulsive behavior, interpersonal sensitivity, depression, anxiety, hostility, phobic anxiety, paranoid ideation, and psychoticism), and 3 global indices (a Global Severity Index to measure overall psychological distress level, a Positive Symptom Distress Index to measure the intensity of symptoms, and the Positive Symptom Total to report the number of self-reported symptoms). The BSI is a shortened version of the Symptom Check List-90 (SCL-90 [27]), a widely used scale assessing current psychological distress and symptoms in both patient and non-patient populations. The BSI measures the experience of symptoms in the preceding seven days including the day the BSI was completed. Responses are on a 5-point scale, from 0 = "not at all" to 4 = "extremely." Raw scores are converted to T-scores using age- and sex-appropriate non-patient norms. The BSI has high scale-by-scale correlations with the SCL-90. The BSI also has high internal consistency (Cronbach's alpha: 0.71–0.85), test–retest reliability, and convergent, discriminant, and construct validity.

Conclusion

The suggested test battery, consisting of a set of psychophysical tests, tests for assessing performance in everyday activities and questionnaires to evaluate personality and psychological condition as well as novel sensation, while using an AVD allows for standardized estimation of the efficacy of the AVD under investigation. The entire test battery is needed to get a meaningful estimation of patient benefit and visual functions, using well-controlled prospective study designs.

References

1. Ricci F, Cedrone C, and Cerulli L: "Standardized measurement of visual acuity" Ophthalmic Epidemiol. Mar 1998; 5(1):41–53
2. Kniestedt C and Stamper RL: "Visual acuity and its measurement" Ophthalmol Clin North Am. Jun 2003; 16(2):155–170
3. Tamai M, Kunikata H, Itabashi T, Kawamura M, Saigo Y, Sato H, Wada Y, and Nakagawa Y.: "An instrument capable of grading visual function: results from patients with retinitis pigmentosa." Tohoku J Exp Med. 2004 Aug; 203(4):305–312.

4. Marmor MF, Holder GE, Seeliger MW, and Yamamoto S: "Standard for clinical electroretinography" Doc Ophthalmol 2004; 108:107–114

5. Bach M, Hawlina M, Holder GE, Marmor M, Meigen T, Vaegan, and Miyake Y: "Standard for Pattern Electroretinography" Doc Ophthalmol 2000; 101:11–18

6. Odom JV, Bach M, Barber C, Brigell M, Marmor MF, Tormene AP, Holder GE, and Vaegan: "Visual Evoked Potentials Standard" Doc Ophthalmol 2004; 108:115–123

7. Kersten D and Yuille A: "Bayesian models of object perception" Curr Opin Neurobiol. Apr 2003; 13(2):150–158

8. Albright TD and Stoner GR: "Contextual influences on visual processing" Annu Rev Neurosci. 2002; 25:339–379

9. Hayes JS, Yin VT, Piyathaisere D, Weiland JD, Humayun MS, and Dagnelie G: "Visually guided performance of simple tasks using simulated prosthetic vision" Artif Organs. Nov 2003; 27(11):1016–1028

10. Thompson RW Jr, Barnett GD, Humayun MS, and Dagnelie G: "Facial recognition using simulated prosthetic pixelized vision." Invest Ophthalmol Vis Sci. Nov 2003; 44(11):5035–5042

11. Hossain P, Seetho IW, Browning AC, and Amoaku WM: "Artificial means for restoring vision" BMJ. 1 Jan 2005; 330(7481):30–33

12. Alteheld N, Roessler G, Vobig M, and Walter P: "The retina implant–new approach to a visual prosthesis" Biomed Tech (Berl). Apr 2004; 49(4):99–103

13. Loewenstein JI, Montezuma SR, and Rizzo JF 3rd: "Outer retinal degeneration: an electronic retinal prosthesis as a treatment strategy" Arch Ophthalmol. Apr 2004; 122(4):587–596

14. Chow AY, Chow VY, Packo KH, Pollack JS, Peyman GA, and Schuchard R: "The artificial silicon retina microchip for the treatment of vision loss from retinitis pigmentosa" Arch Ophthalmol. Apr 2004; 122(4):460–469.

15. Zrenner E: "Will retinal implants restore vision?" Science 8 Feb 2002;295(5557):1022–1025.

16. Stanley OH: "Cortical development and visual function" Eye. 1991; 5 (Pt 1):27–30

17. Gwiazda JE: "Detection of amblyopia and development of binocular vision in infants and children" Curr Opin Ophthalmol. Dec 1992;3(6):735–740

18. Hess RF: "Assessment of stimulus field size for strabismic amblyopes" Am J Optom Physiol Opt. May 1977; 54(5):292–299.

19. Bach M: "The 'Freiburg Visual Acuity Test' – Automatic measurement of visual acuity" Optometry and Vision Science 1996; 73:49–53

20. Dennis RJ, Beer JM, Baldwin JB, Ivan DJ, Lorusso FJ, and Thompson WT: "Using the Freiburg Acuity and Contrast Test to measure visual performance in USAF personnel after PRK" Optom Vis Sci 2004; 81:516–524

21. Schulze-Bonsel K, Feltgen N, Burau H, Hansen LL, and Bach M: "Visual acuities 'Hand Motion' and 'Counting Fingers' can be quantified using the Freiburg Visual Acuity Test." Invest Ophthalmol Vis Sci 2006; 47:1236–1240

22. Green DM andSwets JA (1966) Signal detection theory and psychophysics. New York, Wiley

23. Harvey LO: "Efficient estimation of sensory thresholds" Behavior Research Methods, Instruments, Computers 1986; 18:623–632

24. Alagappan K, Steinberg M, MancherjeN, Pollack S, and Carpenter K: "The psychological effects of a four-week emergency medicine rotation on residents in training" Academic Emergency Medicine 1996; 3(12): 1131–1135

25. Aroian KJ and Patsdaughter CA: "Multiple-method, cross-cultural assessment of psychological distress" Image – the Journal of Nursing Scholarship 1989; 21(2): 90–93

26. Boulet J: "Reliability and validity of the Brief Symptom Inventory" Psychological Assessment 1991; 3(3):433–437

27. Vonk ME and Thyer BA: "Evaluating the effectiveness of short-term treatment at a university counseling center" Journal of Clinical Psychology, 1999; 55(9), 1095–1106

6
The IMI Retinal Implant System

Ralf Hornig[1], Thomas Zehnder[2], Michaela Velikay-Parel[3], Thomas Laube[4], Matthias Feucht[5] and Gisbert Richard[5]

[1]*IMI Intelligent Medical Implants GmbH, Bonn, Germany*
[2]*IMI Intelligent Medical Implants AG, Zug, Switzerland*
[3]*Division of Ophthalmology, University Hospital Graz*
[4]*Division of Ophthalmology, University Hospital Essen, Essen, Germany*
[5]*Division of Ophthalmology, University Medical Center Hamburg-Eppendorf, Hamburg, Germany*

Introduction

The IMI Retinal Implant System currently under development is a new therapeutic approach for restoring vision in patients with retinal degeneration displaying a diverse pathogenesis such as retinitis pigmentosa (RP) and age-related macular degeneration (AMD).

It is known that, in most cases of retinal degeneration, the photoreceptors degenerate while a large number of the other nerve cells in the retina remain intact [1–4]. This is also true for the ganglion cells which form the output cells of the retina. If the output cells of the retina are intact and can be stimulated electrically, artificial excitation of visual sensations is possible. Electrical stimulation has been extensively investigated since the time of Galvani and Volta. Today functional electrical stimulation is performed by several other medical devices, e.g. cochlear implants and pacemakers.

The activities of Intelligent Medical Implants (IMI) go back to 1998, when the company Intelligent Implants GmbH was established. The goal of Intelligent Implants was to bundle the available know-how on retinal implants in order to develop the first commercially available device that could be used by blind patients. Information technology for the preprocessing of image and stimulation signals is an area of special focus at IMI. Computer-aided learning processes are used to adapt the retinal implant to the needs of individual patients.

IMI is currently developing an epiretinal Learning Retinal Implant. Its operation is based on the transformation of image signals from an extraocular camera into sequences of current pulses applied by implanted microelectrodes. The microelectrodes are placed epiretinally in the area of the macula. The images from the camera are processed by a signal processor into stimulation signals that are conveyed via wireless transmission to the implanted Retina Stimulator.

Functional models of the Retina Implant system have now been developed and the first tests have been performed with animals and humans. The safety of acute electrical stimulation has been demonstrated in a preclinical study. Following these tests, IMI performed an acute human trial involving 20 subjects suffering from RP.

Retinal Implant Technology

Overview

The aim of the Learning Retinal Implant System is to restore vision in blind people. This is accomplished by capturing images of the environment with a digital camera, processing these images into electronic data signals, and then sending them by wireless transmission to a receiver implanted in the eye. This receiver translates these data signals into electric stimulation currents which are applied through microelectrodes to the epiretinal side of the eye. This leads to an activation of the underlying nerve cells that eventually elicits visual perceptions in the brain.

The Retinal Implant device consists of three main components: the Retina Stimulator, the Visual Interface, and the Pocket Processor. Only the Retina Stimulator is implanted in the eye. The Visual Interface and the Pocket Processor are external components. Additionally, a special software running on a standard PC is provided. The software allows the user to tune certain parameters to meet the needs of individual patients.

External Components

The Visual Interface consists of several electronic components (camera, data, and energy transmitter) mounted in the frame of eyeglasses; it serves to record visual information and to send data and energy to the implanted Retina Stimulator. The Visual Interface is connected via a cable to a Pocket Processor that can be carried on a waist belt or a strap and is responsible for image processing and power supply.

The Visual Interface (Figure 6.1) carries the digital camera; it records images of the environment and sends the data via cable to the Pocket Processor. The processed data is transmitted back to the Visual Interface via the same cable. The Visual Interface contains a wireless communication unit that transmits data and energy to the Retina Stimulator (Figure 6.2). For this energy transmission an electromagnetic approach is preferred. For the transmission of data an optical channel is possible. The transmission of both energy and data has, as far as possible, to be independent of eye movement.

The camera in the Visual Interface features a dynamic range which allows for vision both indoor and outdoor without additional light sources.

The Pocket Processor contains proprietary software that processes the camera images and translates this information into data signals for the Retina Stimulator.

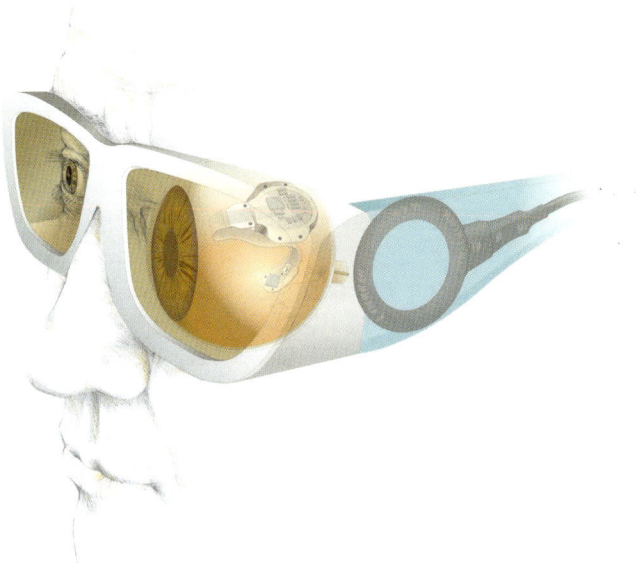

FIGURE 6.1. Retinal Implant System.

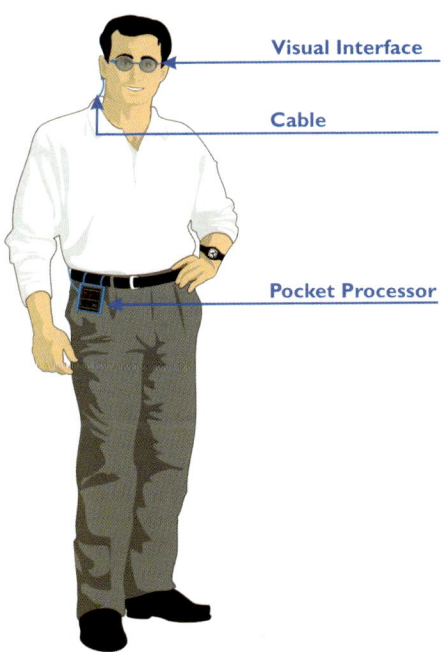

FIGURE 6.2. Visual Processor components of the Learning Retinal Implant System.

In addition, the Pocket Processor contains a rechargeable battery providing an energy supply for the entire system and has connections for a battery charger and a PC. It also has control keys and is able to send acoustic signals. The battery pack for the Pocket Processor is connected to a metal coil, located in the frame of the eyeglasses, which creates a high-frequency alternating electromagnetic field. This allows wireless transmission of the energy to the implant in the eye.

The processed visual information is sent from the Pocket Processor to an infrared (IR) data transmitter inside the Visual Interface, located in front of the eye, which delivers the entire data stream for retinal stimulation to the IR receiver on the implant in the eye via IR light-emitting diodes.

The Visual Interface and the Pocket Processor are connected via a cable that is placed behind the ear and possibly under the clothes (see Figure 6.2).

Retina Encoder

In a healthy retina, image processing is performed by several types of cells. The photoreceptor layer with its rods and cones constitutes the functional input layer. The representation of the visual information at the level of the photoreceptors is comparable to that of a photosensor array in a technical image-processing system. In the natural retina, the image information is processed with the network of horizontal, bipolar, amacrine, and ganglion cells. Photoreceptors, horizontal cells, and bipolar cells process the information with electrotonic potentials. Ganglion cells have a spiking information process whereby the information is coded in the pulse frequency and timing of the action potentials. Amacrine cells process information in both ways, electrotonic and spiking. Retinal implants with epiretinal electrodes stimulate surviving nerve cells at the output of the retinal network. The information processing of the degenerated input network has to be simulated. Information processing in the primate retina has been the subject of extensive investigation [5–7]. The Retina Encoder algorithm can be developed on the basis of these measurements [8, 9]. The first implementation of the Retina Encoder simulated the ganglion cell output of single ganglion cells. However, single cell activation with non penetrating electrodes appeared to be unrealistic since different cells lay close together and the stimulation electrodes have to be relatively large to provide the necessary current. It has to be assumed that microelectrodes in the order of $100\,\mu m$ and larger stimulate up to several hundreds of cells. Therefore the Retina Encoder has to adapt its algorithm to a summary of the information processes in all the stimulated cells. It is possible that one information processing type is dominant in the stimulated cell group.

It is known that the degeneration of the light-sensitive photoreceptor layer triggers a pathological process [10, 11]. As a result of this process, the structure of the retina is irreversibly altered. Consequently, the Retina Encoder has to be even more flexible than is assumed in the literature on the retinal information process.

The Retina Encoder can be implemented as a set of spatiotemporal filters (Figure 6.3). Every filter has a spatial input area within the input images. For temporal processing a series of input images is taken into account. The output of each filter constitutes the stimulation commands for individual electrodes.

The Retina Encoder has a large number of intrinsic parameters. The effect of some parameters, such as the size and location of receptive fields, can be easily understood. Some other parameters (e.g. parameters of spatial or temporal filters) are more complex and their impact on the resulting image quality is not obvious. In addition, the parameters of the encoder are often not independent of each other.

For the tuning of the Retina Encoder, algorithms are developed that permit efficient parameter optimization. This tuning procedure will be carried out during a rehabilitation period subsequent to the implantation of the Retina Simulator.

Figure 6.4 shows the procedure used to tune the Retina Encoder. This data is processed via simulation using subjects with normal vision [12, 13]. Therefore, the Retina Encoder output was translated back into an image using an inverter module. This inverter module has the inverse function of the Retina Encoder. The inverter module is valid for a certain set of Retina Encoder parameters. Subsequently, the Retina Encoder parameters were untuned. The input image of the Retina Encoder and the output image of the inverter module are shown on a screen. Simple geometric moving figures (e.g. moving circles) were used as the input images. A test subject looking at a screen can see the input image (on the left in Figure 6.4) and the output image (one of the images on the right in Figure 6.4). The output image is distorted because the Retina Encoder has been untuned. During the tuning process, the Retina Encoder parameters have to be adjusted so that the output image matches the input image. At the beginning of the tuning process the system randomly generates several sets of parameters for the Retina Encoder. The test subject has to choose the sets that result in output images which are most comparable with the input image. Once the test subject has made his or her choice, the system generates new sets of parameters

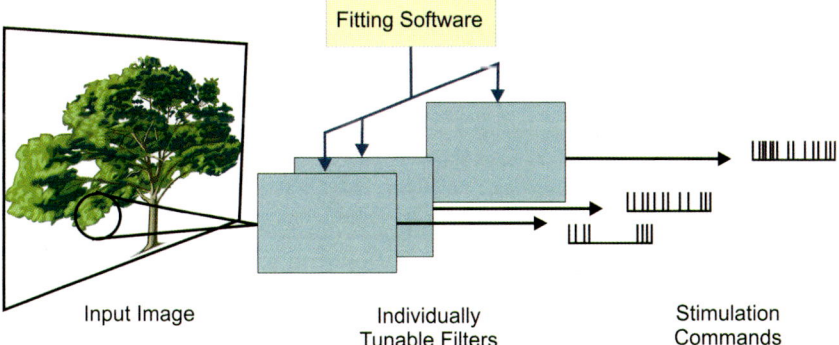

FIGURE 6.3. Retina Encoder information processing.

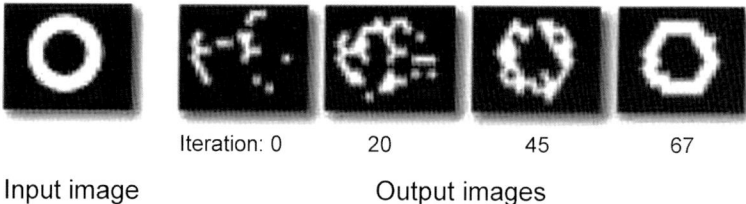

Iteration: 0   20   45   67

Input image   Output images

FIGURE 6.4. Learning procedure (input image, output image after 0, 20, 45, 67 iterations).

for the Retina Encoder. While doing this, the system takes consideration of the previously chosen parameter sets. Afterward, the test subject appraises the new parameters in the same way as the first. This procedure is repeated until the parameters are good enough to perceive the original image at the output of the inverter module. The experiments carried out in the laboratory showed that this is possible after fewer than 100 iterations (see Figure 6.4).

Retina Stimulator

The implantable Retina Stimulator consists of a flexible plastic carrier onto which the various microelectronic components are mounted. The electronic components are used on the one hand to receive data and energy and on the other hand to electrically stimulate the retina with microelectrodes. The Retina Stimulator is implanted into one eye only and has no wire connection to the other parts of the system. The Retina Stimulator has no batteries or energy storage capabilities.

Figure 6.5 shows a functional model with 49 electrodes which has been implemented for animal tests. The carrier of the Retina Stimulator is a flexible circuit board made of polyimide with built-in conducting paths in gold. For long-term hermeticity an additional encapsulation is planned.

The Retina Stimulator has an extraocular part that consists of the transceiver unit containing a high-frequency receiver coil that receives the electromagnetic field from the external sending coil and provides the energy required for the entire stimulation electronics. The extraocular part is fixed with sutures at the sclera. It is connected via a flexible film with the intraocular part.

The intraocular part contains an IR receiver which receives data containing visual information from the external transmitter via an IR optical link permitting high data transmission rates. The IR receiver translates the optical signals into electrical impulses and sends them to the stimulator electronics. Eyelid closing will cause an interruption of data transfer and thereby prevent sight – exactly as would happen in individuals with normal vision.

The stimulator electronics processes data and controls the stimulation currents for the electrode array placed on the retina. The electrode array is composed of stimulation electrodes and is positioned at the epiretinal side of the retina in

FIGURE 6.5. Functional model of the Retina Stimulator with 49 electrodes.

the area of the macula. The electrodes are in direct contact with the retina and activate the retinal nerve cells via electrical currents.

To provide a reference potential, a common electrode is positioned at some distance to the electrode array.

The Retina Stimulator can be fastened to the retina by driving one or several surgical retina tacks into the posterior part of the Retina Stimulator [14–16].

The application of electric current is performed by the small electrodes of the microcontact array. To attain stimulation with a good selectivity – which means that only a few cells are selectively stimulated – the electrodes should be as small as possible. With small electrodes, however, only low charges can be applied because the electrodes have limited charge capacities that are a function of the electrode materials.

The charge capacity of an electrode material is the maximum amount of charge per unit area that can be passed in a biphasic pulse without causing electrode damage. Robblee and Rose [17] give a good overview of electrode materials for microelectrodes. The charges that can be safely applied are proportional to the surface of the electrode. IrO_x, which has a large charge capacity, is a useful electrode material [28]. Figure 6.6 shows electrode test structures coated with IrO_x. The electrode surface is electroplated with IrO_x and is electrically activated.

For patients with high thresholds, the stimulation charges have to be increased significantly over 100 nC [3, 18, 19]. Consequently, the electrodes have to be relatively large even though small electrodes have a better selectivity. The optimal electrode size can be defined as the one that is able to apply stimulation charges above the threshold value in nearly all patients. To identify the optimal size a clinical study was performed; this study is described in the Section "Clinical Study".

Figure 6.6 shows two test structures with IrO_x electrodes that have been designed for preclinical and clinical acute trials. The electrode arrangement and the outer geometry will be different in a later chronic retinal implant. The

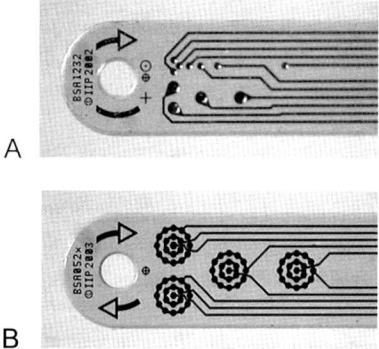

A

B

FIGURE 6.6. Microcontact films for electrical stimulation of the retina.

electrodes of film A have the diameters 50 and 100 μm. Film B has segmented electrodes with effective electrode diameters of 50, 200 and 360 μm.

The electrodes of the microcontact film were tested by impedance spectroscopy at frequencies ranging from 1 to 100 kHz. Figure 6.7 shows the impedance plot of a single 360 μm electrode. The impedance at 1 kHz is 4 kΩ.

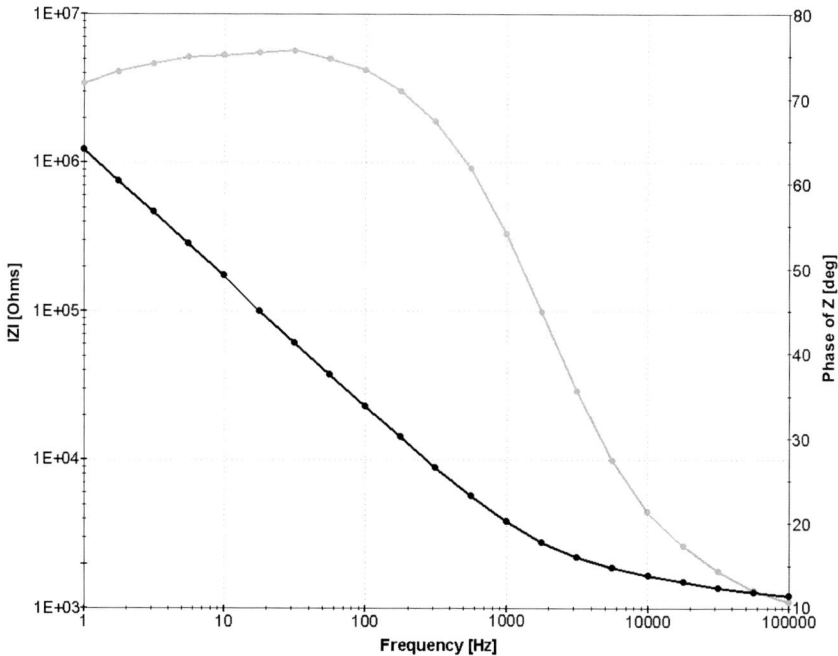

FIGURE 6.7. Impedance plot of a single 360 μm electrode in a 0.9% NaCl solution. The impedance magnitude is shown in black and the phase angle in gray.

Preclinical Studies

The safety and biocompatibility of electrical stimulation of the retina was tested in a preclinical study performed with Göttinger mini pigs [20, 21]. An electrode array, as shown in Figure 6.6, connected via a cable to an external current source was temporarily implanted into the eyes of 12 mini pigs. A three-port vitrectomy was carried out under general anesthesia; the microcontact film was inserted through a scleral port and fixed epiretinally in the area of the macula with perfluorodecaline (PFCL). The mini pigs were randomly assigned to four groups. Group 1 was a control group with no stimulation. In groups 2, 3, and 4, continuous stimulation was performed for 4 hours with charge densities of 0.5, 1, and $4\,mC/cm^2$, respectively. After the planned stimulation, the electrode array was removed and the eyes were examined by indirect ophthalmoscopy. The surgical openings were then closed and the animals were followed for an additional 2-week period, after which they were sacrificed for histological examination of the retina.

The following results were obtained: immediately after stimulation, the eyes of two animals in group 4 (stimulation with $4\,mC/cm^2$) developed severe edema in the stimulated area. In groups 1, 2, and 3, no changes were observed. Two weeks after stimulation, the retinas of the animals in groups 1, 2, and 3 exhibited sporadic pigment coating; in two animals belonging to group 4, the retinas showed pale areas locally. Histological evaluation revealed no structural changes in groups 1, 2, and 3. One animal in group 4 displayed a curved structure in the retina which correlated with the observed edema. Based on these results, we concluded that stimulation of the retina with charge densities of up to $1\,mC/cm^2$ for 4 hours is safe and does not induce any tissue damage.

Materials that are in direct contact with the human body can be tested according to a detailed set of standards to determine their biocompatibility (ISO 10993-1 ff). For implantable devices which stay in contact with tissue or bone for > 30 days, the following tests are required.

- Cytotoxicity
- Sensitization
- Irritation
- Acute Toxicity
- Subchronic Toxicity
- Genotoxicity
- Implantation
- Chronic Toxicity
- Carcinogenicity.

In view of the acute human trial planned by the IMI, the processed polyimide films, including gold paths and pads, have already passed the biocompatibility tests for cytotoxicity, sensitization, and irritation.

Clinical Study

Purpose

It is known from previous studies that visual perceptions are elicited when the retina of patients with RP is stimulated electrically. The minimum charge that must be supplied to each electrode is the perceptual threshold charge. This value may differ for individual patients. For blind human beings it may even exceed the charge capacity of modern electrode materials [19, 22]. Knowing that this value has a major influence on the design of a Retinal Implant System, IMI started an acute clinical trial to evaluate the threshold charge for patients with RP [18, 23–25].

Materials and Methods

The study performed to determine the minimal threshold currents in blind patients was designed as an open, acute, multi-center study involving the following clinical centers and investigators: Division of Ophthalmology, University Hospital Hamburg-Eppendorf, Prof. Dr. Gisbert Richard; Division of Ophthalmology, University Hospital Essen, Dr. Thomas Laube; Division of Ophthalmology, RWTH Aachen, Prof. Dr. Peter Walter; and General Hospital of the City of Vienna, Division of Ophthalmology and Optometry, Prof. Dr. Michaela Velikay-Parel.

The study was approved by the institutional review boards concerned and satisfied the requirements of Directive 93/42/EEC and the Declaration of Helsinki.

Twenty human subjects with RP were recruited. The general inclusion criteria were volunteers, aged between 18 and 79 years, suffering from RP with best corrected vision (ETDRS) of the study eye $\leq 4/200$, normal hearing and linguistic understanding and good general condition. The general exclusion criteria were as follows: the study eye was the better or only eye; concomitant ophthalmologic diseases (e.g. optic nerve atrophies with the exception of those related to retinitis pigmentosa, diabetic retinopathy, glaucoma, etc.), serious general illness, cardiac pacemakers or other electronic implants, and pregnancy or breastfeeding. All subjects were briefed comprehensively about the study procedure as well as the associated risks and signed a statement of informed consent.

All participating subjects were prepared psychologically for the test. The procedure was discussed in detail and all subjects attended a special training. It was assumed that the subjects could have visual perceptions that were not caused by electrical stimulation but by the intraocular light source or mechanical irritation. Therefore, one focus of the training was to teach the subjects to distinguish true test signals from interfering signals. This training was performed with tactile signals instead of electrical stimulation signals using a hand-held device. Stimulation events consisted of three mechanical vibrations of 0.5 seconds each accompanied by a sound and spaced at 0.5 second intervals. Interfering audio signals were mixed in and the subjects had the task of detecting the true stimulation signal in a noisy environment. This training was carried out one day before the electrical stimulation of the retina.

The intraocular electrical stimulation of the retina was carried out during regular eye surgery performed under local subconjunctival anesthesia. Four of the eye muscles were fixed (muscle fixation was not performed at one center). A pars plana vitrectomy with focus on complete removal of the vitreous was performed. Afterward, the electrodes were introduced into the eye with a specially developed device (MESE 12 system; Figure 6.8) and placed onto the macula. The "MESE 12" system is a surgical instrument for controlled positioning of a microcontact film onto the retina. It has to be held in the surgeon's hand during the entire stimulation procedure.

Two different microcontact films were used, as shown in Figure 6.6. Electrodes with diameters of 100, 200, and 360 μm were tested. Moreover, each microcontact film had one common electrode with a surface area of $0.25\,\text{mm}^2$, positioned 5.5 mm away from the stimulation electrodes. All electrodes were made of iridium oxide.

The stimulation currents were generated by a neurostimulator, the STE10, developed by IMI. This device has 10 channels that can be programmed to generate currents between 0.5 and 380 μA. It is powered by batteries and permits monitoring of the voltage of one selected electrode. The device has been successfully tested and certified according to the EN 60601-1 standard. The STE10 was connected by a cable to the MESE 12 system.

For each electrode size, five charge density levels were used (0.05, 0.1, 0.2, 0.5, and $1\,\text{mC/cm}^2$). The pulse frequency was always 60 Hz. Anodic-first charge-balanced current pulses were used. The anodic pulse had a duration of 2 ms and the cathodic pulse had a duration of 1 ms. Between the anodic and the cathodic pulses, the electrode was open-circuited for 100 μs.

To reduce the risk of mechanically damaging the retina, the subjects were not allowed to talk during the stimulation time. They responded to questions posed by the study investigator by pressing pushbuttons held in their right and left hands.

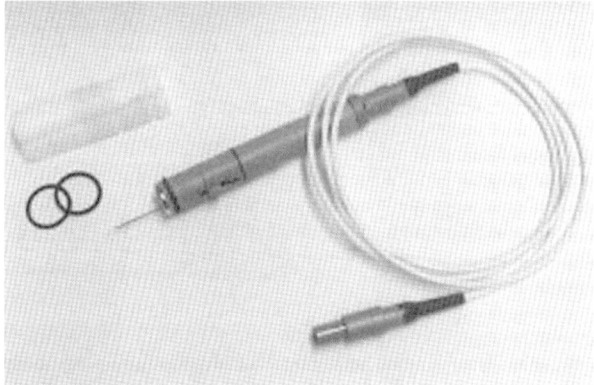

FIGURE 6.8. MESE 12 system with integrated microcontact film.

Threshold determinations were carried out using three consecutive current pulse trains. Each pulse train had a duration of 0.5 s. Consisting of 30 pulses thereby allowing a synchronous triggering of 30 ganglion cell spikes. Synchronously with each electric pulse train, a sound was presented to the subject to indicate the stimulation process. This method allowed the subjects to discriminate visual perceptions generated by true electric stimulation from visual perceptions produced by other sources. As indicated above, the subjects were trained to react only to visual sensations in combination with a sound.

A computer with special software was used as a control and data log system. The computer was equipped with an interface card for recording of electrode voltages and subject answers. All data was continuously logged in a file for subsequent analysis. For immediate analysis, the data was presented on a display. The computer also performed speech synthesis for the subject's answers.

The threshold determinations provided information on the minimum charge necessary to elicit a visual perception by electrical stimulation of the retina. We defined the perceptual threshold as the lowest charge applied to the retina that reliably evoked a visual perception for a given electrode. Every threshold determination consisted of two parts. In the first part, the threshold was determined; in the second part, the threshold was verified. The verification was realized by applying stimulation currents slightly above the determined threshold mixed with randomly inserted "placebo" stimulations. For placebo stimulations no electric current was used but only a sound was presented to indicate stimulation. The verification was deemed successful if no more than 2 errors out of 10 events (5 stimulations and 5 placebos) occurred.

The threshold determinations always began with stimulation at the lowest intensity level (Figure 6.9). After each stimulation event, the subject was asked

Intensity Level												
5						1	1					
4				1	0			1	1		1	1
3			0						0			
2		0										
1	0											
	1	2	3	4	5	6	7	8	9	10	11	12
Pass												

FIGURE 6.9. Hypothetical sequence of a threshold determination. 0 = Subject answers "No," 1 = Subject answers "Yes." In the example shown, the threshold was found at intensity level 4.

if he or she had seen something while hearing the sound. If the subject answered with "no," the current was increased. If the subject answered with "yes," the stimulation event was repeated with the same current. If the subject answered a second time with "yes," the current was decreased until the answer was "no." After that, the current was increased again and the tests were repeated until the subject answered "yes" twice in a row in response to the same stimulation current.

The entire stimulation procedures were carried out within a maximum time of 45 minutes. At the end of the tests, the subjects' eyes were examined for possible damage. The surgical openings were then closed and the subjects were treated according to the hospital's usual standard of care. Immediately after the operation, the subjects were asked to briefly describe their visual sensations; a detailed interview was conducted approximately one hour after stimulation. The subjects were followed up by performing ophthalmologic examinations one day, one week, and three months, respectively, after surgery.

Results

Twenty subjects were tested during the time period between October 2003 and July 2004. The follow-up examinations were completed in October 2004. One subject had no perception after electrical stimulation. Four subjects reported visual perception but it was not possible to verify their thresholds. All other subjects had perceptions with verified threshold measurements.

Within the limited time of maximum 45 minutes, approximately 6 threshold measurements were possible; the exact number is determined by the surgeon after evaluating the patient's condition. According to the definition of the threshold, the lowest value was taken into account for the analysis. For the 15 subjects with verified threshold measurements, the mean threshold after taking into account all used electrode diameters using single and multiple electrodes was 191.5 nC with a standard deviation of 189.7 nC. The minimal threshold was 20 nC measured in a late stage RP patient who had a visual acuity of light localization.

In the standardized postoperative interview the subjects reported light perception of different shapes, colors, and brightness. If the subjects reported more than one type of object, every type of object was given the same weighting in the analysis performed for the individual subject. For the overall analysis every subject had the same weighting.

The perceptions of the subjects covered a wide spectrum. The subjects described the sensations as little stars, points, circles, triangles, rectangles, a half moon, a solar eclipse, a hash (pound) symbol, etc. even if only one electrode was activated.

The sizes of the objects are shown in Figure 6.10. To describe the size of the objects, the subjects had to imagine that the objects were at a distance of 1 m. One subject was not able to imagine the object at this distance; this subject

Size of objects

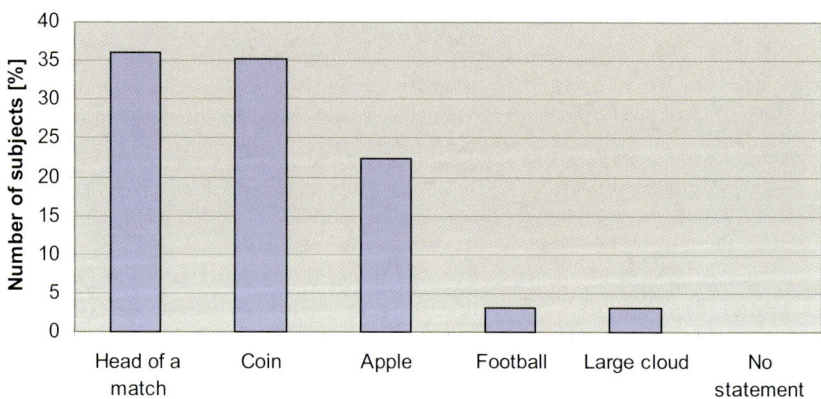

FIGURE 6.10. Size of perceived objects (imagined at a distance of 1 meter).

perceived the object immediately in the eye. Based on the description and the gestures of the subject, the interviewer assumed that the size at a distance of 1 m was about as large as an apple. Most of the subjects had perceptions in the size of a coin or the head of a match.

It has to be mentioned that the object characteristics described in the postoperative interview provide an idea of the overall impression of the subjects. This includes stimulation at threshold level and also stimulation above threshold. The stimulation above threshold may be significant as only five intensity levels from 20 to 380 nC were used. Between one level and the next, the intensity is approximately doubled.

The brightness of the perceived objects is shown in Figure 6.11. Most of the subjects had perceptions with brightnesses comparable to a candle or a light bulb. Objects with very low brightness and extremely bright objects were rare. One subject saw dark shadows instead of light. The threshold of this subject was extremely low and stimulation was not applied significantly above threshold. It is unclear whether the subject would perceive brighter objects if larger stimulation currents were applied.

Objects appeared in a wide range of colors (Figure 6.12). The predominant colors were white, yellow, and blue (see Figure 6.12). Red and green were rare. The perception "black" was reported by the subject who saw shadows instead of light. In general, one perceived object has one color. Different objects perceived by the same subject can have different colors. One subject saw an object that was filled with a mosaic of different colors.

The subjects were asked to categorize the sharpness of the perceived objects. The results are shown in Figure 6.13. It can be seen that most objects were seen sharply and clearly.

Brightness of objects

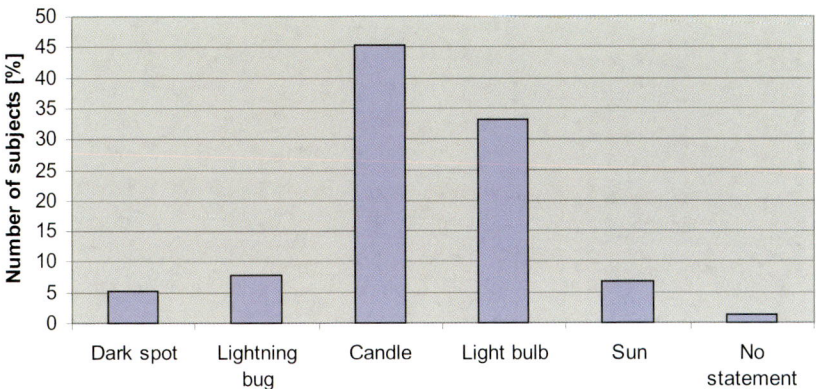

FIGURE 6.11. Brightness of perceived objects.

The subjects were asked whether their visual experience had been pleasant, unpleasant, or painful. Almost all subjects described the objects perceived during the threshold measurement as pleasant. Only two of the subjects said the perceived objects were neither pleasant nor unpleasant or painful.

Except for a peripheral retinal detachment in one subject, there were no significant complications or adverse events during the three-month follow-up period. The detached retina was subsequently reattached.

Color of objects

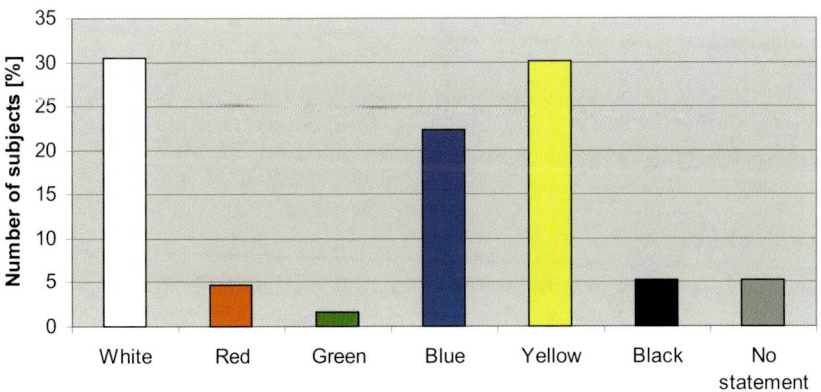

FIGURE 6.12. Color of perceived objects.

Sharpness of objects

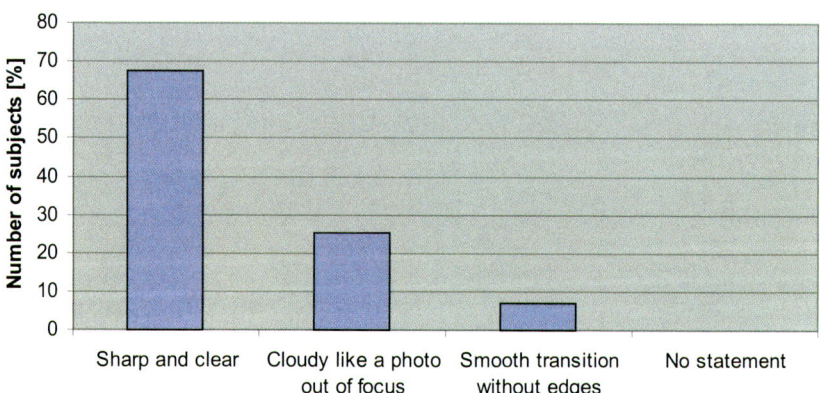

FIGURE 6.13. Sharpness of perceived objects.

Conclusions

Clinical trials have demonstrated, in a limited number of patients, that it is possible to elicit visual perceptions by electrical stimulation of the retina using the epiretinal approach [3, 4, 18, 19, 22, 23, 26, 27].

The collected data show that functional electrical stimulation of the retina is possible in patients with retinitis pigmentosa, i.e. in totally blind persons. The perceptions described by the subjects were typically small and are useful for a chronic retinal implant. The measurement of thresholds showed a mean threshold of 191.5 nC with a significant standard deviation of 189.7 nC. Given the requirement of an electrochemical biocompatible charge limit of $1 \, mC/cm^2$ and an electrode diameter of $360 \, \mu m$, 19 out of 20 subjects experienced visual perception evoked by electrical stimulation.

The perception of the majority of subjects in the acute clinical trial carried out with RP patients has proven to be valuable for the realization of a chronic retina implant.

Acknowledgment. This work is supported by IMI Intelligent Medical Implants, Bonn, Germany, and by the European Community (Healthy Aims Project, Contract No. 001837).

References

1. Stone JL, Barlow WE, Humayun MS, de Juan EJ, and Milam AH (1992) Morphometric analysis of macular photoreceptor and ganglion cells in retinas with retinitis pigmentosa. *Arch. Ophthalmol.* **110**: 1534–1639
2. Santos A, Humayun MS, de Juan EJr, Greenberg RJ, Marsh MJ, and Milam AH (1997) Inner retinal preservation in retinitis pigmentosa: a morphometric analysis. *Arch. Ophthalmol.* **115**: 511–515

3. Humayun MS, de Juan EJ, Weiland JD, Dagnelie G, Katona S, Greenberg R, and Suzuki S (1999) Pattern electrical stimulation of the human retina. *Vision Research* **39**: 2569–2576

4. Humayun MS, Weiland JD, Fujii GY, Greenberg R, Williamson R, Little J, Mech B, Cimmarusti V, van Boemel G, Dagnelie G, and de Juan EJ (2003) Visual perception in a blind subject with a chronic microelectronic retinal prosthesis. *Vision Research* **43**: 2473–2581

5. Lee BB (1996) Receptive field structure of the primate retina. *Vision Research* **36**: 631–644

6. Dacey DM and Peterson MR (1992) Dendritic field size and morphology of midget and parasol ganglion cells of the human retina. *Proc. Natl Acad. Sci.* **89**: 9666–9670

7. Watanabe M and Rodieck RW (1989) Parasol and midget ganglion cells of the primate retina. *J. Comp. Neruol.* **289**: 434–454

8. Eckmiller R (1997a) Learning retina implants with epiretinal contacts. *Ophthalmic Research* **29**: 281–289

9. Eckmiller R, Neumann D, and Baruth O, (2005) Tunable retina encoders for retina implants: why and how. *J Neural Eng* **2**: S91–S104

10. Jones BW, Watt CB, Frederick JM, Baehr W, Chen CK, Levine EM, Milam AH, LaVail MM, and Marc RE (2003) Retinal remodeling triggered by photoreceptor degenerations. *J. Comp. Neurol.* **464**: 1–16

11. Marc RE and Jones BW (2003) Retinal remodeling in inherited photoreceptor degenerations. *Mol. Neurobiol.* **28**(2): 139–147

12. Eckmiller R, Becker M, and Hünermann R (1997b) Dialog Concepts for Learning Retina Encoders. In *Proceedings ICNN 97, June 1997, Houston Texas*

13. Becker M, Hünermann R, and Eckmiller R (1999) Testing perception-based retina implant tuning with normal vision. *In Proceedings ARVO Annual Meeting Ft. Lauderdale*, Abstract 3880

14. Walter P, Szurman P, Vobig M, Berk H, Lüdtke-Handjery HC, Richter H, Mittermayer C, Heimann K, and Sellhaus B (1999) Successful long-term implantation of electrically inactive epiretinal microelectrode arrays in rabbits. *Retina, The Journal of Retinal and Vitreous Diseases* **19**(6): 546–552

15. Szurman P, Walter P, Berk H, and Heimann K (1998) *Investigative Ophthalmology & Visual Science* **39** Abstract 4576

16. Majji AB, Humayun MS, Weiland JD, Suzuki S, D'Anna SA, and de Juan E Jr. (1999) Long-term histological and electrophysiological results of an inactive epiretinal electrode array implantation in dogs. *Invest. Ophthalmol Vis. Sci.* **40**(9): 2073–2081

17. Robblee LS and Rose TL (1990) Electrochemical guidelines for selction of protocols and electrode materials for neural stimulation. *In Agnew WF, McCreery DB Neural Prostheses* Prentice-Hall, Englewood Cliffs 25–66

18. Richard G, Feucht M, Bornfeld N, Laube T, Rössler G, Velikay-Parel M, and Hornig R (2005) Multicenter Study on Acute Electrical Stimulation of the Human Retina With an Epiretinal Implant: Clinical Results in 20 Patients. *In Proceedings ARVO Annual Meeting Ft. Lauderdale*, Abstract 1143

19. Rizzo JF, Wyatt J, Loewenstein J, Kelly S, and Shire D (2003a) Methods and perceptual threshold for short-term electrical stimulation of human retina with microelectrode arrays. *Investigative Ophthalmology & Visual Science* **44**(12): 5355–5361

20. Laube T, Akguel H, Brockmann C, Bolle I, Bornfeld N, Schilling H, and Luedtke-Handjery HC (2004) Verifikation von Schwellenwert-Parametern und Gewebeverträglichkeit bei epiretinaler elektrischer Stimulation der Minipig-Retina mit IrOx-Film-Elektroden. *In Proceedings DOG 23.-26.9.2004 Berlin* Abstract

21. Akguel H, Laube T, Brockmann C, Bolle I, Bornfeld N, Schilling H, and Luedtke-Handjery HC (2004) Verification of threshold parameters and tissue compatibility of IrOx film-electrodes on epiretinal electrical stimulation of retina in mini pigs. *In Proceedings ARVO Annual Meeting Ft. Lauderdale*, Abstract 4184

22. Humayun MS, de Juan EJ, Dagnelie G, Greenberg RJ, Probst RH, and Phillips DH (1996) Visual perception elicited by electrical stimulation of retina in blind humans. *Arch. Ophthalmol.* **114**: 40–46

23. Richard G, Feucht M, Laube T, Bornfeld N, Walter P, Velikay-Parel M, and Hornig R (2004) Visual Perceptions in an Acute Human Trial for Retina Implant Technology. *In Proceedings ARVO Annual Meeting Ft. Lauderdale*, Abstract 3400

24. Hornig R, Laube T, Walter P, Velikay-Parel M, Bornfeld N, Feucht M, Akguel H, Rössler G, Alteheld N, Lütke Notarp D, Wyatt J, and Richard G (2005) A method and technical equipment for an acute human trial to evaluate retinal implant technology. *J Neural Eng* **2**: S129–S134

25. Feucht M, Laube T, Bornfeld N, Walter P, Velikay-Parel M, Hornig R, and Richard G (2005) Entwicklung einer epiretinalen Prothese zur Stimulation der humanen Netzhaut. *Ophthalmologe* (in press)

26. Humayun MS and de Juan EJ (1998) Artificial Vision. *Eye* **12**: 605–607

27. Rizzo JF, Wyatt J, Loewenstein J, Kelly S, and Shire D (2003b) Perceptual efficacy of electrical stimulation of human retina with microelectrode array during short-term surgical trials. *Investigative Ophthalmology & Visual Science* 44(12) 5362–5369

28. Beebe X and Rose TL (1988) Charge injection limits of activated iridium oxide electrodes with 0.2 ms pulses in bicarbonate buffered saline. *IEEE Transaction on Biomedical Engineering* **35**(6): 494–495

7
Challenges in Realizing a Chronic High-Resolution Retinal Prosthesis

Wentai Liu[1], Mohanasankar Sivaprakasam[1], Guoxing Wang[1], Mingcui Zhou[1], James D. Weiland[2], and Mark S. Humayun[2]

[1]Department of Electrical Engineering, University of California
[2]Department of Ophthalmology, University of Southern California

Introduction

Electrical stimulation has been proposed by several research works as a means of restoring vision in blind patients whose vision is impaired due to retinitis pigmentosa (RP) and age-related macular degeneration (AMD). RP affects the rods (used in night vision) first and then the cones (used in ambient daylight levels). AMD results from abnormal aging of the retinal pigment epithelium and retina. Persons with AMD will start to have distorted vision and eventually lose most of the vision in the central 30°. RP is a collective name for a number of genetic defects that result in photoreceptor loss. In both diseases, the vision is impaired due to the damage to the photoreceptors that convert photons to neural signals [1]. Post-mortem evaluations of retina with RP or AMD have shown that a large number of cells remain healthy in the inner retina compared to the outer retina [2]. Further, electrical stimulation of humans with RP and AMD results in the perception of light; so the neural cells can be activated, providing the hope of restoring lost vision in blind persons [3]. A chronic implant with 16 electrode sites on the retina in three blind patients has yielded promising results [4]. After being implanted with the prosthetic device, the patients were able to detect motion of a white bar (up, down, left, or right), detect a rectangular object, count objects, discriminate the orientation of two white bars in an 'L' configuration as to where the corner of the L was positioned, and discriminate between a dessert plate, a coffee cup, and a plastic knife [5]. The success rate of these simple visual tasks differed between subjects and the difference is attributed possibly to the age and the number of years of blindness. The results of acute and chronic studies have encouraged several research and development efforts for realizing chronically implantable, high-resolution retinal prostheses [6].

One implementation of such a retinal prosthesis is shown in Figure 7.1. An external camera, which preferably can be worn on the glasses, captures the image in the field of vision of the patient. This image is processed through

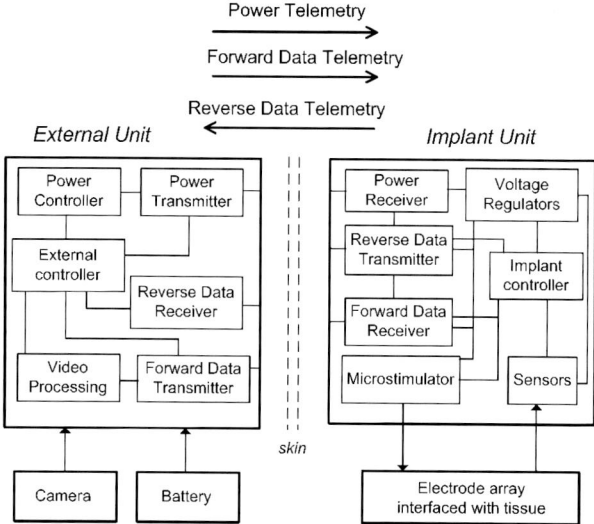

FIGURE 7.1. System diagram of a chronic high-resolution retinal prosthesis.

a video-processing unit for transforming the video data to stimulation data for the electrical pulses delivered to the retina. This data is sent through a wireless link to avoid any penetrating wires that would cause discomfort, open up the possibility of infection to the patient, and require frequent care after the implantation. The stimulation data is received at the implant unit by a data receiver. This data acts as an input to a stimulator that generates the electrical pulses to the retina. A flexible cable transmits the electrical outputs to an electrode array that interfaces the outputs of the stimulator to the retina. All the electronics inside the eye receive power wirelessly, through a power receiver unit. The electronics in the external unit operate from a battery. In an ideal case, the stimulator, data recovery, and power recovery units will be integrated into a single integrated circuit (IC) leading to a system-on-chip (SoC). All the electronics have to be shielded from the biological fluid inside the eye, and vice versa. This requires a hermetic, biocompatible package that protects the electronics. Also, all the implant components inside the eye such as coil, cable should be biocompatible for realizing a chronic implant. The device should function safely and reliably for many years to avoid frequent surgeries which would be needed if there are catastrophic failures. The reliability of electronics refers to the breakdown of any device due to constant stress over a long period of time. The breakdown of devices can occur if the package fails and direct contact is made with biological fluid.

The quality of vision restored through the retinal prosthesis is highly dependent on the resolution of the image, which translates to the number of stimulation sites on the retina. Simulations of prosthetic vision suggest that 600–1000 electrodes will be required to restore visual function to a level that would allow reading vision, independent mobility, and facial recognition [7–10]. A retinal prosthesis with 1000 stimulation outputs is currently under development [11–14]. Many challenges exist in the design of the electronics in order to realize such a chronic high-resolution retinal prosthesis. This article will examine the challenges, and describe ways to overcome these challenges, especially in design and technology areas. We hope that the challenges will also serve as motivation for the researchers in the field to develop new solutions. We observe three basic requirements as the challenges for design of electronics for a chronic high-resolution retinal prosthesis.

1. Increase flexibility – This refers to the amount and variety of functions of the device. More functionality is needed for the patients to improve the quality of the vision, and for the physician to monitor the implant after the implantation.
2. Minimize power – This refers to the power efficiency of the device. In other words, it refers to the amount of power consumed during the operation. Less power consumption translates to less heat dissipation from the electronics and longer battery life, both of which are desirable in any electronic prosthetic device.
3. Reduce size – This refers to the footprint of the device. Miniaturizing the electronics will ease the surgery procedure and will occupy a small volume inside the eye.

As it often happens in engineering design, the three requirements form three corners of a triangle as shown in Figure 7.2. A great deal of flexibility needs to be built in the prosthetic devices for several reasons. The exact requirements for the prosthesis (e.g. stimulation threshold) vary between different patients. Furthermore, these requirements can vary even after the implantation and willl need to be tuned for optimal performance. This requires the device to be programmable externally by both the patient and the physician. Also in a chronic implant, periodic monitoring of the implant is essential, especially in the initial period after the implantation. For the monitoring to be efficient, there needs to be test patterns built in, which the physician can activate using an external interface. This flexibility requires additional circuits to be built which consumes area and power during periodic operation along with standby power consumption. Simply stated, a flexible system, which is naturally complex, results in larger size and consumes more power, compared to a less flexible device. In a wireless system, increasing the size of the receiving element usually reduces the effort (power) to receive the wireless signal. So, a receiving system consuming large area requires smaller power than the system consuming small area. The future sections will describe several challenges related to the three corners of the triangle. The inevitable tradeoffs that have to be made while overcoming one of the challenges, while increasing the other, will also be discussed.

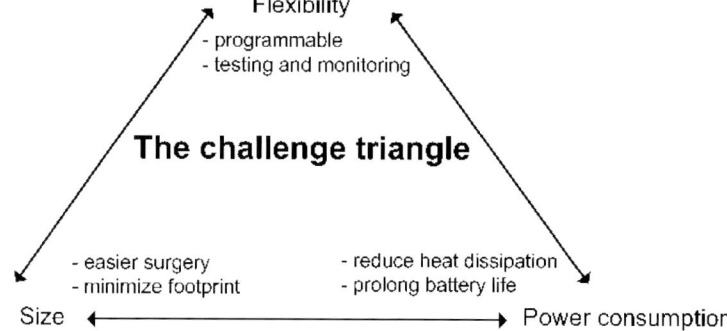

FIGURE 7.2. Challenges for a chronic high-resolution retinal prosthesis.

External Video Processing Unit

From the first look at Figure 7.1, it might seem ideal that the retinal prosthesis consisting of electronics would have all components fully implanted such that communication with any external devices is minimized or even eliminated. But migrating some of the functionality from the implant to the external unit is advantageous in many ways.

- It reduces the area and power consumption of the electronics that would be part of the implant unit.
- It allows for upgrades of functionality by programmable hardware or software, or even by attaching additional hardware to the external unit.
- It allows the image information to be processed and compressed so that the forward telemetry data rate can be reduced.
- It is possible to enhance the quality of the prosthetic vision which requires significant power.

It is well known that the human retina is not a mere receptor of photonic infor-mation, but performs significant image processing due to its layered neural network structure. Since retinal prosthesis can stimulate only a limited number of neurons, it is crucial to enhance this limited perception by means of image processing. These enhancement techniques include edge detection and enhancement, zoom, contrast and brightness adjustment. An external image processing also provides the users the option of tuning their individual devices based on their visual experience after the implantation. The external unit can also supplement the visual sensation provided by the retinal prosthesis with audio information to enhance the quality of the vision. Such a system is described in Ref. [15]. A digital camera mounted on the patient's eyeglasses or head captures images. The contents of the captured image are determined through real-time image-processing algorithms. Following the image processing, a sentence is constructed describing the object's attributes. This method can be used to train the system with the images of critical objects that the patient frequently comes across, so that it can enhance the performance of identification of those objects used in day-to-day life. The challenge faced in designing the external video-processing unit is to reduce the size and weight so as to make it wearable. The unit should also have an easily accessible user interface that can allow the patient to perform the necessary image enhancement adjustments. This user interface can be a visual, tactile, or audio or a combination of these. The interface between the output of this sensing system and the hardware that performs the function should be highly reliable since any malfunction could confuse the patient. An error detection system that can reset the system to the standard mode or prompt the patient to see assistance should be built in. The external unit can also be used by the physician during diagnosis to apply test patterns through software system in the clinic. Thus the unit should also be able to interface with known software.

Large Stimulation Voltage

The typical biphasic pulse used for stimulating the retina is shown in Figure 7.3. The time interval between the two phases is called the interphase interval. Usually the cathodic phase is the leading one which depolarizes the cell membrane and elicits a neural response. This is followed by the anodic phase which is used for balancing the first phase so that no net charge accumulates at the electrode site. The interphase interval separates the pulses slightly so that the second pulse does not reverse the physiological effect of the first pulse. The amount of charge generated at the stimulation site, which should be above the threshold for creating a response from the stimulation, is equal to the product of the amplitude of the current pulse and the pulse width. The rate at which the stimulation pulses are delivered is called the stimulation rate and is expressed in pulses per second. Experiments have shown that non-flickering perception can be achieved with stimulation rates of 40 to 50 Hz [3]. The biphasic pulse can be generated by using one or two supply voltages. In Figure 7.4a, two supply voltages are used. Two switches control the flow of current by turning 'on' the required path in a mutually exclusive fashion. In Figure 7.4b, only one supply voltage is used. A careful observation will show that while in Figure 7.4a, only one connection lead is needed from the electrode site to the biphasic current generator, two connection leads are needed from the electrode site as in Figure 7.4b.

In a high-resolution device with 1000 electrodes, Figs. 7.4a and 7.4b need 1001 and 2000 connection leads, respectively, from the electrodes to the stimulator IC. The additional one electrode is needed for the common ground connection. Since most retinal prostheses are aiming to increase the number of electrodes and thus the resolution, Figure 7.4a is highly preferred. Also, considering that it is a great challenge to place 1000 electrodes, it is important to convert every interconnection lead from the electrodes on the retina to the stimulator IC to a stimulation site. But using two supply voltages (usually equal) presents a challenge of accommodating a large voltage in the chip. For example, if the individual supply voltage is 5 V, then the maximum voltage in the chip is 10 V. In reality, this

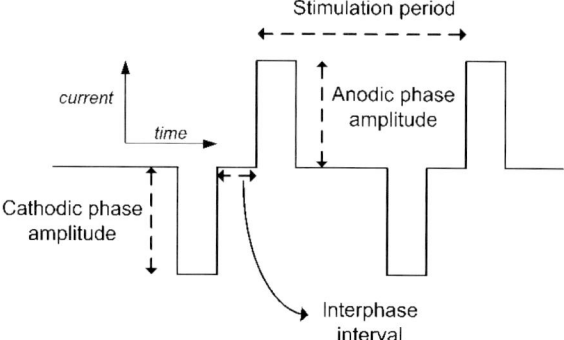

FIGURE 7.3. Biphasic stimulus pulse.

voltage is even higher since additional headroom of 1 to 2 V is consumed by
the power recovery circuits to convert the AC voltage to the required regulated
DC levels. In semiconductor device technology, operating supply voltage levels
are scaled down along with reduction in device dimensions. This is necessary to
avoid the breakdown of the gate oxide and the impact ionization that occurs at
the drain due to high fields if voltage levels are not reduced. Table 7.1 shows the
feature sizes, and the number of metal layers (used for interconnecting circuits)
available for some process technologies from different vendors. This data was
derived from MOSIS, an IC fabrication service [16]. As it can be seen from the
table, except for some special processes, most CMOS processes operate at less
than 5 V. In general, smaller feature size and more metal layers are available for
lower operating voltages. Smaller feature size helps in reducing the area of the
circuits (also referred to as active area) and more metal layers, the interconnect

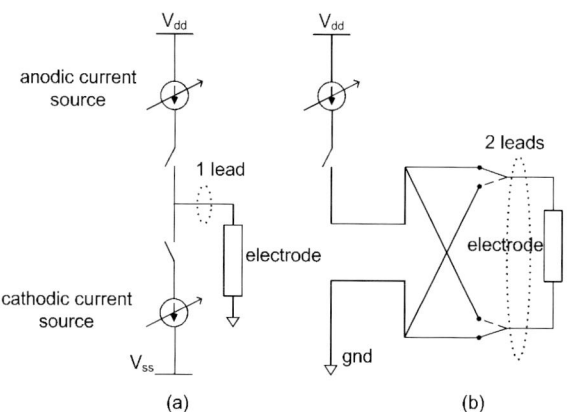

FIGURE 7.4. Biphasic stimulus generation using (a) one supply voltage (b) two supply
voltages.

TABLE 7.1. Process technologies and their characteristics.

Vendor	CMOS Feature size (micrometers)	Number of metal layers	Operating voltage
AMIS	0.35	4	3.3
	0.5	3	5
	0.7	3	100
	1.5	2	5
TSMC	0.18	6	1.8, 3.3
	0.25	5	2.5, 3.3
	0.35	4	3.3, 5
IBM	0.13	8	1.2, 2.5
	0.18	6	1.8, 3.3
	0.25	5	2.5, 3.3
austriamicrosystems	0.35	4	2.5, 50
	0.8	2	5, 50

area. Operating at low voltages not only helps in choosing a smaller and denser process technology, but also reduces the power consumption of the circuits.

It is obvious that the supply voltages are directly determined by the maximum voltage appearing at the electrode during stimulation. Usually additional voltage of around 0.5 V over the electrode voltage is required for the current generators to function properly. From the circuit design point of view, using a supply voltage higher than the specified operating voltage introduces the problem of long-term reliability of the circuit components. This requires careful design to avoid any high voltages appearing across the devices at any point of time during circuit operation. Circuit techniques that stack transistors to divide the high voltage into smaller voltages can be used to alleviate this problem [14]. It is clear that reducing the electrode voltage will help the design of low power, miniaturized IC along with ensuring the reliability of the circuit components. Reducing the electrode voltage is a challenge since this requires reduction of the stimulation current and/or reduction of the electrode impedance. For a given surface area of the retina to place the electrodes, migrating from a low-resolution device to a high-resolution device requires reducing the size and area of the electrodes. A reduction in size is often accompanied with an increase in the impedance of the electrodes. So making low-impedance electrodes even when reducing the electrode size is one of the major challenges. Placing the electrodes closer to the retina has been shown to reduce the stimulus current threshold [17]. These efforts are necessary in order to overcome the challenges of power consumption and area.

Stimulation Flexibility

There are two ways of connecting the current generators and the electrodes. One biphasic current generator can be dedicated to one electrode or a number of electrodes can be grouped together and they can be assigned to one current generator (also called demultiplexing). Figure 7.5 shows these two options. Each biphasic current generator has a digital-to-analog converter (DAC), which allows for digital control of the stimulation current. The option in Figure 7.5b exploits the large stimulation period compared to the pulse widths. For example, if we consider stimulation pulses of with cathodic phase width, anodic phase width and interphase interval of 1 ms each, and a stimulation period of 50 Hz, it is possible for one biphasic current generator to serve 6 electrodes in a serial fashion, still ensuring non-flickering perception. While the advantage of this option is the reduction in the number of DACs and the current generators, it also reduces the flexibility of stimulation. For example, in this scheme, two electrodes cannot be stimulated simultaneously. This tradeoff should be considered during the stimulator design to reduce area if possible. This requires prior experiments to determine the required sequence of stimulation. If the

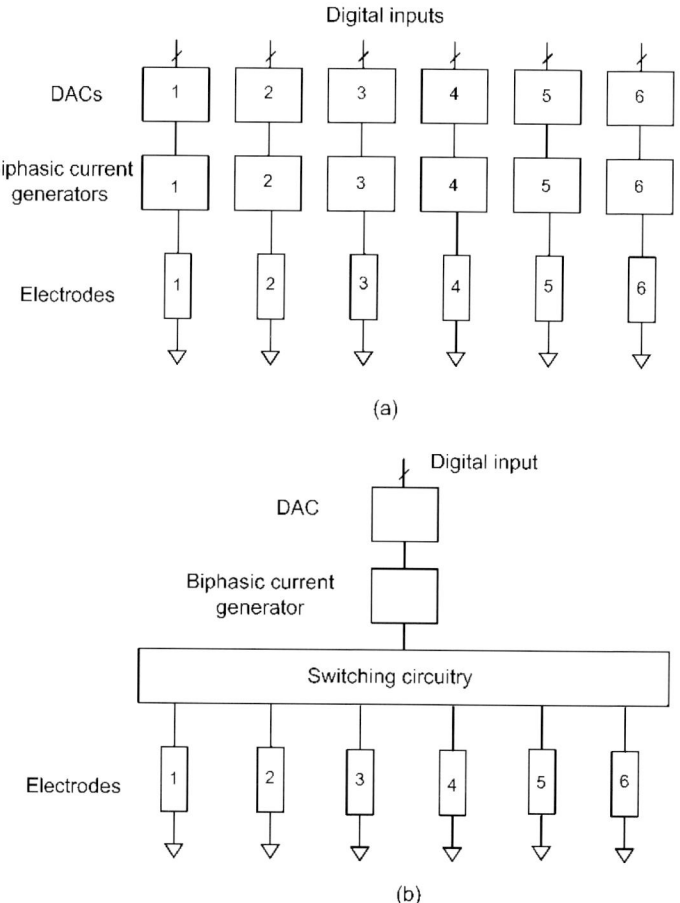

FIGURE 7.5. Electrode assignments (a) no demultiplexing, (b) with demultiplexing.

demultiplexing scheme is used, the electrodes which do not require simultaneous stimulation should be grouped together.

Another challenge for high-density retinal prosthesis is the requirement of a large data rate for forward telemetry. The data rate depends on the number of electrodes, number of bits per stimulation, and stimulation rate. The stimulation data consists of information on cathodic and anodic pulse amplitudes, pulse widths, whether anodic or cathodic is the leading pulse, and interphase interval. Also additional information for connecting the electrodes to the common ground potential to remove any residual charge (this is required for ensuring tissue safety) need to be transmitted. If we consider 20 bits for stimulation with a stimulation rate of 50 Hz, for 1000 electrodes, a data rate of 1 Mbps is required. This calculation assumes that it is not necessary to address each electrode separately. In this way no address bits are required. But if the flexibility of arbitrary stimulation is required, each electrode needs to be addressable with a 10-bit address. This increases the data rate by 50%, to 1.5 Mbps. If additional data such as error detection and configuration data are included, then the required data rate can go up to 2 Mbps. An increase in data rate is usually accompanied with increase in power consumption of the data transmitter and/or receiver. The challenge of achieving high data rate will be addressed in the coming sections. At this point it is not clear how much flexibility would be needed for a 1000 electrode retinal prosthesis in terms of stimulation sequence, delay between two arbitrary stimulations. So for a test device, the implant controller has to be designed to provide the maximum flexibility to enable the physician to conduct various experiments on different parameters.

Powering of the Retinal Implant

Both the external unit and the implant unit need electrical power for their operation. For the external unit, battery is a natural choice for the energy source and it should be rechargeable. Currently, lithium-ion battery is the most widely used rechargeable battery due to its high energy density (0.2 Watt hours/cm^3) and long shelf-time (> 10 years) [18]. With the size of a cell phone battery, the capacity is about 5 Watt hours. If 20 hours of continuous operation is required, the maximum power it can support is 0.25 Watts. This includes the power dissipated by the camera, external unit, and the implant unit. If more is needed, obviously the size (and weight) of the battery has to be increased. The overall power dissipation of the system has to be maintained at a low level to implement an external unit which is wearable. In order to minimize the battery size, the implant powered by the external battery should consume low power and the power transfer from the external battery to the implant should be highly efficient.

The retinal implant needs power for the stimulator to stimulate the tissue. In addition, other circuits such as the voltage regulation, data telemetry, and implant controller also need power for operation. Also, for continuous real-time imaging, a continuous supply of power is necessary. The amount of power needed depends on the stimulation current requirements. At this time, the optimal stimulation parameters (amplitude, frequency, and pattern) are not clearly known. The total power dissipation of the implant with 1000 electrodes is estimated to be less than 100 mW. In order to provide the full-scale functionality, the system should be able to transmit the maximum power. In addition, it would be ideal if the power transmission adaptively adjusts the transmitted power by sensing the required implant power.

Powering the retinal implant through an implantable battery is not an option due to the stringent size limitation and relatively large power requirement. With 100 mW power capacity and continuous operation for 20 hours, the minimum size of the lithium-ion battery would be around 10 cm^3, which is larger than the size of an eyeball. In addition, the number of rechargeable cycles is also limited (about 500), which makes it impossible for the requirement of more than 10-year lifetime. This shows that an implantable battery is not feasible for chronic high-resolution retinal prosthesis, unless the power requirement is dramatically reduced through technology advancements.

A conductive tethering wire connecting a battery in the external unit to the implant unit is another straightforward way to provide power. But wire penetrating through the skin may require continued medical supervision to guard against any infection. In retinal prosthesis, rapid eye movement can break any penetrating wires passing through the sclera (tissue envelope covering the eyeball except the cornea). In addition to the technical difficulties, it also poses inconvenience to the patients. Considering the difficulties of dealing with a percutaneous wire in an implantable device, especially for long-term implantation, wireless methods of power delivery are preferred. Inductively coupled coil pair, infrared power link, and thermal-to-electric conversion are some of the methods of obtaining power wirelessly. Of the wireless methods of power transfer, inductive link is the widely used one. It can continuously provide relatively large amounts of power (up to several Watts) with reasonable efficiency. The challenges associated with an efficient inductive power transmission design are addressed in the next section.

Wireless Power Transmission

Coil, the Key Component for High Power Efficiency

A coil is formed when a conducting wire is wound into one or several loops [Figure 7.6a]. The magnetic field lines generated by each loop of the coil combine with the lines generated by other loops to produce a concentrated field at the center of the coil. When another (receiving) coil is placed close to the transmitting coil, the magnetic field generated by the transmitting coil passes through the windings of the receiving coil. Now, these two coils are magnetically (inductively) coupled and energy transfer between the two coils is possible.

For a coil, there are several important properties such as self-inductance (L), resistance (R), and self-resonating frequency ω_0. For a given size of a coil, the

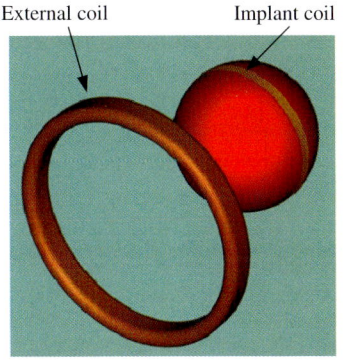
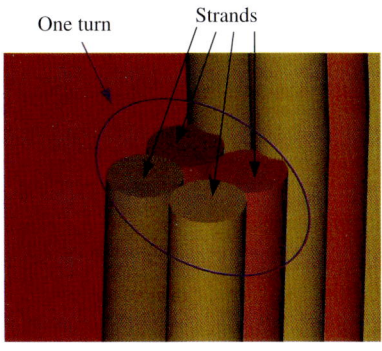

FIGURE 7.6. Coil (a) conceptual inductive coupling between two coils (b) turns and strands of a coil.

inductance is roughly proportional to the square of the number of turns (N) [19]. The effective resistance of the coil is a function of the coil geometry and increases with the frequency of the electrical current passing through it [20]. The self-resonating frequency is determined by the coil inductance and the capacitances between the windings of the coil. In general, the coil has to be operated at a frequency below its self-resonating frequency. Beyond the self-resonating frequency, the coil loses its inductive property and acts like a capacitor.

A coil is generally characterized by its quality factor, Q, which is defined as:

$$Q = \frac{\omega L}{R} \tag{1}$$

In Eq. (1) ω is the angular frequency of the current, L is the inductance, and R is the resistance of the coil. For high efficiency power transfer, the Q of the coils should be maximized. From Eq. (1), the quality factor can be increased by decreasing the resistance of the coil, increasing the inductance, or increasing the frequency. To decrease the resistance, a phenomenon called skin effect should be considered. Skin effect refers to the phenomenon of increase in the resistance of a conductor when AC current passes through it. The current tends to concentrate on the surface of the cross section of the conductor (also called current crowding), thus increasing the equivalent resistance. Higher the frequency, more severe is the skin effect. The wire diameter of the coil should be chosen according to the frequency of the electrical current to minimize the skin effect. One effective way to reduce the skin effect is to construct the coil with several insulated wires (strands) twisted together (e.g. Litz wire) to increase the equivalent cross section conducting area while keeping the individual wire small. However, the maximum number of strands is limited by the physical size of the coil. For a given size, the number of strands trades off with the number of turns since the total number of wires that a coil can hold is fixed. Increasing the number of strands per turn will reduce the number of turns, and thus the inductance is reduced. The number of strands needs to be optimized for a given design to achieve high power efficiency. To increase the Q of the coil, L and ω can be increased. Increasing the inductance L can be done by adding more turns to the coil. However, as the number of turns increases, the resistance increases as the equivalent wire length increases. Increasing the quality factor by increasing the frequency is limited by two factors. One is the self-resonating frequency of the coil. Another factor is that when the frequency is high, the eddy current in the wire induced by the magnetic field causes additional power dissipation and thus increases the equivalent resistance of the coil [21].

The design of a coil for power transmission involves the above factors in a non-trivial way. Given the physical size, the coil should be optimized through the right choice of diameter of wire, number of turns, number of strands, and operating frequency. Unfortunately, calculation of the inductance and resistance of a given coil involves magnetic field distribution and there are no simple analytical equations, which makes the optimization of the coils difficult [21]. A formal design procedure was developed in Ref. [12], which formulates a semi-analytical relationship between the inductance and the resistance of a coil and finds an optimal coil structure within a given size limitation to achieve high power transfer efficiency.

Power Transmission and Recovery

The transmitting coil needs alternating current to generate an alternating magnetic field so that the required voltage can be induced across the receiving coil. This is done by a power amplifier that converts the DC energy from the battery to AC and drives the transmitting coil. Class-E power amplifier is a good choice due to its high power conversion efficiency (90–97%) and ability to generate high-amplitude output signals across the inductor while operating at a small supply voltage. The power coupled to the receiving coil is AC and needs to be converted to DC for the operation of implant electronics. As mentioned before, for biphasic stimulus of large number of electrodes, stimulation through two supply voltages is preferred. To convert the AC voltage into dual DC supply, two forms of rectification can be employed, with one favoring the power efficiency and the other favoring the size of the implant.

As shown in Figure 7.7a, the diode D_1 and capacitor C_S rectifies the AC voltage V_2 to DC voltage V_S. The series regulator provides the first stage voltage regulation followed by a shunt regulator. The shunt regulator provides the required supply voltages V_{dd} (positive) and V_{ss} (negative) with a common ground (zero potential). However, for biphasic stimulation, the anodic stimulation does not occur at the same time as the cathodic stimulation. In this case, the anodic current (i_a) flows through the electrode and the cathodic shunt regulator, as shown in Figure 7.7a. This results in additional power loss in the cathodic shunt regulator. The power dissipated on the cathodic shunt regulator ($V_{ss} \cdot i_a$) is wasted, which is about half the total power delivered by the series regulator (($V_{dd} + V_{ss}) \cdot i_a$). In other words, if the required stimulation power is P_{load}, the series regulator needs to supply twice the power of that ($2P_{load}$).

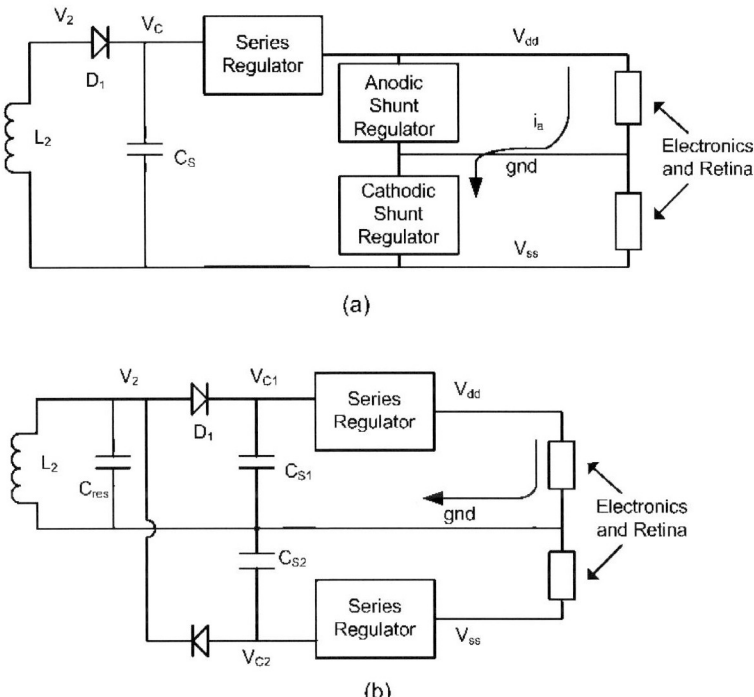

FIGURE 7.7. Rectification Topologies (a) single diode, (b) dual diode.

The power dissipation in the cathodic shunt regulator can be avoided if a return path is provided for the anodic current, bypassing the cathodic regulator. The rectification topology shown in Figure 7.7b provides this return path. The two voltages are generated through a dual-diode rectification with the series regulators without any shunt regulators. This topology provides a low impedance return path for the stimulation current to the storage capacitors. Thus even when only anodic/cathodic stimulation is activated the current will not flow through the cathodic/anodic regulator as in the case of Figure 7.7a. This allows the regulated power to stimulate the tissue without any additional loss. However, this topology requires an additional diode and a capacitor compared to the topology in Figure 7.7a, which increases the size of the implant especially when off-chip components are used.

The choice between the two rectification topologies has to be made from the system design perspective. If the stimulation of different electrodes can be scheduled in a way such that at a given time, the combined anodic current of one set of stimulated electrodes is equal to the cathodic current of the rest of the stimulated electrodes, then the current passing through shunt regulators can be reduced. However, at the time of this writing, the effective stimulation pattern is not known. Thus the dual-diode topology is preferred to maintain the power efficiency.

Choice of Storage Capacitor

The capacitor(s) C_S shown in Figure 7.7 stores the charge that is to be delivered to the tissue through electrodes. It also acts as a filter to remove the ripple from the output of the diode. Due to the varying magnitude of the stimulus current, the capacitor needs to be large enough to store charge to avoid unacceptable voltage drop when charge is delivered to the electrodes. Figure 7.8 illustrates this problem for the dual-diode topology case. For the single diode topology, the analysis is similar but the requirement of the capacitance is even higher since the capacitor has to supply both anodic and cathodic currents. If N is number of electrodes stimulated, the total charge delivered to the tissue by the stimulator $Q_{stim} = I_{stim} \cdot N \cdot PW_{stim}$. The average current from source is $I_{s\_avg} = Q_{stim}/T_{stim}$. The voltage drop ΔV_{cap} during the stimulation can be calculated as $\Delta V_{cap} = (1/C_S) \cdot (I_{stim} \cdot N - I_{s\_avg})^* PW_{stim}$. If $N = 1000$, $I_{stim} = 100\,\mu A$, $PW_{stim} = 1$ millisecond, $T_{stim} = 16$ milliseconds and the maximum voltage drop the system can tolerate is 1 V, then the required capacitance is $93.75\,\mu F$. This is far too large a capacitance to be implemented in an IC. Even an off-chip capacitor of this size will consume a large area for an implant. Fortunately, this is an extreme condition, when all the 1000 electrodes are stimulated at the same time, which is unlikely. In addition, if the stimulation of electrodes can be

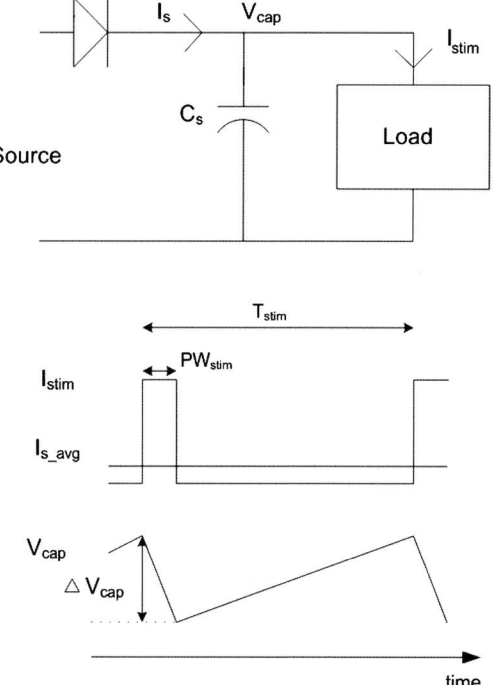

FIGURE 7.8. Storage capacitor and related waveforms.

scheduled in a sequence such that the current consumption of the electronics does not vary much over time, the capacitance can be reduced. However, there is not enough knowledge at this point, about the stimulation pattern and the interactions between electrodes. While clinical experiments should focus on obtaining more insights on these parameters, the test device for these experiments should be designed with enough flexibility to facilitate the experiments. So, the capacitance should be chosen as large as possible which can be integrated within the available area.

Coil Coupling and Loading Variations

For the patient's comfort, the external unit should have some freedom of movement instead of a rigid fixture, which may cause the change of distance and orientation angle between the transmitter and the receiver coils. As a result, the coupling between the two coils varies and the coupling coefficient is estimated to change from 0.08 to 0.24 for a distance variation from 15 to 5 mm. Furthermore, the power consumption in the implant unit may change significantly with stimulation data pattern through changes in stimulation currents. For example, during standby operation, less power is needed. Both the coupling and the loading variations may cause one of the following: (1) increased implant heating due to excess transmitted power, causing a detrimental effect on the tissue over long term; (2) decreased implant supply voltage/current due to insufficient transmitted power, causing improper device operation or shutdown. This poses challenges to electronic design to transmit the optimal power required by the implant.

Research efforts have been directed for compensating the loading/coupling variations. The work in Ref. [22] proposed to estimate the coupling coefficient by detecting the primary coil voltage and current. This method somewhat reduces the impact of coupling change, although it is only applicable to systems with constant loading. Furthermore, without knowing the exact information on received power at the secondary side, it cannot achieve optimal compensation. In Ref. [23], we implemented an adaptive wireless power telemetry system where the transmission power is controlled according to the needs of the implant. By detecting the power level inside the implant and wirelessly transmitting this information back to the external unit, a closed loop control was realized to achieve immunity to coupling and/or loading variations. This system requires reverse telemetry, communication from the implant to external unit, which will be discussed in the next section.

Wireless Data Communication

The data communication between the external unit and the retinal implant needs to be bi-directional. The one from external unit to implant is termed "forward data telemetry" and the one from implant to external unit is termed "reverse data telemetry." Forward data telemetry transfers the parameters of the stimulation (converted from image information by signal processing unit) to the implant. Besides the image information, forward data telemetry also needs to relay configuration information to the implant. Reverse data telemetry transfers information to the external unit from the implant. The information includes implant status such as temperature and pH, received power level, impedance measurement data, etc.

Forward Data Telemetry

To achieve maximum flexibility of stimulation, forward data rate needs to be as high as possible. For high-density (1000 pixels) retinal implant, the estimated forward data rate is about 1 to 2 Mbps, as mentioned earlier. There are several ways to transmit information to the implant. The simplest method of data transmission to and from the implant is a percutaneous wire connecting the two units. The disadvantages of using a tethering wire as discussed in the power telemetry section apply here as well. Optical telemetry is another option for transmitting data, where a modulated beam of infrared light is used to encode information. This information would then be decoded at the receiving end using optoelectronic detectors and further processed using electronics. However, optical transmission efficiency decreases rapidly with skin thickness and it is more sensitive to relative dislocation between the transmission element and the receiving element compared to magnetic coupling with coils [24] as well as the ambient light sources. The challenges associated with inductive method of data telemetry will be discussed below.

To transmit data into implant, one way is to utilize existing power transmission link by modulating the data onto the power carrier. Another way is to use an RF link with frequency different from the power carrier. In the first case, the power carrier is modulated with the data, thus employing the power carrier as the data carrier. This type of telemetry system is termed 'single band telemetry' referring to the fact that the same carrier frequency is used for transmitting both power and data through the same physical means. The method has the obvious advantage of not needing a dedicated coil pair for data telemetry. However, it faces the challenge of high data rate required by high-density stimulation. Normally the power carrier frequency is less than 10 MHz to penetrate the human body without significant absorption and achieve high efficiency of power transmission. Achieving high data rates (1 to 2 Mbps) at a relatively low carrier frequency is a great engineering challenge. Another approach to high data rate and high efficiency power transmission, termed 'dual-band telemetry', has been proposed in Ref. [25]. Shown in Figure 7.9, the power and data are transmitted through separate pairs of coils using two different carrier frequencies, forming a hybrid dual-frequency link which allows optimization of efficiency power telemetry and data rate of the forward telemetry through allocation of different frequencies. The power carrier frequency f_L is lower than the data carrier frequency f_H.

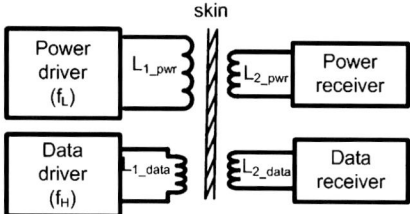

FIGURE 7.9. Dual-band telemetry.

However, the dual-band approach faces the challenge of interference between the power and the data link. As shown in Figure 7.10, the four coils have six coupling coefficients, which make the design complicated. The power transmitter generates a strong magnetic field in order to transmit power to the implant. This magnetic field can generate a fairly high voltage on the data receiver coil, which may corrupt the data signal completely. One way to tackle this problem is to filter out the power interference before passing the data signal to the recovery circuits. This filter may consume additional power in the implant, which is not desirable. Another way is to minimize the coupling by relative positioning of the power and data coils. For example, the data coils (L_{1_data} and L_{2_data}) can be

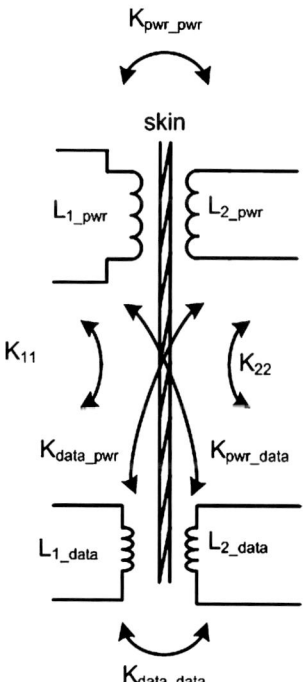

FIGURE 7.10. Interference between power and data telemetry.

placed orthogonal to the power coils (L_{1_pwr} and L_{2_pwr}), which ideally reduce the coupling coefficients of K_{11} and K_{22} to zero. However, winding L_{2_data} around L_{2_pwr} (to make the zero coupling of K_{22}) creates a bump on coil L_{2_pwr}, which is undesirable from the surgical perspective.

Compared to the single-band approach, the dual-band approach requires one more coil to be implanted, which increases the size of the implant. On the other hand, it can save power since it allows power link to be optimized without sacrificing for data transmission. Design of a dual-band telemetry system should try to minimize the size increase due to the extra coil.

Reverse Data Telemetry Design

Reverse telemetry may use another frequency as data carrier, similar to forward data telemetry. This approach is called 'active reverse telemetry'. The data carrier may use the existing coil pair or another coil pair as physical link. In general, this approach requires an active driver to drive the coil/antenna. Another approach is called "passive reverse telemetry," as it does not require a dedicated data driver. Figure 7.11 shows the principle of using passive telemetry to transmit data from the implant to the external unit. It should be noted that in a dual-band telemetry system, the reverse telemetry can use the power telemetry as its data carrier, since it does not require high data rate and the low frequency power carrier is a good candidate. As shown in Figure 7.11, the coil is not driven by any power amplifier. Instead, a change in the loading of the coil is introduced by turning on and off the switch S. This loading change at the secondary coil creates a detectable change in the coil current (hence called load shift keying [22]). A receiver in the external unit connected to the primary coil detects this change and recovers the data. The load shift keying shown in Figure 7.11 may generate a high voltage stress on the switch and the diode, when the switch is open, since interrupting current flow through an inductor produces a large voltage to counteract the change. This high voltage in turn makes it difficult to implement the switch in the IC, if it is above the operating voltage of the chosen process technology, and may require discrete components, which increases the size of the implant. Thus the reverse telemetry should be designed to maintain the advantage of low power dissipation while minimizing the discrete components and consuming as less space as possible.

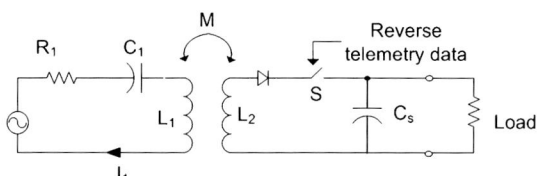

FIGURE 7.11. Passive reverse telemetry.

Conclusions

We have examined several challenges in the electronics design toward the implementation of chronic high-resolution retinal prosthesis. The key challenges are to reduce power consumption, provide more flexibility, and reduce the size. As seen through the examples in the above sections, solving these challenges often involve direct tradeoffs among themselves. The stimulator circuit faces the challenge of providing maximum flexibility, which increases the power and area. The power telemetry faces the challenge of increase in area in the form of off-chip components like coil, diode and capacitor, while maintaining a high efficiency and long battery life. The data telemetry faces the challenge of supporting high data rate for maximum flexibility hence requiring an additional coil pair. A system design is highly critical in realizing a chronic high-resolution retinal prosthesis, during which the above tradeoffs should be considered to make the suitable choices. In addition to the design of the electronics, the choice of technology for implementing the circuits deserves careful attention. A small feature size technology with more metal layers is preferred to reduce the size of the IC. This in turn requires the stimulation voltages to be small enough to allow the use of such a process technology. This also poses challenge for circuit design to accommodate high voltages. Passive components such as the storage capacitor, coils, and active components such as the back telemetry switch are difficult to implement in an IC. Off-chip versions of these components, which can lend themselves to be integrated with the ICs through post-processing on the surface of the IC, are highly preferred over components that require to be connected using wires.

References

1. Margalit E and Sadda SR (2003) Retinal and optic nerve diseases. Artificial Organs 27:963–974
2. Humayun MS, Prince M, De Juan Jr E, Barron Y, Moskowitz M, Klock IB, and Milam AH (1999) Morphometric analysis of the extramacular retina from postmortem eyes with retinitis pigmentosa. Investigative Ophthalmology and Visual Sciences 40:143–148
3. Humayun MS, De Juan Jr E, Weiland JD, Dagnelie G, Katona S, Greenberg R, and Suzuki S (1999) Pattern electrical stimulation of the human retina. Vision Research 39:2569–2576
4. Humayun MS, Greenberg RJ, Mech BV, Yanai D, Mahadevappa M, van Boemel G, Fujii GY, Weiland JD, and De Juan Jr E (2003) Chronically implanted intraocular retinal prosthesis in two blind subjects. In: Proceedings of ARVO Annual Meeting, April 2003.
5. Weiland JD, Yanai D, Mahadevappa M, Williamson R, Mech BV, Fujii GY, Little J, Greenberg RJ, De Juan Jr E, and Humayun MS (2004) Visual task performance in blind humans with retinal prosthetic implants. In: Proceedings of the 26th Annual International Conference of the IEEE EMBS

6. Weiland JD, Liu W, and Humayun MS (2005) Retinal prosthesis. Annual Review of Biomedical Engineering 7:361–401

7. Cha K, Horch K, and Normann RA (1992) Simulation of a phosphene-based visual field: visual acuity in a pixelized vision system. Annals of Biomedical Engineering 20:439–449

8. Cha K, Horch KW, Normann RA, and Boman DK (1992) Reading speed with a pixelized vision system. Journal of Optical Society of America A: Optics, Image Science and Vision 9:673–677

9. Cha K, Horch KW, and Normann RA (1992) Mobility performance with a pixelized vision system. Vision Research 32:1367–1372

10. Hayes JS, Yin JT, Piyathaisere DV, Weiland JD, Humayun MS, and Dagnelie G (2003) Visually guided performance of simple tasks using simulated prosthetic vision. Artificial Organs 27:1016–1028

11. Liu W and Humayun MS (2004) Retinal prosthesis. In: IEEE International Solid-State Circuits Conference Digest of Technical Papers, pp 218–219

12. Kendir GA, Liu W, Wang G, Sivaprakasam M, Bashirullah R, Humayun MS, and Weiland JD (2005) An optimal design methodology for inductive power link with class-E amplifier. IEEE Transactions on Circuits and Systems – I 52:857–866

13. Wang G, Liu W, Bashirullah R, Sivaprakasam M, Kendir GA, Ji Y, Humayun MS, and Weiland JD (2004) A closed loop transcutaneous power transfer system for implantable devices with enhanced stability. In: Proceedings of the IEEE International Symposium on Circuits and Systems. pp 17–20

14. Sivaprakasam M, Liu W, Humayun MS, and Weiland JD (2005) A variable range bi-phasic current stimulus driver circuitry for an implantable retinal prosthetic device. IEEE Journal of Solid-State Circuits 41:763–771

15. Fink W, Tarbell M, Weiland JD, and Humayun MS (2004) DORA: Digital Object Recognition Audio-Assistant for the visually impaired. Investigative Ophthalmology and Visual Science 45:4201

16. http://www.mosis.org/products/fab/vendors/

17. Mahadevappa M, Weiland JD, Yanai D, Fine I, Greenberg RJ, and Humayun MS (2005) Perceptual thresholds and electrode impedance in 3 retinal prosthesis subjects. IEEE Transactions on Neural Systems and Rehabilitation Engineering 13:201–206

18. Soykan O (2002) Power sources for implantable medical devices. Medical Device Manufacturing and Technology

19. Grover FW (1973) Inductance Calculation Working Formulas and Tables. Instrument Society of America, Research Triangle Park, North Carolina.

20. Lee TH (1998) The Design of CMOS Radio-Frequency Integrated Circuits. Cambridge University Press, Cambridge, United Kingdom

21. Sullivan CR (1999) Optimal choice for number of strands in a litz-wire transformer winding. IEEE Transactions on Power Electronics 14:283–291

22. Tang Z, Smith B, Schild JH, and Peckham PH (1995) Data transmission from an implantable biotelemeter by load-shift keying using circuit configuration modulator. IEEE Transactions on Biomedical Engineering 42:524–528

23. Wang G, Liu W, Sivaprakasam M, and Kendir GA (2005) Design and analysis of an adaptive transcutaneous power telemetry for biomedical implants. IEEE Transactions on Circuits and Systems-I 52:2109–2117

24. Takahashi M, Watanabe K, Sato F, and Matsuki H (2001) Signal transmission system for high frequency magnetic telemetry for an artificial heart. IEEE transactions on Magnetics 372921–372924

25. Bashirullah R, Liu W, Ji Y, Kendir GA, Sivaprakasam M, Wang G, and Pundi B (2003) A smart bi-directional telemetry unit for retinal prosthetic device. In: Proceedings of International Symposium on Circuits and Systems. pp 5–8

8
Large-scale Integration–Based Stimulus Electrodes for Retinal Prosthesis

Jun Ohta[1], Takashi Tokuda[1], Keiichiro Kagawa[1], Yasuo Terasawa[2], Motoki Ozawa[2], Takashi Fujikado[3] and Yasuo Tano[4]

[1]*Graduate School of Materials Science, Nara Institute of Science and Technology (NAIST)*
[2] *Vision Institute, R&D Div., NIDEK Co., Ltd.*
[3]*Department of Applied Visual Science, Osaka University Medical School*
[4]*Department of Ophthalmology, Osaka University Medical School*

Introduction

Since large-scale integration (LSI) technologies allow integration of smart functions such as image sensing, control, amplification, and signal processing with stimulators, LSI-based retinal prosthesis devices are attracting significant interest [1, 2]. We have been developing retinal prosthesis devices based on Complementary Metal Oxide Semiconductor (CMOS) LSI technology [3–11]. In our project, the implanted stimulator is placed underneath the retina (subretinal implantation) or upon the suprachoroid; this stimulation method is called suprachoroidal transretinal stimulation (STS) [12, 13].

In subretinal implantation, a photosensor is integrated with a stimulus electrode in order to substitute for photoreceptor cells. Thus far, a simple photodiode array without any bias voltage, i.e. the solar cell mode, has been used as a photosensor mainly due to its simple configuration [14, 15]. The photocurrent is directly used as the stimulus current into retinal cells. In order to realize sufficient stimulus current using a photosensor in a daylight environment, we first propose a pulse frequency modulation (PFM) photosensor, which is fabricated using standard CMOS LSI technology, although a PFM photosensor needs an external power supply, which can be provided through RF coupling coils. A PFM photosensor converts input light intensity into an output pulse train, the frequency of which is proportional to the light intensity [16, 17]. Recently, another group has proposed that a PFM photosensor or a pulse-based photosensor might be applied to retinal implantation [18, 19].

The PFM appears to be suitable as a retinal prosthesis device in subretinal implantation for the following reasons. First, PFM produces an output of pulse streams, which would be suitable for stimulating the cells. In general, pulse stimulation is effective for evoking cell potentials. In addition, such pulse form is compatible with logic circuits which enable highly versatile functions. Second, PFM can operate at a very low voltage without decreasing the signal-to-noise ratio. This is suitable for an implantable device. Finally, its photosensitivity is sufficiently high for detection in normal lighting conditions and its dynamic range is relatively large. These characteristics are very advantageous for the replacement of photoreceptors. Although the PFM photosensor is essentially suitable for application to a retinal prosthesis device, some modifications are required and these will be described herein.

There are many technical challenges to overcome when applying LSI-based stimulator devices to a retinal prosthesis. First, the LSI-based interface must be biocompatible. The standard LSI structure is unsuitable for a biological environment; silicon nitride is conventionally used as a protective top layer in standard LSIs, but will be damaged in a biological environment in the case of long-time implantation. Second, stimulus electrodes must be compatible with the standard LSI structure. Wire-bonding pads, which are made of aluminum, are usually input–output interfaces in standard LSIs, but are completely inadequate as stimulus electrodes for retinal cells, because aluminum dissolves in a biological environment. Finally, in addition to electrode materials, the shape of the electrode affects the efficiency of stimulation. A convex shape is suitable for efficient stimulation, but the electrodes in the LSI are flat.

This manuscript is organized as follows. First, we describe a PFM photosensor for a retinal prosthesis device. Some modifications of the PFM photosensor are discussed and demonstrated. In addition to these modifications, we implement image processing such as edge enhancement in the PFM photosensor. Such image processing would be effective for blind patients if the stimulus device has a low resolution. The design and fundamental characteristics of the fabricated device are presented.

In order to verify the effectiveness of the PFM photosensor as a retinal prosthesis device, the fabricated stimulator device based on PFM photosensors is applied to the simulation of a retina that has been detached from the eye of a frog. For the in vitro experiment, we have developed a new electrode and packaging technology that is suitable to standard LSI structure. We describe this technology in detail, and then demonstrate the in vitro experimental results. Finally, we describe the flexible and extendible LSI-based electrode array, which is suitable to the STS device.

The PFM Photosensor as Subretinal Implantable Device

Operation Principle and Fundamental Characteristics of the PFM Photosensor

In this section, we briefly describe the operation principle and fundamental characteristics of the PFM photosensor. The detail analysis is described in

Ref. [6]. The PFM is an output representation in which an analog output is converted into a pulse frequency. In order to employ a PFM, relaxation oscillation circuits are constructed as shown in Figure 8.1. The photodiode PD acts as a variable current source controlled by the input light intensity with capacitance C_{PD} associated with the pn-junction. The capacitance is charged through the reset transistor M_r, and then electrically floated. The photocurrent discharges the charged capacitance and thus the voltage at the cathode node of the PD, V_{PD}, decreases from the reset voltage. When V_{PD} reaches the threshold voltage of the inverter V_{th}, the output pulse is produced after a delay time t_d and simultaneously M_r is reset. Repeating the process, the output pulse train is generated according to the input light intensity. In such a configuration, the stronger the light intensity, the higher the pulse frequency. The analog value of the light intensity is consequently converted into a pulse train as a digital signal. The output pulse frequency is approximately expressed as:

$$f_{out} \approx \left[\frac{C_{PD}(V_{DD} - V_{th})}{R_{ph}P + I_d} + 2t_d \right]^{-1} \tag{1}$$

where R_{ph} is the photosensitivity, I_d is a dark current of PD, P is the input light power, t_d is the delay time of feedback circuitry, respectively. From Eq. (1), the output frequency is proportional to the input light intensity if the delay time is negligible.

Figure 8.2 shows the experimental result of output pulse frequency as a function of input light intensity [6]. A wide dynamic range of over 120 dB is

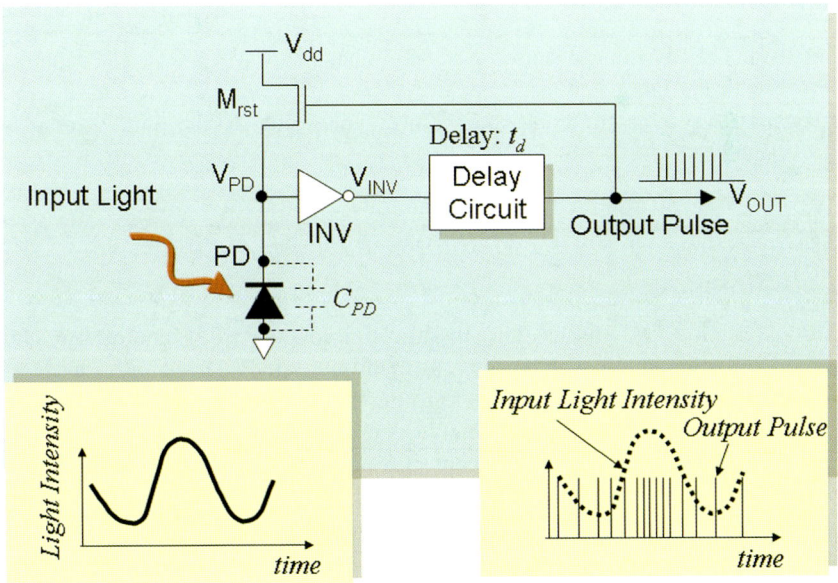

FIGURE 8.1. Block diagram of the PFM photosensor.

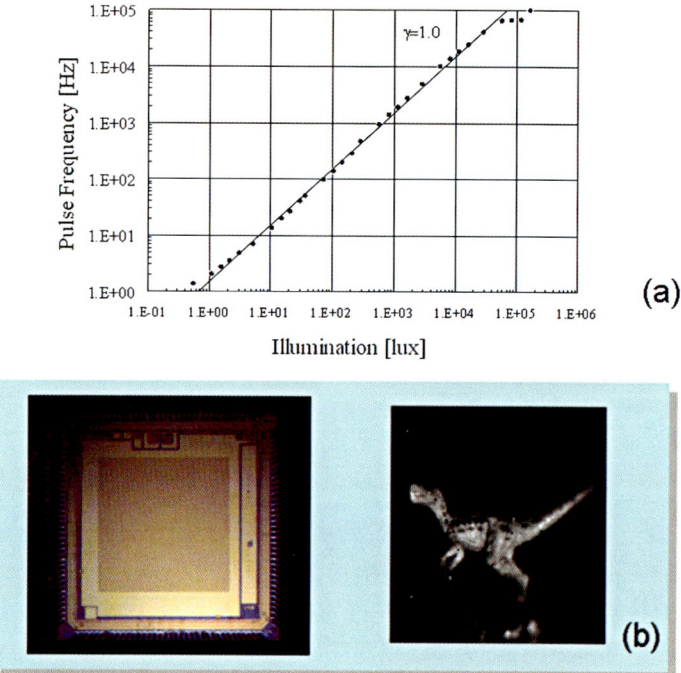

FIGURE 8.2. Experimental results of the PFM image sensor (a) output pulse frequency of the PFM photosensor as a function of input light intensity, (b) image captured by the image sensor with 128×128 pixels.

obtained. The PFM image sensor used herein has 128×128 pixels. The captured image with the fabricated sensor is also shown in Figure 8.2. The output pulse frequency was converted into the voltage value outside the chip to display the image.

Modification of the PFM Photosensor for Retinal Cell Stimulation

In this section, we discuss the modification of the PFM photosensor for retinal cell stimulation. The reasons for modifying the PFM photosensor are as follows [4]. First, the output from a PFM photosensor is in the form of a voltage pulse waveform, whereas current output is preferable for injecting charges into retinal cells constantly, even if the contact resistances between the electrodes and the cells are changed. Second, biphasic output, i.e. positive and negative pulses, is preferable for charge balance. Third, output frequency limitation is needed because an excessively high frequency may cause damage to retinal cells. The output pulse frequency of the original PFM, as shown in Figure 8.2, is generally too high (approximately 1 MHz) for stimulating retinal cells.

TABLE 8.1. Modifications of the PFM photosensor for stimulating retinal cells.

	Requirements	Our work
Photosensor	–	Pulse frequency modulation
Photo sensitivity	High	High
Stimulation	Current	Voltage
	Pulse	Pulse
	Biphasic	Monophasic
Frequency	< 100 Hz	< 1 MHz
Power supply	–	Required
Injection charge	< 1 mC/cm^2	Sufficient

The frequency limitation, however, causes a reduction in the range of input light intensity. We have alleviated this problem by introducing a variable sensitivity: the output frequency is divided into 2^{-n} with a frequency divider, where n is the number of divisions. Note that the digital output of the PFM is suitable for the introduction of such a logic function of the frequency divider. Table 8.1 summarizes the photosensor requirements for a retinal prosthesis.

Based on the above modifications, we have designed and fabricated a pixel circuitry using standard 0.6-μm CMOS technology. Figure 8.3 shows a block diagram of the pixel. The frequency limitation is achieved by a low pass filter using switched capacitors. The biphasic current pulse is implemented by switching the current source and sink alternatively.

Figure 8.4 demonstrates the experimental result of the variable photosensitivity using the chip. The original output curve has a dynamic range of over 6-Log

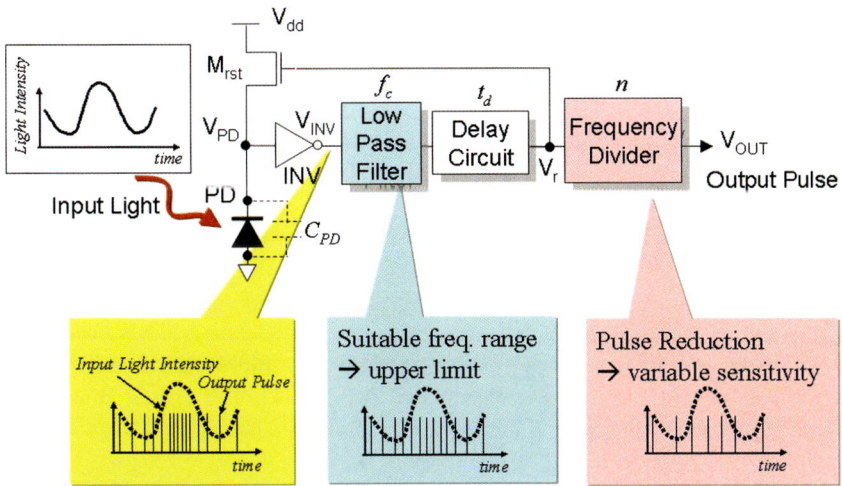

FIGURE 8.3. Block diagram of the PFM photosensor with variable photosensitivity.

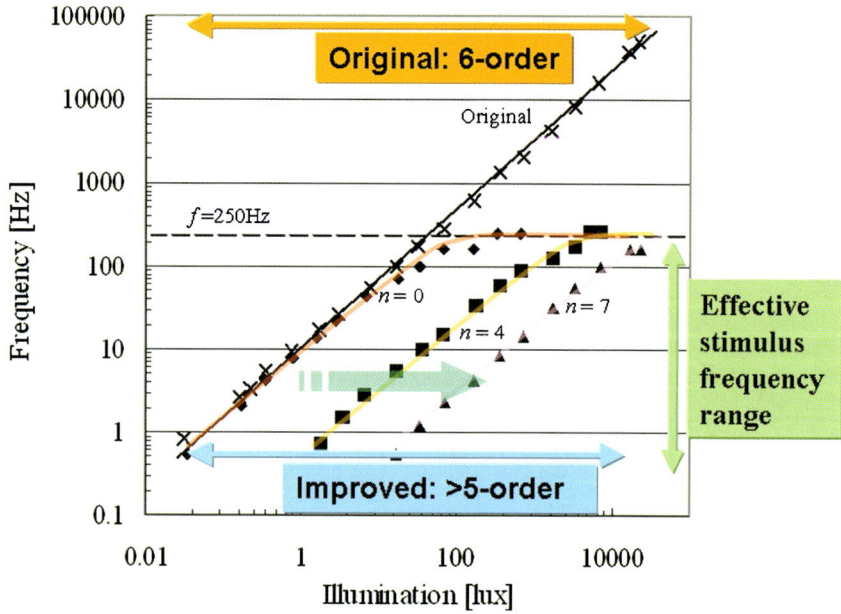

FIGURE 8.4. Output pulse frequency of the PFM photosensor with variable photosensitivity as a function of input light intensity. Reprinted from Ref. [4] with permission from IEEE.

(6th-order range of input light intensity), but is reduced to around 2-Log to be limited at 250 Hz when the low pass filter is turned on. By introducing the variable sensitivity, the total coverage of input light intensity becomes 5-Log between $n = 0$ and $n = 7$. The other functions in Table 8.1 have also been demonstrated.

Image Preprocessing Using the PFM Photosensor

When the PFM-based stimulator device is applied to a retinal prosthesis, the resolution is less than approximately 30×30. This limitation arises because the electrode pitch is larger than $100\,\mu$m, according to the electrophysiological experiments, and the width of the chip is less than approximately 3 mm according to the implantation operation. In order to obtain a rough but clear-shaped image with such a low resolution, image processing, such as edge enhancement, is preferable.

In order to implement image processing, we have proposed a new principle of spatial filtering in the pulse frequency domain [7]. The spatial filtering is generally based on the spatial correlation operation using a kernel h as

$$g(x, y) = \sum_{x'} \sum_{y'} h(x', y') f(x + x', y + y').\qquad(2)$$

Here, $f(x, y)$ and $g(x, y)$ indicate the pixel values at (x, y) of input and output images, respectively; $h(x, y)$ is a kernel weight.

Usually, f, g, and h are analog values for analog image processing or integers for digital image processing. In our scheme, f and g are represented as pulse frequency. Thus, for our implementation, we consider a method by which to represent the kernel weight in the pulse domain.

We introduce the interaction with the neighboring pixel as the gate control of the pulse stream from the neighboring pixel. This concept is illustrated in Figure 8.5. The absolute value of the kernel weight, $|h|$, is expressed as the on–off frequency of the gate control. The sign is expressed as follows. In order to realize negative weights in the spatial filtering kernel, the pulses from a pixel are collided with those from its neighboring pixels in order to make them disappear. For positive weights, the pulses from the pixel are merged with those from the neighbors. These mechanisms can be achieved by simple digital circuitry. In the architecture, a 1-bit pulse-buffering memory is implemented to absorb the phase mismatch between the pulses to be interacted.

The proposed architecture enables us to execute fundamental types of image processing, such as edge enhancement, blurring, and edge detection. The advantage of our architecture over straightforwardly implemented digital spatial

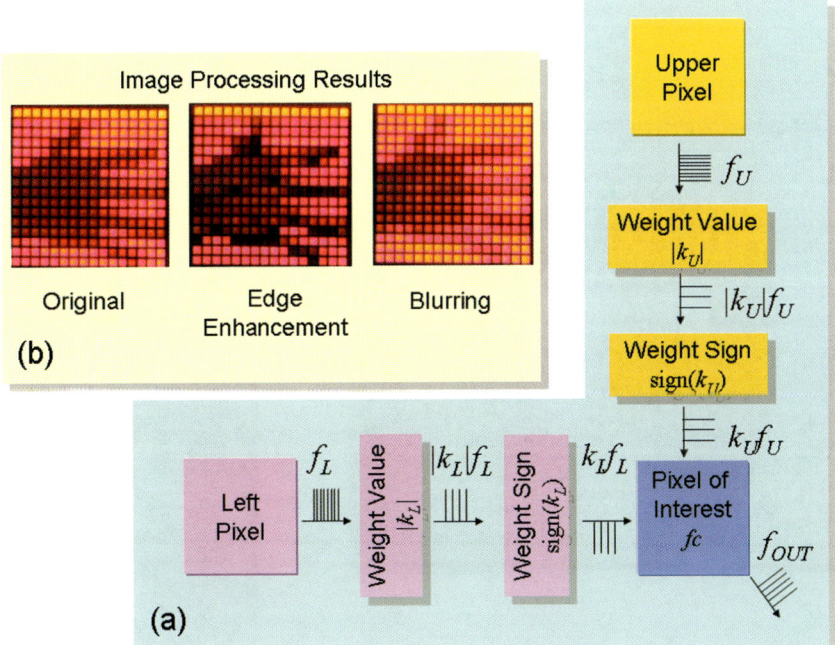

FIGURE 8.5. Concept of (a) image processing in the PFM-based image sensor, and its image processing results, (b) original captured image, edge enhancement, and blurring.

filtering is that the number of required logic gates is small because there is no need for adders or multipliers.

We have also implemented a binarization circuit based on an asynchronous counter with N-bit D-FF. The input of the D-FF of the MSB is fixed to HIGH. When 2^{N-1} pulses are input, the output of the counter turns HIGH from LOW. This means that the counter works as a digital comparator with a fixed threshold of 2^{N-1}.

According to the proposed architecture, we have designed and fabricated a PFM-based retinal prosthesis device with 16×16 pixels. Figure 8.6 shows microphotographs of the fabricated chip, peripheral circuits, and its pixel. As shown in Figure 8.7, each pixel has a PFM photosensor, the above-described image processing circuits, stimulating circuits, and a stimulus electrode, i.e. this chip can stimulate retinal cells. We used this chip for in vitro electrophysiological experiments of stimulating retinal cells, as described in the following section. Here, we present only the image processing results. Figure 8.5b shows the experimental results of image processing using the chip: an original image, edge enhancement, and blurring. These results clearly demonstrate that the proposed architecture works properly.

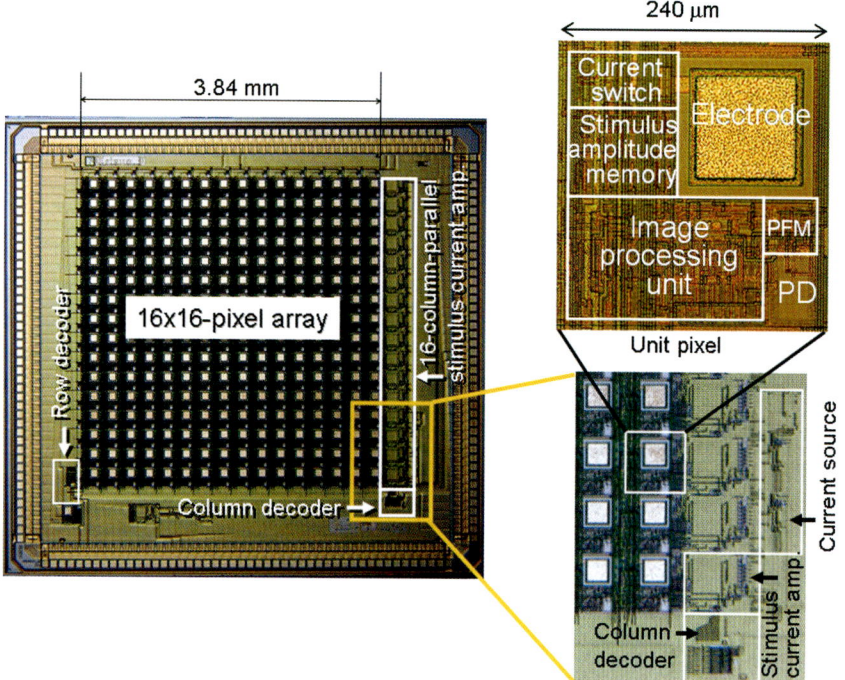

FIGURE 8.6. Microphotographs of the fabricated chip, peripheral circuits, and its pixel.

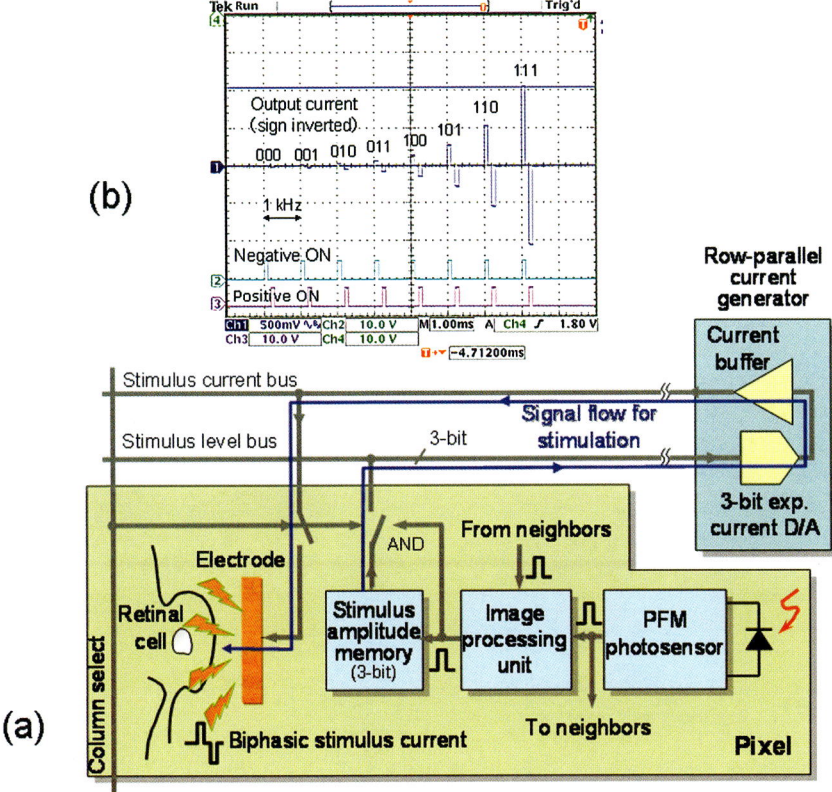

FIGURE 8.7. Block diagram of the chip (a) and its output biphasic pulses with 3-bit amplitude resolution (b).

Application of PFM Photosensor to the Stimulation of Retinal Cells

Electrodes and Packaging for Biological Environment

In this section, we demonstrate that the proposed PFM-based stimulator described in the previous section is effective in stimulating retinal cells. In order to apply the Si-LSI chip to electrophysiological experiments, we must protect the chip against the biological environment, and make an effective stimulus electrode that is compatible with the standard LSI structure. In order to meet these requirements, we have developed a Pt-/Au-stacked bump electrode process.

Figure 8.8a shows the fabrication process of our electrode [8, 10]. First, a Pt-/Au-stacked bump structure is formed on an Al-bonding I/O pad in a standard LSI chip. Although aluminum cannot be used in the biological environment, platinum (Pt) is a suitable electrode material due to its excellent biocompatibility and charge injection efficiency. The Pt electrode juts out of the top surface of the

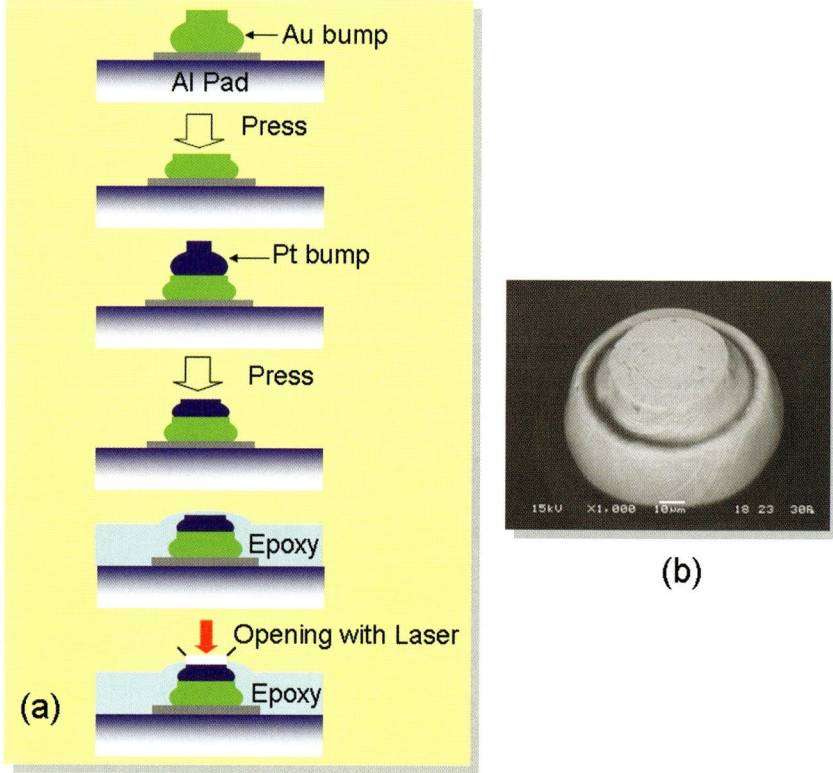

FIGURE 8.8. Fabrication process of stacked bump electrodes (a) and SEM of the stacked bump electrode (b).

chip, enabling close contact with retinal cells. In order to construct this jutting Pt electrode, we have formed the stacked bump in two steps, in which a gold bump is initially formed, on top of which the Pt bump is formed. The gold bump acts as a cushion for the hard Pt bump. Direct formation of the Pt bump sometimes breaks the LSI I/O pad; I/O pad is usually formed on an Si substrate and insulators such as silicon dioxide and silicon nitride films are sandwiched in between.

After the fabrication of the stacked Pt/Au bump, the entire LSI chip, including bonding wires, is covered with a biocompatible epoxy resin. Finally, the resin on top of the electrodes is completely removed using a high-power Ar ion laser, and thus the entire LSI chip except for the top of the Pt electrodes is covered with resin. Parylene coating can be additionally applied to the resin to ensure durability in the biological environment and can be removed by Ar ion laser.

Figure 8.8b shows a scanning electron micrograph (SEM) of the stacked Pt/Au bump electrode. The diameter of the structure is approximately $100\,\mu m$ and can be controlled by changing the Au and Pt wire diameter. Note that other biocompatible and efficient materials such as TiN and IrO$_x$ can be deposited on top of the Pt electrode.

In Vitro Electrophysiological Experiment Using the PFM Photosensor

In order to verify the operation of the PFM photosensor chip, we have conducted in vitro experiments using detached bullfrog retinas. The chip acts as a stimulator that is controlled by input light intensity as is the case in photoreceptor cells. A current source and pulse shape circuits are integrated onto the chip. The details have been reported in [11]. The Pt/Au stacked bump electrode and chip molding processes were performed as described in the previous section. Before molding, the chip was bonded onto a printed circuit board for handling. A reservoir of Ringer solution was placed surrounding the chip.

A piece of the bullfrog retina was placed, with the retinal ganglion cell (RGC) side face up, on the surface of the packaged chip. Figure 8.9 shows the experimental setup. Electrical stimulation was performed using the chip at a selected single pixel. A tungsten counter electrode with a tip diameter of 5 μm was placed on the retina, producing a transretinal current between the counter electrode and the chip electrode. A cathodic-first biphasic current pulse was used as the stimulation [20]. The pulse parameter is described in the inset of Figure 8.10a. Note that near-infrared (NIR) light does not excite the retinal cells, but does excite the PFM photosensor cells. A typical response curve is shown in Figure 8.10a. The arrow indicates the response from a RGC. Figure 8.10b demonstrates the experimental results of evoking retinal cells with the PFM photosensor, which is illuminated by input NIR light. We have confirmed that the firing rate increases in proportion to the input NIR light intensity. This demonstrates that the PFM photosensor activates the retinal cells through the input of NIR light, and suggests that it can be applied to human retinal prostheses. Details of the experimental setup are given in [11].

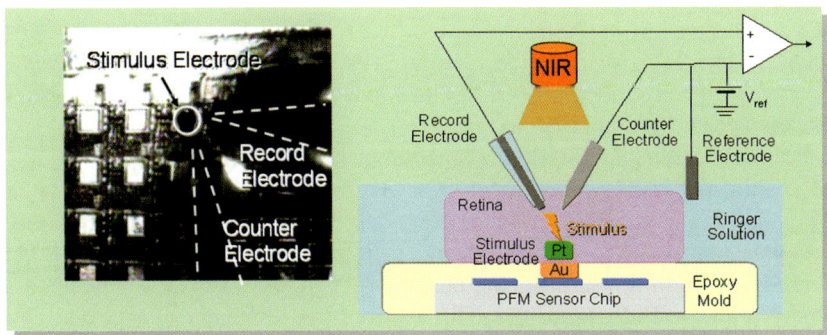

FIGURE 8.9. Experimental setup for in vitro stimulation of detached frog retina with the PFM photosensor. Reprinted from Ref. [11] with permission from Elsevier.

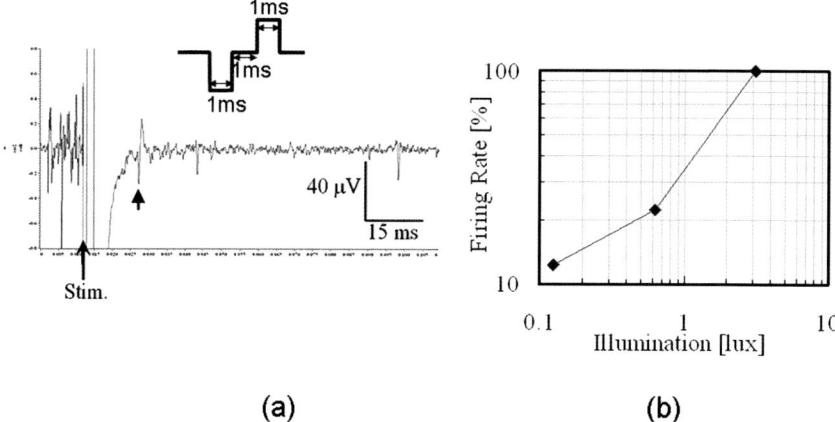

(a)   (b)

FIGURE 8.10. Experimental results of evoking retinal cells with the PFM photosensor. Response data (a) and firing rate as a function of input light intensity (b). Reprinted from Ref. [11] with permission from Elsevier.

Implantation of LSI-based Retinal Prosthesis Devices

In the previous sections, we have demonstrated the possibility of application of the PFM photosensor in subretinal stimulation. In this section, we focus on the implantation of the LSI-based device into the eye. Using the packaging and electrode technologies described in the previous sections, a thin Si dummy chip with bump electrodes was implanted on the suprachoroid+ of a rabbit. A 3-mm-wide and 4-mm-long chip was mounted on a polyimide substrate. Although the chip was successfully implanted, such a thin LSI chip must be handled very carefully. Another issue is how bending affects the device characteristics [5].

   In order to overcome the issue of mechanical rigidity and realize a feasible LSI-based device, we have proposed a device architecture consisting of small microchips that work under a single set of control signals [8–10]. Figure 8.11 shows the concept of the proposed smart distributed stimulator. The array consists of a number of LSI-based microchips, each of which is approximately 500-μm square in size. Each microchip has several Pt-/Au-stacked bump electrodes and is covered by the process described in a previous section. The detailed fabrication process is described in [10].

   In order to implant the device smoothly, the thickness of the device must be as thin as possible. The proposed device has a thickness of approximately 200 μm, which is acceptable. A microchip-based stimulator is fabricated in order to demonstrate proof-of-concept. The microchips are dummy Si chips with Pt/Au bumps. The entire image of the stimulator with Pt wires covered with silicone tubing is shown in Figure 8.12a. The width of the stimulator is approximately 3 mm.

   The device was implanted into a pocket that was made in the sclera of a rabbit eye to stimulate retinal cells by the STS method. Figure 8.12b shows the

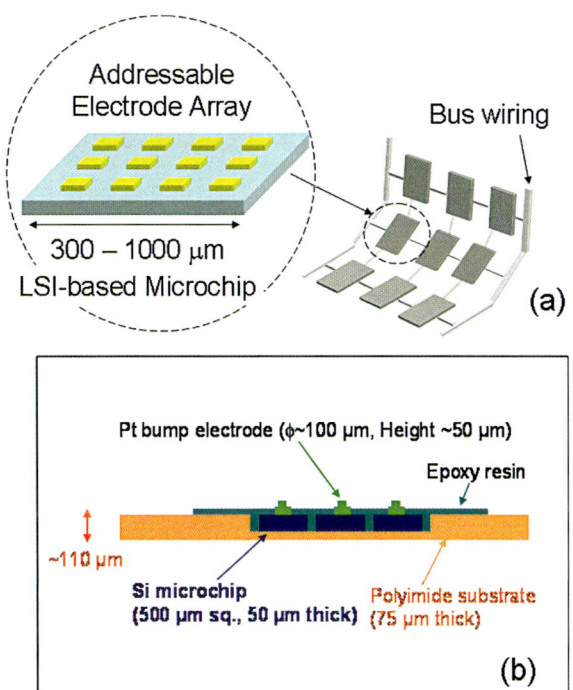

FIGURE 8.11. The concept and structure of the distributed microchip-based stimulator. Reprinted from Ref. [10] with permission from Elsevier.

extracted rabbit eye into which the device was implanted in 2 weeks. The device is bent along the eye curvature. The short-time stimulation experiment, in which only the small change of electrically evoked potentials (EEPs) was observed between the first and last stages of the implantation, demonstrate that the device can work properly during the implantation.

In the next step, we develop an advanced architecture of the distributed microchip-based stimulator device. In this device, the microchips are connected to each other via two wires, not including the power supply lines, and are placed on a flexible polyimide substrate. The features of the device are as follows. First, the device is thin and flexible so that it can contact neural cells more closely when implanted, and thus is suitable for simulating neural cells. Second, the introduction of LSI to the microchip makes it possible to introduce, for example, a PFM photosensor. This provides the device with high performance and versatility. In addition, the LSI reduces the number of input/output pads needed in the device. Apart from the power supply lines, only two signal lines of stimulation/record and control are required. The control line operates the entire set of microchips. Each microchip includes enough circuitry to decode the control signal. Third, the device can be connected to another device. Such a daisy chain could combine a large number of electrodes, e.g. over 1000 electrodes.

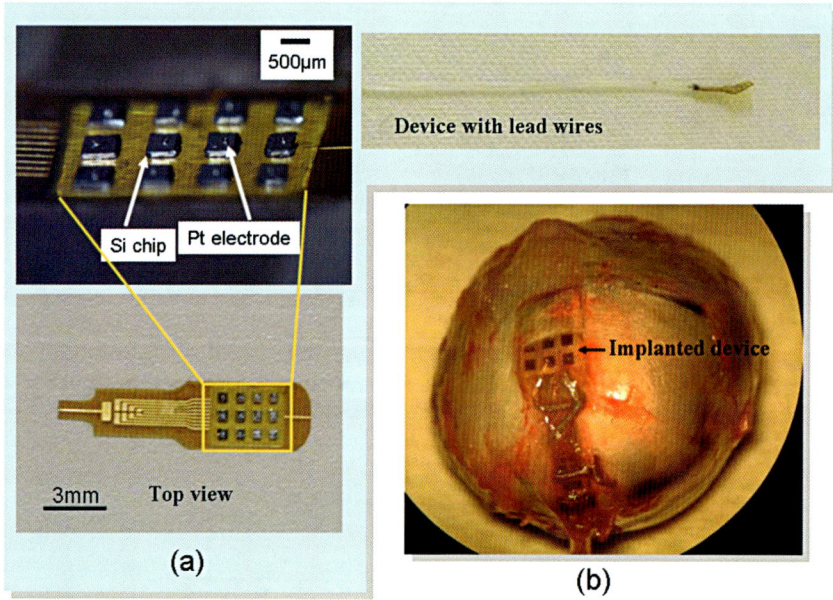

FIGURE 8.12. The photograph of the distributed microchip-based stimulator device (a) and an extracted eyeball implanted device in the sclera pocket (b).

Figure 8.13 shows the circuit block diagram and the microphotograph of the microchip [8, 11]. The chip is fabricated using 0.6-μm 2-poly 3-metal standard CMOS technology. The microchip is so small, at $600\,\mu m \times 600\,\mu m$, that it can be thinned down to less than $100\,\mu m$ without risk of breakage. The fabricated microchip has nine stimulation/recording electrodes and control circuits with

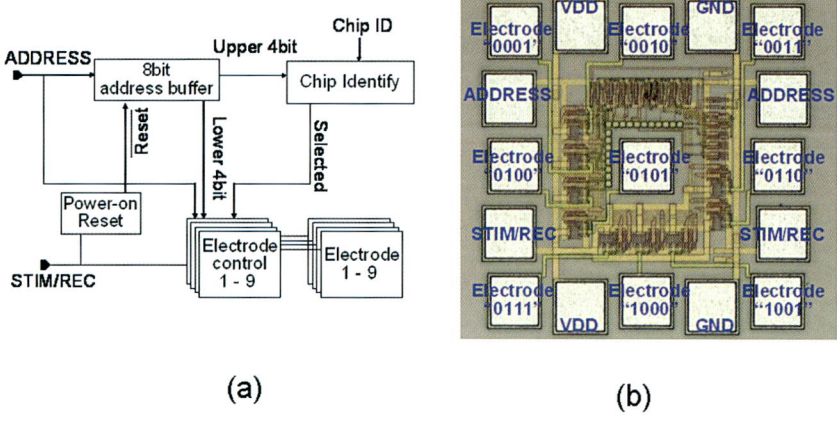

FIGURE 8.13. Circuit block diagram (a) and the layout of the microchip (b). Reprinted from Ref. [8] with permission from Institute of Pure and Applied Physics.

four I/O pads for addressing (ADDRESS) and stimulation/record (STIM/REC) and four pads for power supply (VDD and GND), as shown in Figure 8.13a. Note that each microchip relays ADDRESS and STIM/REC lines in the vertical direction and VDD and GND lines in the horizontal direction, as shown in Figure 8.13. This wiring architecture reduces the wiring area on the substrate.

We use a broadcast topology to assign one electrode to be activated. This consumes only a small area of circuitry, sufficiently small for the size of the microchip. An external controller broadcasts a control signal to all of the microchips. Each microchip has its own identification (ID). The microchip has an 8-bit asynchronous counter as an address buffer. The addressing counter counts the digital pulses applied to the ADDRESS line, and the microchip interprets the value in the counter as the address of the selected electrode. The upper 4 bits and the lower 4 bits represent the addresses of the selected chip and the selected electrode, respectively. Only the selected electrode on the selected chip is connected to the STIM/REC line. Once selected, the neural stimulation/recording can be activated at the selected electrode via the STIM/REC line. Figure 8.14

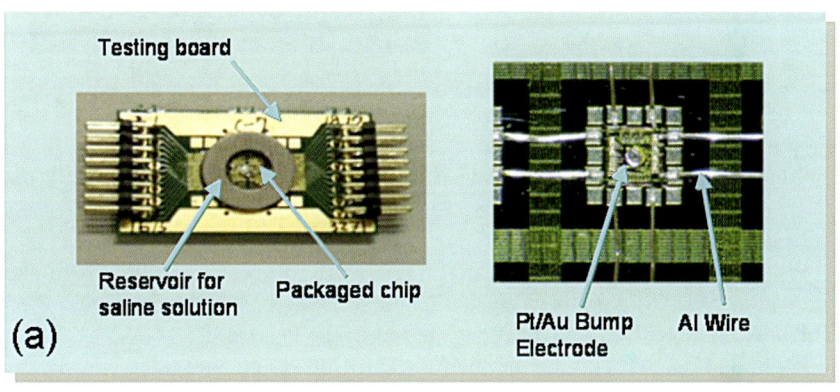

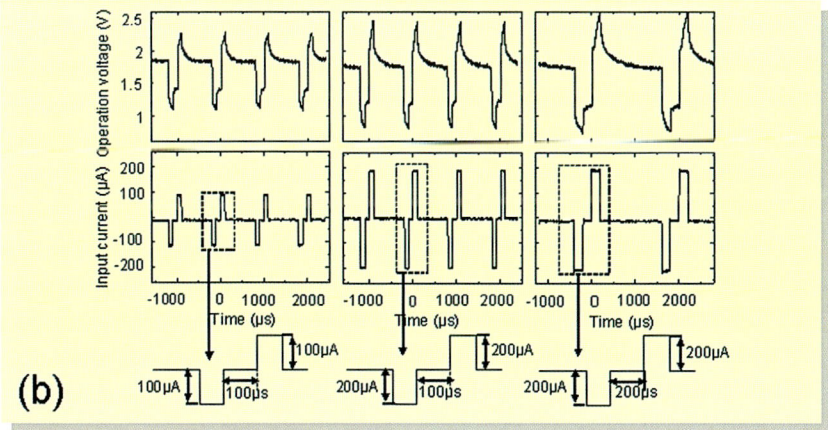

FIGURE 8.14. Experimental result of the distributed microchip-based stimulator device under the saline environment (a) testing board, (b) experimental results.

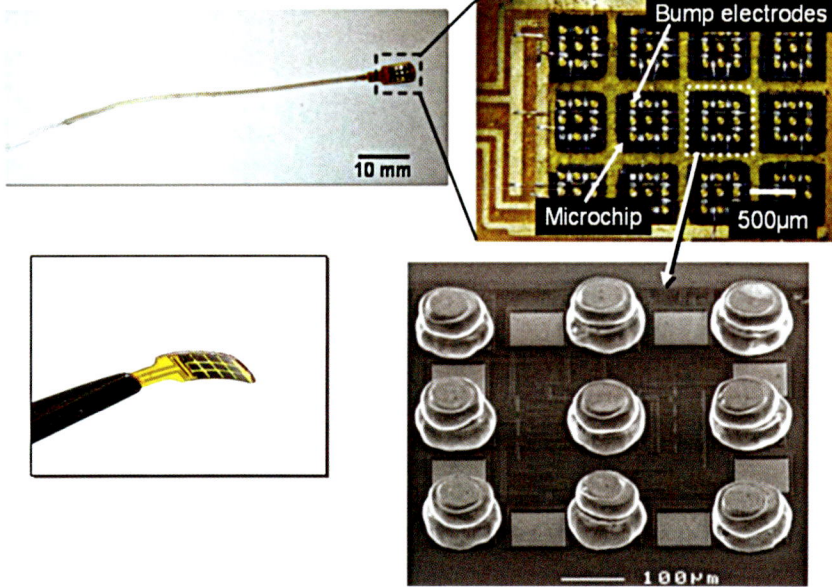

FIGURE 8.15. Photograph of the fabricated microchip-based stimulator device.

shows the packaged chip and the experimental result in saline solution of the fabricated stimulator that is packaged as described. For each timing, only one electrode is activated. This demonstrates that we can select an arbitrary electrode to be activated.

We have developed a dedicated fabrication process for the distributed microchip-based stimulator. The fabricated stimulator is shown in Figure 8.15. The stimulator will be implanted and tested in the near future.

It should be noted that the number of microchips to be controlled is restricted with the asynchronous counter design, and that this number could easily be increased by designing the counter to have a greater number of bits. A stimulator that consists of 4×4 microchips with 9 electrodes (total 144 electrodes) is connected with another stimulator in a daisy chain, and thus a stimulator with 288 electrodes is realized.

The distributed microchip-based device we have fabricated has a broadcast architecture. Only a single electrode can be activated. In order to activate multiple electrodes as well as different pulse parameters in each electrode, we have developed serial bus architecture and confirmed its fundamental functions [9]. In the future, we will develop a more sophisticated device in order to introduce such serial architecture.

Summary

We have demonstrated the PFM photosensor that is suitable for application as a subretinal implantation device. The fabricated PFM-based image sensor with 128×128 pixels demonstrated a wide dynamic range of $60\,dB$. In order to effectively apply the PFM photosensor to the stimulation of retinal cells, we modified the original PFM photosensor. The output was converted into biphasic current pulse and the photosensitivity was variable. These features can be easily introduced thanks to the nature of the PFM photosensor, which is compatible with logic circuits. The PFM photosensor has been applied to the stimulation of detached retinas from a frog eye. In vitro electrophysiological experiments demonstrated that the PFM photosensor evokes retinal cells under the illumination of IR light. Finally, we proposed a distributed microchip-based stimulator device that can be bent easily and can realize a large number of electrodes. The short-time implantation via the STS method was performed using the proposed implantation device and showed that the device can stimulate retinal cells. Further development is required for applying the device to clinical implantation.

Acknowledgments. The authors would like to thank Professor Tetsuya Yagi for valuable advices on the electrophysiological experiments, and Dr. Shigeru Nishimura and Naoko Tsunematsu for their continuous encouragement and valuable discussion. The authors also would like to thank Dr. Uehara of Nidek and Dr. Furumiya of NAIST for the experimental data used here. This work was supported by the New Energy Development Organization (NEDO) of Japan "Artificial Vision System" Project and Health and Labor Sciences Research Grants, Japan.

References

1. W. Liu, P. Singh, C. DeMarco, R. Bashirullah, M. Humayun, and J. Weiland, "Semiconductor-based implantable microsystems," Chapter 6, in *Handbook of Neuroprosthetic Methods* (edited by W. Finn, and P. LoPresti), CRC Publishing Company, 2002, pp. 127–161.
2. W. Liu, and M. S. Humayun, "Retinal prosthesis," *IEEE Int'l Solid-State Circuits Conf. Dig. Technical Papers*, pp. 218–219, San Francisco, CA, 2004.
3. J. Ohta, N. Yoshida, K. Kagawa, and M. Nunoshita, "Proposal of application of pulsed vision chip for retinal prosthesis," Jpn. J. Appl. Phys. **41** (4B), 2322–2325, 2002.
4. K. Kagawa, K. Isakari, T. Furumiya, A. Uehara, T. Tokuda, J. Ohta, and M. Nunoshita, "Pixel design of a pulsed CMOS image sensor for retinal prosthesis with digital photosensitivity control," Electron. Lett. **39** (5), 419–421, 2003.
5. D. C. Ng, K. Isakari, A. Uehara, K. Kagawa, T. Tokuda, J. Ohta, and M. Nunoshita, "A study of bending effect on pulsed frequency modulation based photosensor for retinal prosthesis," Jpn. J. Appl. Phys. **42** (12), 7621–7624, 2003.
6. K. Kagawa, N. Yoshida, T. Tokuda, J. Ohta, and M. Nunoshita, "Building a simple model of A pulse-frequency-modulation photosensor and demonstration of a 128×128-pixel pulse-frequency-modulation image sensor fabricated in a standard 0.35-μm complementary metal-oxide semiconductor technology," Opt. Rev. **11** (3), 176–181, 2004.
7. K. Kagawa, K. Yasuoka, D. C. Ng, T. Furumiya, T. Tokuda, J. Ohta, and M. Nunoshita, "Pulse-domain digital image processing for vision chips employing low-voltage operation in deep-submicron technologies," IEEE Selected Topics Quantum Electron **10** (4), 816–828, 2004.

8. Y.-L. Pan, T. Tokuda, A. Uehara, K. Kagawa, J. Ohta, and M. Nunoshita, "A flexible and extendible neural stimulation device with distributed multi-chip architecture for retinal prosthesis," Jpn. J. Appl. Phys. **44** (4B), 2322–2325, 2005.

9. A. Uehara, Y.-L. Pan, K. Kagawa, T. Tokuda, J. Ohta, and M. Nunoshita, "Micro-sized photo detecting stimulator array for retinal prosthesis by distributed sensor network approach," Sensors & Actuators A **120** (1), 78–87, 2005.

10. T. Tokuda, Y.-Li Pan, A. Uehara, K. Kagawa, M. Nunoshita, and J. Ohta, "Flexible and extendible neural interface device based on cooperative multi-chip CMOS LSI architecture," Sensors & Actuators A **122** (1), 88–98, 2005.

11. T. Furumiya, D. C. Ng, K. Yasuoka, K. Kagawa, T. Tokuda, M. Nunoshita, and J. Ohta, "Functional verification of pulse frequency modulation-based image sensor for retinal prosthesis by in vitro electrophysiological experiments using frog retina," Biosensors and Bioelectronics **21** (7), 1059–1068, 2006.

12. H. Kanda, T. Morimoto, T. Fujikado, Y. Tano, Y. Fukuda, and H. Sawa, "Electro-physiological studies of the feasibility of suprachoroidal-transretinal stimulation for artificial vision in normal and RCS rats," Invest. Ophthalmol. Vis. Sci. **45** (2), 560–566, 2004.

13. K. Nakauchi, T. Fujikado, H. Kanda, T. Morimoto, J.S. Choi, Y. Ikuno, H. Sakaguchi, M. Kamei, M. Ohji, T. Yagi, S. Nishimura, H. Sawai, Y. Fukuda, and Y. Tano, "Transretinal electrical stimulation by an in-trascleral multichannel electrode array in rabbit eyes," Graefes Arch. Clin. Exp. Ophthalmol. **243** (2), 169–174, 2005.

14. A. Y. Chow, M. T. Pardue, V. Y. Chow, G. A. Peyman, C. Liang, J. I. Perlman, and N. S. Peachey, "Implantation of silicon chip microphotodiode arrays into the cat subretinal space," IEEE Trans. Neural Syst. Rehab. Eng. **9**, 86–95, 2001.

15. E. Zrenner, "Will retinal implants restore vision?" Science, 295, 1022–1025, 2002.

16. W. Yang, "A wide-dynamic-range, low-power photosensor array," *IEEE Int'l Solid-State Circuits Conf. Dig. Technical Papers*, pp. 230–231, San Francisco, CA, 1994.

17. F. Andoh, H.Shiammoto, and Y. Fujita, "A digital pixel image sensor for real-time readout," IEEE Trans. Electron. Device **47**, 2123–2127, 2000.

18. M. Mazza, P. Renaud, D.C. Bertrand, and A.M. Ionescu, "CMOS Pixels for Subretinal Implantable Prosthesis," IEEE Sensors Journal **5**, 32–37, 2005.

19. J. Deguchi, T. Watanabe, T. Nakamura, Y. Nakagawa, T. Fukushima, S. J-Chill, H. Kurino, T. Abe, M. Tamai, and M. Koyanagi, Jpn. J. Appl. Phys. **43** (4B), 1685–1689, 2004.

20. L. Li, Y. Hayashida, and T. Yagi, "Temporal properties of retinal ganglion cell responses to local trans retinal current stimuli in the frog retina," Vision Res. **45** (2), 263–273, 2005.

9
Development of a Wireless High-Frequency Microarray Implant for Retinal Stimulation

G.W. Auner[1,4], R. You[1], P. Siy[1], J.P. McAllister[3,4], M. Talukder[1] and G.W. Abrams[2,4]

[1]*SSIM/Biomedical Engineering/Elec and Comp Engineering, Wayne State University*
[2]*Kresge Eye Institute, Wayne State University*
[3]*Dept of Neurosurgery, Wayne State University*
[4]*Ligon Research Center of Vision, Wayne State University*

Abstract: We have developed an electrical stimulator and diagnostic research microarray with wireless power and communications to facilitate spatial stimulation of retinal tissue. A third generation 32×32 prototype of this retinal neural implant array has been developed. Integrated into the microarray is a functionally graded Ti/IrO_2 microbump electrode system for interface with neural tissue with decreased impedance for stimulation. The microarray is designed for basic research to determine retinal tissue stimulation thresholds and spatial effects. The array is connected to a telemetry chip, which uses magnetic induction for wireless power with a digital overlay for communication. In our design, changes in the induced current in the telemetry coil are used to send information to the reading coil. Since the reading and telemetry coil are magnetically coupled, the current change can be sensed for bidirectional communication. Combined, this chip set provides a 1024 array that can stimulate neural tissue spatially, can sense neural signals spatially, and has wireless power and communication in a package of less than 2 mm size.

Introduction

This paper outlines a retinal prosthesis device designed to take external data such as a visual field, processes it, and finally reproduces it on the retina to create a visional form by means. A conceptual system of retinal prosthesis is given in Figure 9.1. It consists of *extraocular* and *intraocular* units. Extraocular unit is built on the spectacle, while the intraocular unit is implanted inside the eye. These two units are linked together through RF coupling. The block diagram

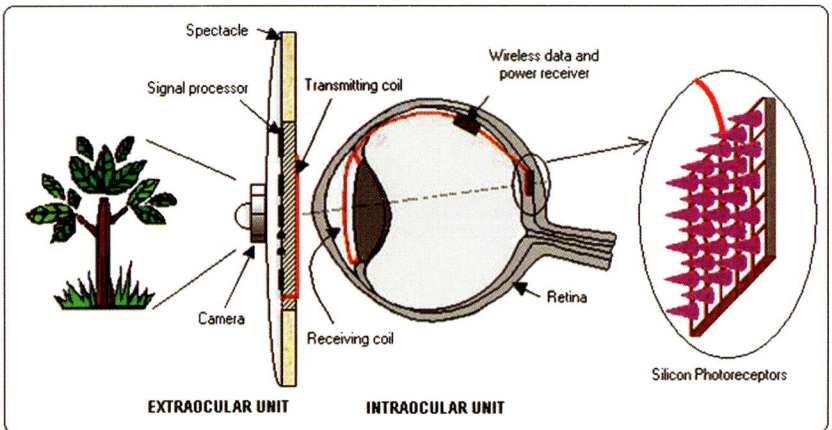

FIGURE 9.1. Complete system of retinal prosthesis.

of different functional units of a retinal prosthesis is shown in Figure 9.2. In extraocular unit, image from a camera is reduced to lower resolution by an image processor, and transmitted after time multiplexing. In intraocular unit, a wireless receiver recovers serial data of each pixel and sends them to the stimulator array for further processing and stimulation. This chapter outlines the current research and development of a retinal prothesis.

In general, to excite the retina the signal should be a biphasic current waveform (Figure 9.3) [1]. There are four parameters associated with this waveform. They are amplitude, width, interphase delay, and frequency. Amplitude represents the gray level of the pixel, width and interphase delay depend on the degree of a patient's retinal damage, while frequency is the frame rate. The characteristics of these waveform parameters will depend on patient feedback on visual perception. Typical values of the waveform reported in different papers are: amplitude $10–600\,\mu A$, width $100\,\mu s–2\,ms$, delay $0–1\,ms$, and frequency $10–125\,Hz$. The typical impedance of the retinal tissue is $10\,K\Omega$ for retinitis pigmentosa (RP) and age-related macular degeneration (AMD) patient. However, variations of stimulation waveforms such as high-frequency stimulation have

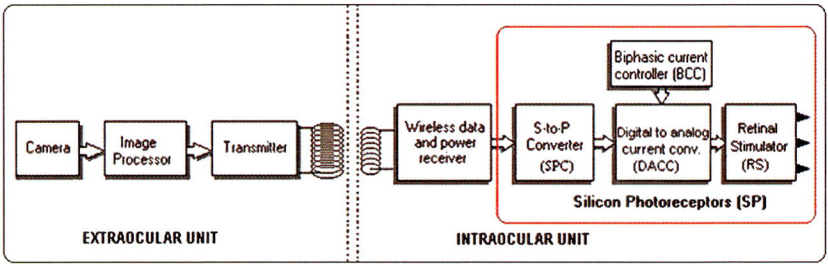

FIGURE 9.2. Block diagram of the retinal prosthesis.

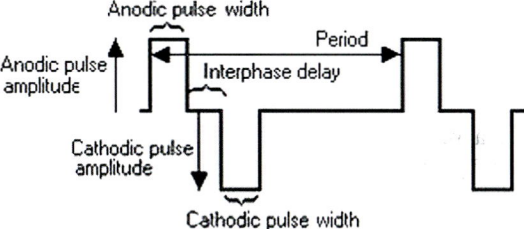

FIGURE 9.3. Biphasic waveform.

not been extensively investigated in a retinal implant. Therefore, the design and fabrication of a device that allows for an array of stimulation waveforms and electrode size and spacing variation would be very useful for understanding the effects on neural stimulation. Further, if the device can sense neural signals as feedback, a greater understanding of neural stimulation effects could be obtained.

There are two approaches to implanting the stimulator array. They are epiretinal implant and subretinal implant (Figure 9.4). In both the cases, the stimulation is done by microelectrodes on the stimulator array arranged as a two-dimensional array. We are pursuing a subretinal implant based on an RF link for data and power transfer [2, 3]. As the subretinal implant goes underneath the

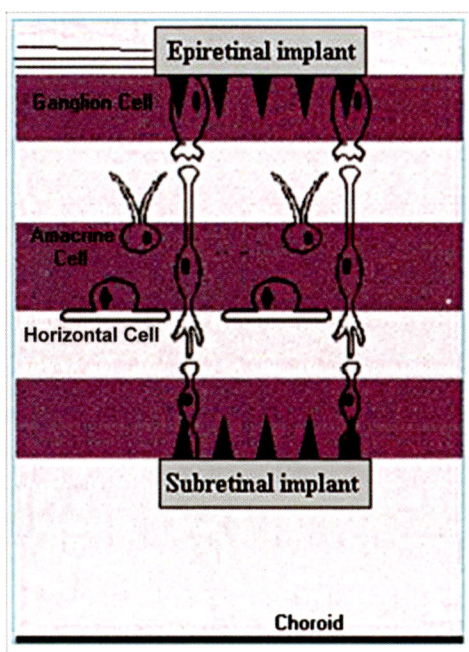

FIGURE 9.4. Location of implants.

outer layer, relevant criteria include compactness, on-chip electrodes, wireless or should have fewer wires if the wireless receiver is not on board of the stimulator array, power consumption, radiated heat in retinal tissue, and biocompatible packaged reliably. Further, given the unknowns in implant stimulation, it is desirable to have a versatile experimental system that contains a high density of electrodes. Ideally, these electrode stimulation points can be programmed to activate in blocks to mimic larger electrodes or programmed with inactive electrodes separating active ones to investigate optimal electrode spacing. Further, the ability to sense neural activity by switching between stimulation and sensing modes would add information on diagnosing the implant effectiveness and the way the retina responds to stimulation parameters. With these views in mind we have designed a stimulator array applying multiplexing technique. In this technique, stimulator array communicates using only six wires, a substantial reduction in the number of wires compared to flexible epiretinal implants [4]. Utilizing this technique we are able to design a 32×32 array stimulator array with only one serial to parallel converter (SPC), one digital to analog current converter (DACC), and one biphasic current controller (BCC). Serial data of each pixel is converted to parallel data by SPC. DACC converts the digital pixel to equivalent analog current, which is fed to the retinal stimulator (RS). The stimulator decodes the time-multiplexed analog current and stimulates the neurons. It does not need address to stimulate the neurons. This device is a bidirectional device that can also retrieve responses from retina as a reader and sends out multiplexed readings through reverse telemetry. In theory the stimulator array is capable of generating visual perception in almost real time eliminating the needs of memory unit. As a result, power consumed by the stimulator array is so small that the generated heat is ultra low. We are also capable of generating a large array of experimental stimulation waveforms. This implant chip is linked to our wireless power and communications chip.

Wireless Implantable Bio-Device Interface (WIBI)

Wireless implantable bio-device interface (WIBI) system provides power and bidirectional communication capability to the retina-stimulating chip. The WIBI is a multipurpose system that can support various implantable devices applications. It replaces the batteries and metal wires by two separated coils, one implanted and another placed outside, to extend the device lifetime and reduce the danger of infection (Figure 9.5).

Operation Principles

The WIBI uses the mutual inductance between two coils to transmit power and data. The physics of power transmission is the same as transformers. However,

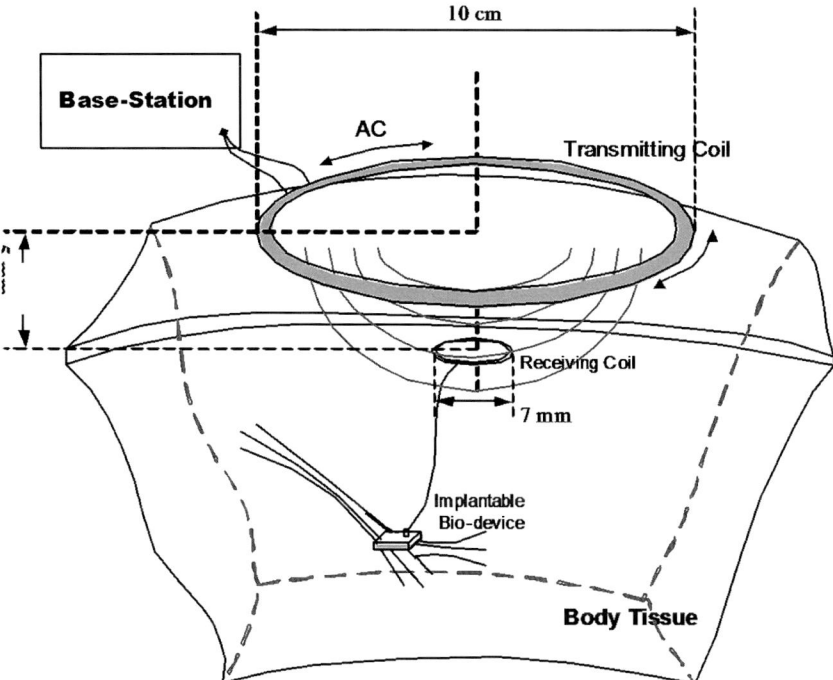

FIGURE 9.5. Implantable devices with WIBI.

in a regular transformer, two coils are coupled using an iron core, which significantly reduces the magnetic leakage and increases the coupling coefficient k to nearly 1. In the WIBI system, k is typically less than 0.05. We must design a converting circuit sensitive enough to pick up the low induced voltage in the implanted coil. Data transmission is overlaid on the continuous power transmission. Two directions of communication are named as download (from the base station to the WIBI) and upload (from the WIBI to the base station). Download transmission utilizes amplitude modulate with 100% modulate rate. The base station shuts the current in the base station coil shortly to send a pulse, and a series of pulses carries the data. Since the WIBI has limited power, the upload transmission uses a "reflecting" phenomenon. The WIBI shorts the implanted coil promptly to generate a current pulse, and this pulse is received by the base station coil.

The WIBI System Architecture

The purpose of developing the WIBI system is to support various applications with different requirements. The flexibility of WIBI is one of the important design targets. Table 9.1 lists the requirement of the WIBI design.

TABLE 9.1. Design requirements of the WIBI system.

Parameter	Values
Carrier frequency	13.56 MHz
Power supply	> 10 mW
Data rate	> 50 kbps
Clock frequency	500 kHz ~ 1 MHz

The WIBI system architecture is shown in Figure 9.6. Here, the RF–DC, power-up reset, DC–DC, and transceiver are analog modules, while the encoder, decoder, controller, and data buffer are digital modules. The digital phase-locked loop is a hybrid module with an analog oscillator and a digital phase detector.

There are five wires from the WIBI: Vdd, GND, CLK, DATA, and SYNC. WIBI can support multiple functional modules (sensors, stimulators, and others). All the functional modules are connected to these wires in parallel. Vdd and GND provide 7 V power at a 10 ~ 20 mW level. CLK is a 500 kHz to 1 MHz clock signal. And DATA and SYNC are the two wires in a serial bus with the highest speed of 50 kbps. Each functional module has a distinguishable ID code. Using this ID code, the WIBI can activate a functional module or initiate a communication with it.

Analog Circuits of WIBI

There are five blocks in this portion of the WIBI system: RF–DC, transceiver, power-up reset, DC–DC, and a hybrid module digital PLL.

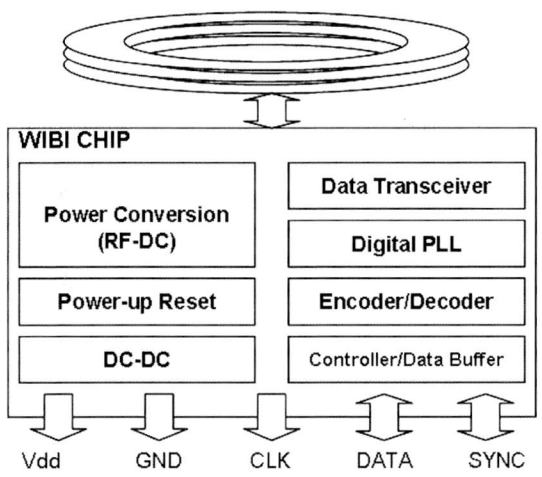

FIGURE 9.6. WIBI architecture.

RF–DC

This module converts the 13.56 MHz current induced in the coil to 7 V DC power. The arrangement of RF–DC circuits is a voltage-double-rectifier with a voltage feedback control. The reason to use these circuits is based on the fact that an MOS transistor has lower pass-resistance when the applied gate voltage is high.

Power-up Reset

It outputs a high voltage only after the Vdd reaches a certain level and is stabilized. The power-up reset is not only for digital circuits to initialize, but also to enable the analog modules as a global start-up signal. Many analog circuits may lock the Vdd at zero with lack of proper start-up.

DC–DC

It further filters Vdd and outputs a 3.3 V power. Since most circuits do not require high voltage, this lower voltage saves 80% power and provides a very clean power source. The DC–DC module has two separated outputs for analog and digital to reduce the noise. One or two 3.3 V wires are optional outputs of WIBI.

Transceiver

For transmitting, it shorts the implanted coil to switch its load; for receiving, it retrieves the envelope of the waveform induced in the coil, and detects it using a threshold. The main design issue is how to avoid any mistransmitting pulse, since during the transmitting a significant power loss will happen.

Digital PLL(DPLL)

It is one of the essential functions of the WIBI system. It provides a clock signal synchronized to the base station. The DPLL uses a novel algorithm that enables large pull-in range with high accuracy based on a simple phase detection method. It can tolerate high circuit mismatch and high noise. Since the working frequency of this new DPLL is equal to its output frequency, the power consumption is very low.

Digital Circuits of WIBI

The digital modules handle the encoding and decoding of the bidirectional communication and control the data flow. They also monitor the working condition of the implantable device and feedback these in formations to the base station. There are five modules: two-wire data bus, decoder, encoder, data buffer, and controller.

Two-Wire Data Bus

It contains two wires: DATA and SYNC. Utilizing wire-OR logic through pull-down resistors, the data bus is a high capable data flow path that can be established between two digital blocks, or between a digital block and a functional module. This arrangement reduces the components along the data flow path, which is important for communication reliability and low-power-consumption strategy.

Decoder and Encoder

They work with the analog transceiver. Decoder translates the incoming Manch-ester coded waveform to a data sequence, and encoder translates data sequence to waveform. Both of the modules are controlled using asynchronous reset and clock-locking-enable. They do not process the data, buffer them, or check the bus for any acknowledgment. So the timing of enable and reset signals is critical for meaningful communication.

Data Buffer

The data buffer has two basic functions: temporarily buffering a small piece of data from the bus and giving a bus access to the controller. It can read or write the bus. In addition, acknowledged signals and functional module IDs are all written by the data buffer.

Controller

For extended flexibility, the controller is designed based on a light-weighted RISC architecture, but not a pre-designed state machine. Currently 16 hardware commands are implemented in the controller. Through them the controller can control the WIBI system as well as the functional modules (using the two-wire-bus protocols). Except the program in ROM, a remote control mode is available. In this mode, the base station sends a series of commands, and then controller runs them.

Simulation

The WIBI analog circuits have been designed using Cadence IC 5.033 custom IC development software and simulated using Cadence Spectre simulator. Figure 9.7 shows the power conversion waveform of RF–DC and DC–DC modules, Vdd is the 7 V RF–DC output, and Vddd/Vdda are the 3.3 V DC–DC outputs. Due to the switching operations of digital circuits, Vddd has some voltage spikes, but Vdda has no effect from them. We can also observe the Vddd and Vdda are only available after Vdd is established. This is controlled by the power-up reset (not shown).

Figure 9.8a shows the downloading scenario: the induced voltage in the implanted coil (/net 148), the threshold detection result of transceiver (RCV),

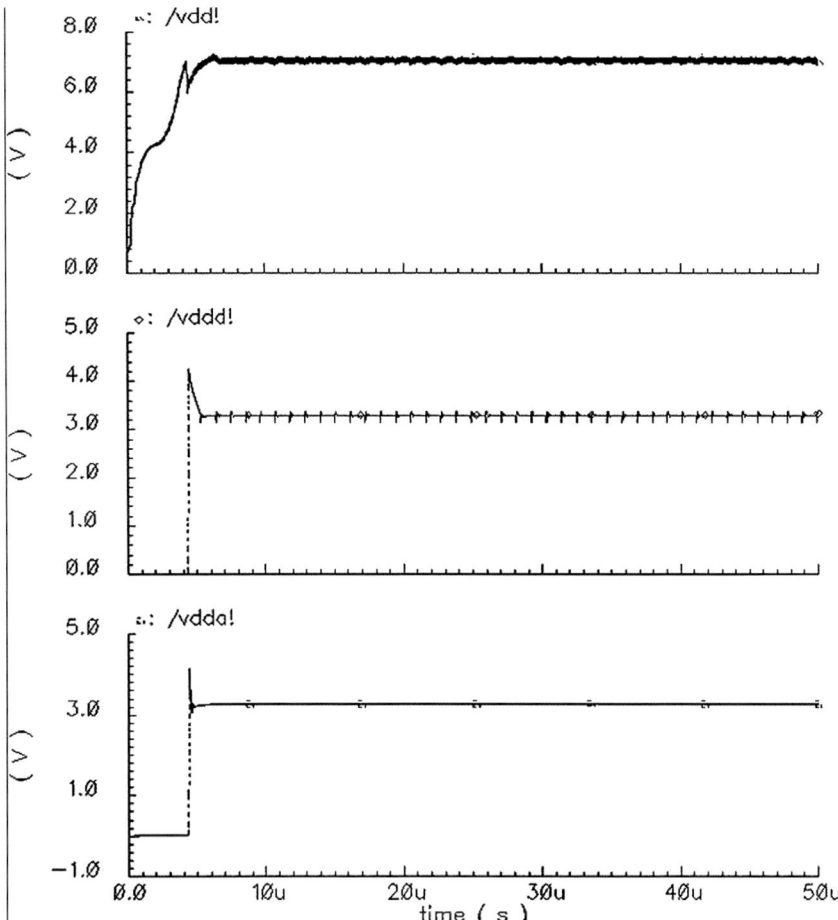

FIGURE 9.7. WIBI power outputs simulation.

and the influence of download communication to the Vdd. The pulses in the implanted coil are clearly retrieved to a binary sequence. Figure 9.8b shows the uploading transmission. (/net 109) is the binary sequence being transmitted. Then transceiver modulates the load from the implanted coil and causes a series of current pulses in the coil (/L1/PLUS). This transmission also has an effect on the output power Vdd.

The output power can reach 10.16 mW, and the corresponding RF–DC converting efficiency is higher than 70%. The bidirectional simplex communication can transmit up to 500,000 pulses per second. If a 10:1 Manchester coding is used, 50 kbps data rate can be achieved with plenty of redundancy. The WIBI digital circuits are designed using Verilog HDL with Cadence BuildGate and Cadence Silicon Encounter design packages, and simulated by Cadence LDV software. Figure 9.9 shows the simulated scenario of a decoding and encoding

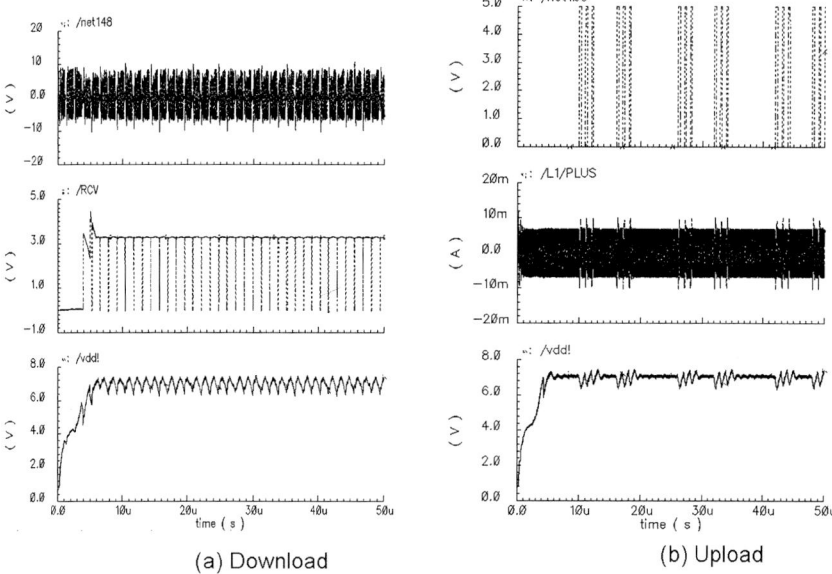

FIGURE 9.8. WIBI bidirectional communication simulation.

operation. First the controller writes a binary number $01100011(0 \times 63)$ to the data buffer (db_reg), then data buffer delivers this number bit by bit to the encoder and generates a binary sequence (coil_out). After the encoding, this sequence is feedback to the decoder (coil_in), and the number is written back to the data buffer (db_reg).

(a) Encoding period
(b) Decoding period

Design of Retinal Prosthesis

The stimulator array is the central part of the retinal prosthesis. The function of the stimulator array is to convert serially received digital pixel from wireless receiver into analog biphasic current and to stimulate the neuron the pixel is intended to. We are presenting a unique design of stimulator array. As shown in Figure 9.10, stimulator array has been designed with one SPC, one DACC, one BCC, and one RS to carry out its function. The design of each unit is described below.

Serial to Parallel Converter (SPC)

We have decided that each pixel would be 8-bits deep. To handle 8-bits pixel, we have designed 8-bits-wide double buffering SPC (Figure 9.10). It has a shift

A

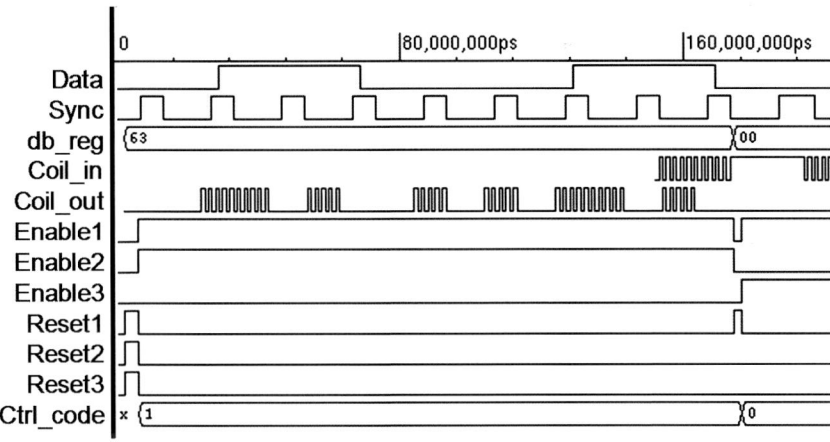

B

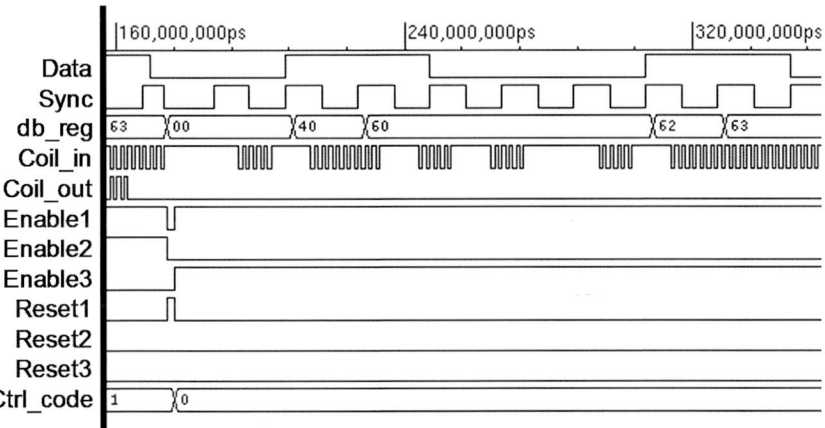

FIGURE 9.9. WIBI digital circuits functional simulation.

register (DFF0-DFF7), a latch (LCH0-LCH7), and a counter (DFF8-DFF10). The time multiplexed serial data from the receiver is shifted in bit by bit; when counter counts 8, the data is latched and available right away for DACC. This parallel data in latch is held for the next 8 cycles, while shift register is receiving another pixel.

Digital to Analog Current Converter (DACC)

For stimulation, we are to apply biphasic analog current. DACC (Figure 9.11a) converts 8-bits pixel into analog biphasic current. It applies binary-weighted

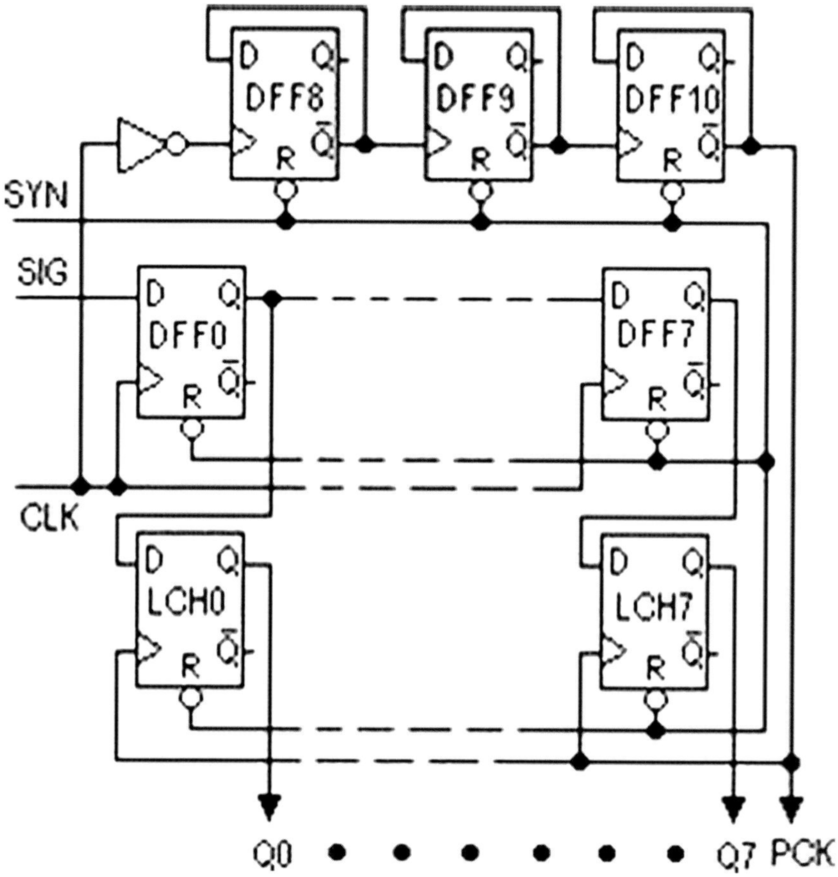

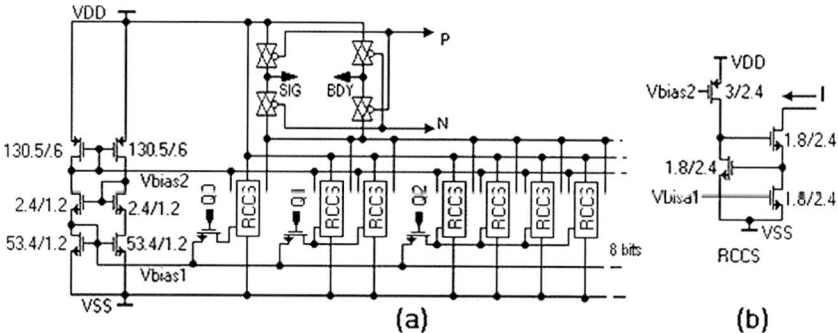

FIGURE 9.10. Serial to parallel converter.

FIGURE 9.11. (a) DACC with bridge to generate biphasic current. (b) Regulated cascade current sink.

data conversion technique. We are to generate 600 μA current at 10 KΩ load from 8-bit pixel. It is very important that the output current is linear with the pixel value to produce the linear gray level in the retina and at the same time the supply voltage requirement does not go up. Unlike some other designs [1, 5, 6], our DACC is capable of delivering 600 μA current at 10 KΩ load linearly for the full range of the pixel. We designed our DACC using regulated cascade current sink (RCCS) (Figure 9.11b). RCCS uses negative feedback to stabilize the output current and increases the output impedance. We checked our design for different voltages and found that it is capable of delivering our required current linearly at 7 V power supply (Figure 9.13). The loading affect on output current is shown in Figure 9.12 for 5 V power supply. From the plot, we see that if we limit supply voltage to 5 V then our DACC is able to supply 600 μA at 7.5 KΩ instead of 10 KΩ. It is possible to reduce the load resistance from 10 KΩ by bringing the reference electrode closer to the stimulating electrode. The resolution of the DACC is 2.5 μA. To generate biphasic waveform, usually one requires midpoint of power supply and use of PMOSs and NMOSs together. Our DACC uses only NMOSs, and we do not need midpoint of supply voltage. A bridge circuit consisting of four transmission gates is incorporated with anodic control voltage P and cathodic control voltage N. By controlling voltages P and N the anodic and cathodic pulses are generated. The linearity problem that arises with the use of PMOS and NMOS together is eliminated by using only NMOS transistors, which is essential for biphasic charge-balance current stimulation (Figure 9.3).

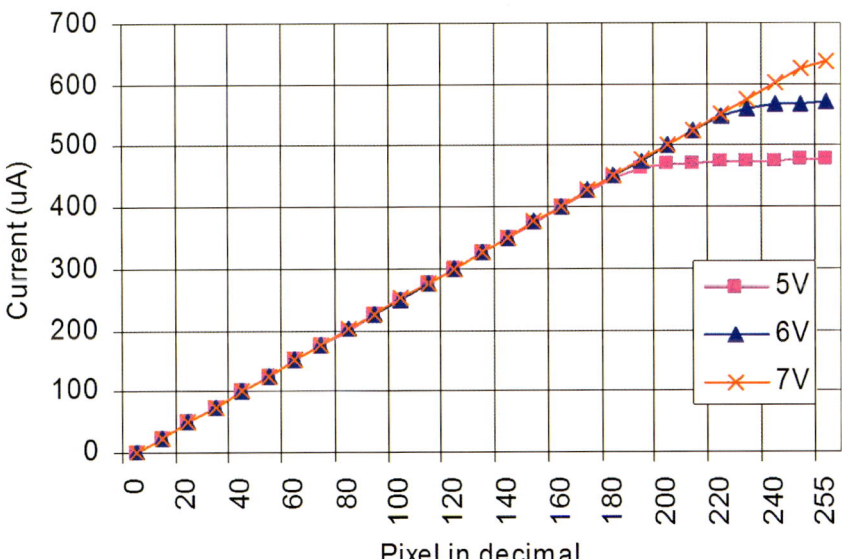

FIGURE 9.12. Output current generated by 8-bit pixel.

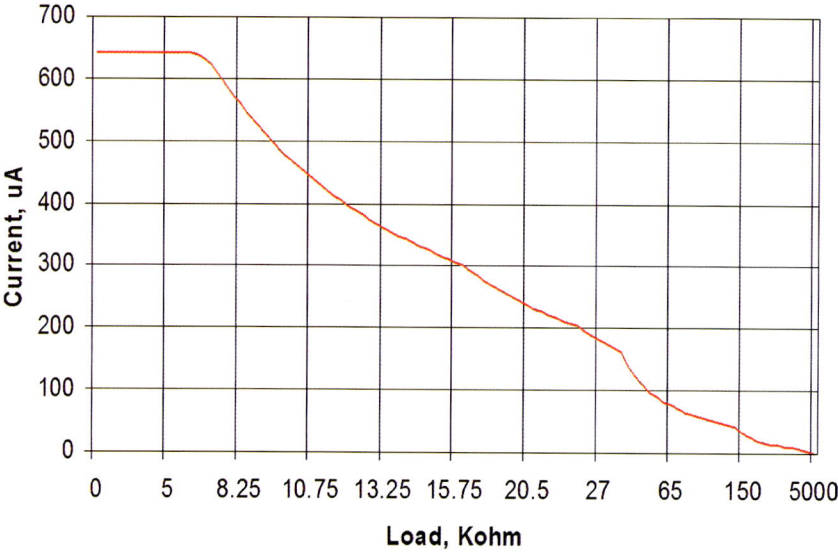

FIGURE 9.13. Loading effect on current.

Biphasic Current Controller (BCC)

As mentioned earlier, the width and interphase delay of biphasic current waveform depend on the degree of a patient's retinal damage. This will be ascertained by the doctor through clinical experiments on each patient. To generate biphasic current waveform of variable width and interphase delay, we have designed BCC (Figure 9.14). It is designed to control the widths of anodic and cathodic pulses and interphase delay independently. The first block of BCC controls P, the second ensures interphase delay, and the third controls N. Each block is composed of three T Flip-Flops (TFF) connected as three-stage synchronous binary counter with three parallel inputs $P0$, $P1$, and $P2$. By programming 9 bits of the three blocks, the variable widths and interface delay are generated. When RST is asserted low, each block is loaded with $P0$, $P1$, and $P2$. Binary counter can count up to 7 from 0 or any number programmed by $P0$, $P1$, and $P2$. On reaching count 7, RC becomes high, which is fed back to the input of a1 after inversion blocking the toggle of TFFs and locking the counter state. On reset P is set high and N low. After the set time of anodic width, P is reset triggering interphase delay. On expiry of interphase delay, N is set high enabling cathodic pulse, which is reset after the set period of time of cathodic width. The beauty of the circuit is that we can generate only anodic or cathodic pulse by setting the other to zero or biphasic pulse without delay by setting interphase delay to zero. For longer width or delay the counter should be set to lower value. This is because the width or delay is the difference between the set value and 7.

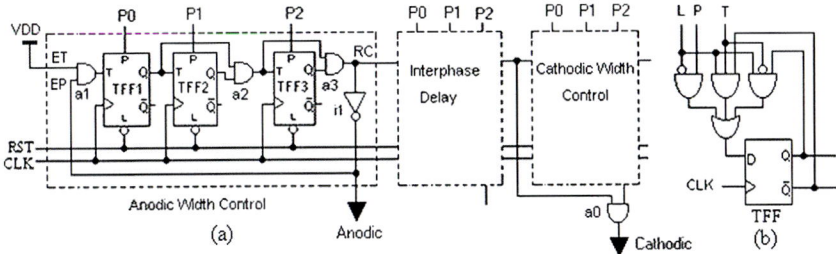

FIGURE 9.14. (a) Biphasic current control circuit, (b) T Flip-Flop.

Retinal Stimulator

The design of a 32×32 array RS is shown in Figure 9.15a. P11-P3232 is the probe matrix and DFF1-DFF64 is the decoder. It has six I/Os (SIG, BDY, PCK, SYN, VDD, and VSS). Figure 9.15b shows the structure of a probe. When the chip is fabricated, the entire top layer (metal-3) is covered with glass. But for creating contact with the neurons, we left an opening of $7.8\,\mu m \times 7.8\,\mu m$ on the metal-3 layer for each probe. Microelectrodes are grown on this opening to reduce the contact resistance with the neurons. Each probe is controlled by two NPN transistors connected in series. These two transistors are again controlled by the decoder. Control of these two transistors helped us reduce the number of DFFs required for decoder significantly. To turn any particular probe on, we are to apply control voltages to both the transistors. At any particular time, only one probe is turned on. Inputs of all the probes are connected together and

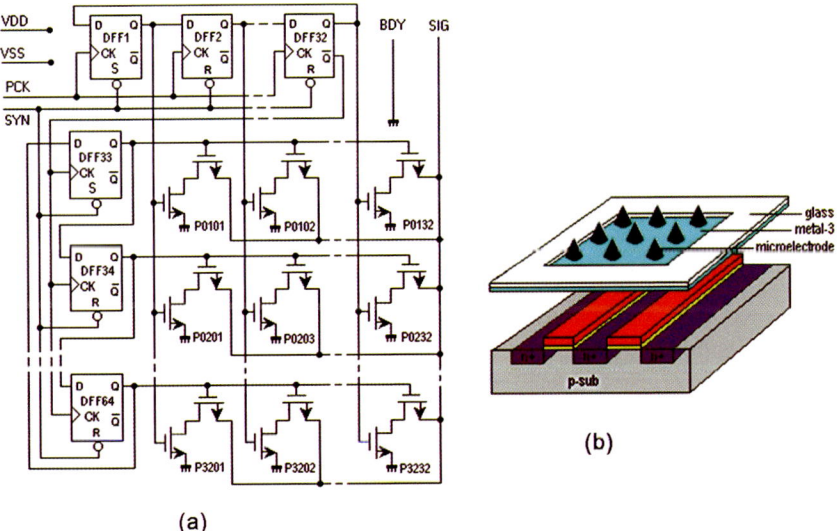

FIGURE 9.15. (a) Design of 32×32 array RS, (b) structure of a probe.

marked as SIG. This common bus carries the time-multiplexed signals of all the probes. PCK is the clock for DFF1 to DFF32 connected in a ring, while Q' of DFF32 is the clock for DFF33 to DFF64 connected in another ring. DFF1 and DFF33 are designed with set input, while the rest with reset input. When SYN is made low, DFF1 and DFF33 are set and the rest of the DFFs are cleared selecting probe P0101 and initiating the stimulation process beginning with P0101. With the rising edge of next clock cycle P0102 is selected, and so on till the 32nd clock cycle. At the rising edge of 33rd clock cycle, Q' of DFF32 switches from low to high clocking row decoder. With that, 2nd row is selected and stimulation proceeds with P0201, P0202, etc. This pattern goes on till the end of 32nd row and 32nd column. So we see the stimulation progresses from left to right and from top to bottom of the array and it progresses with the rising edge of the clock. Each probe is sequentially selected at each rising edge of clock resetting the previous one. After the 32×32nd clock cycle, SYN is made low again to force the decoder to point to the beginning of the array for synchronization. For retinal stimulation, data is sent in multiplexed fashion. Upon sending the SYN pulse, the first data corresponding to P0101 is sent, followed by P0102, in sequence from left to right and from top to bottom of the array. For retinal response monitoring, the multiplexed signal is decoded. Upon sending the SYN pulse, the first decoded data corresponds to P0101, followed by P0102, etc. in the same sequence as in stimulation.

Experimental Results

Figure 9.16 shows the microscopic view of 32×32 array RS fabricated using $0.5\,\mu m$ CMOS technology. For testing, RS is fabricated without package, and tested by microtest probes under microscope. We have also fabricated it in package specially designed for testing. Typical $10\,K\Omega$ neural impedance is used as load. Test result of RS as a stimulator is presented in Figure 9.17a. We have presented here the output of probes P0201, P0202, and P0203 only. These three probes are consecutive probes. We have excited them by biphasic waveforms. Their stimulations are one cycle apart. The result shows that each probe was excited for the duration of one clock cycle by the biphasic waveform. We have tested it for other experimental waveforms and found that it stimulates the neurons as expected. This means that RS is capable of stimulating the neurons sequentially by any kind of analog waveforms.

The RS has also been tested as reader (Figure 9.17b). In this test, we have applied 0.5 V AC signals of different frequencies to the probes simultaneously and monitored the output on output signal (SIG) line. Here we have presented the result for probes P0201 and P0202. P0201 was excited by 1 KHz sinusoidal signal and P0202 by 4 KHz sinusoidal signal. The SIG came out multiplexed. This signal can be sent out through reverse telemetry.

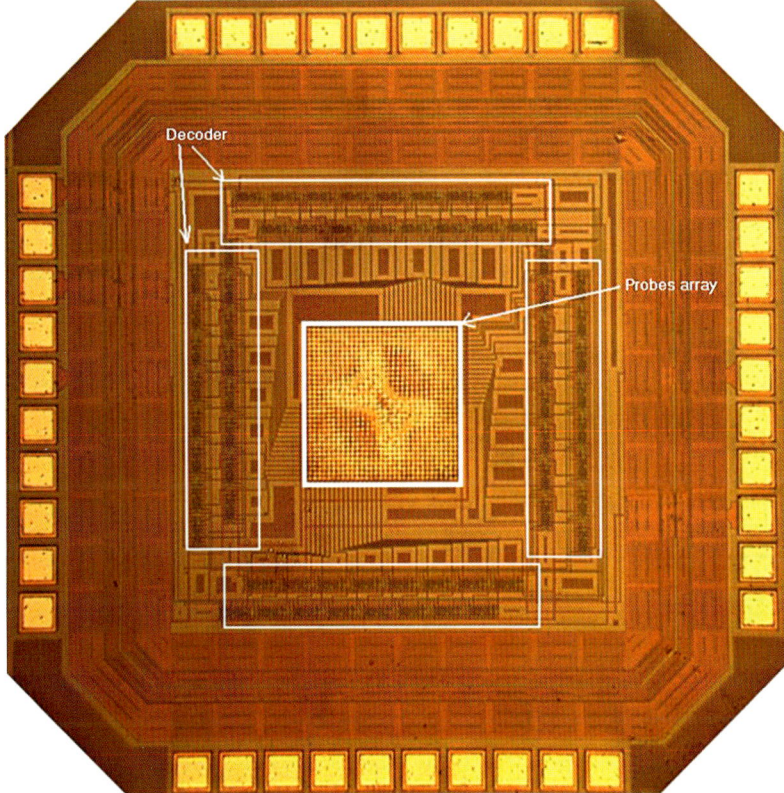

FIGURE 9.16. Microscopic view of 32×32 array RS.

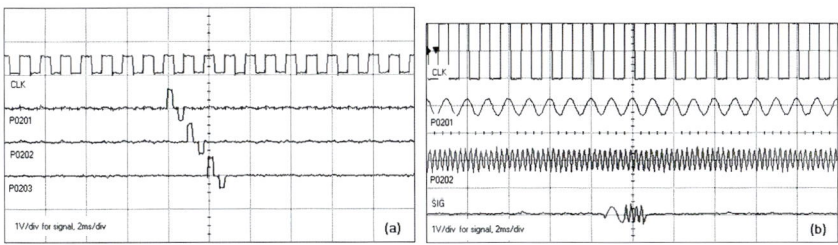

FIGURE 9.17. (a) Test result of RS as stimulator (b) test result of RS as reader.

Conclusion

We have presented the design, fabrication, and testing of a CMOS stimulator microarray for retinal prosthesis. The retinal prosthesis is designed for bidirectional data communication. A multiplexing technique is utilized allowing the use of only one SPC and one DACC. The design eliminates direct wiring requirements and uses only six interface wires for power and control. Furthermore, the prosthesis does not require direct addressing of any probe for active stimulation. Rather, a sequential stimulation by the clock timing eliminates need for wireless address transmission overhead. The system is also capable of stimulating the retina by programming large variety of experimental waveforms.

References

1. W. Liu, K. Vichienchom, M. Clements, S. C. DeMarco, C. Huges, E. McGucken, M. S. Humayun, E. DeJuan, J. D. Weiland, and R. Greenberg, "A Neuro-stimulus Chip with Telemetry Unit for Retinal Prosthetic Device," IEEE Journal of Solid-State Circuits, vol. 35, no. 10, Oct 2000, pp. 1487–1497.
2. E. Zrenner, K. D. Miliczek, V. P. Gabel, H. G. Graf, E. Guenther, H. Haemmerle, B. Hoefflinger, K. Kohler, W. Nisch, M. Schubert, A. Stett, and S. Weiss, "The development of sub-retinal microphotodiodes for replacement of degenerated photoreceptors." Opthalmic Research, 1997, 29, 269–280.
3. G. Peyman, A. Y. Chow, C. Liang, V. Y. Chow, J. I. Perlman, and N. S. Peachey, "Subretinal semiconductor microphotodiode array," Ophthalmic Surg. Lasers, vol. 29, pp. 234–241, 1998.
4. H.K. Trieu, L. Ewe, W. Mokwa, M. Schwarz, and B. J. Hostica, "Flexible Silicon Structures For A Retina Implant", *IEEE Transaction, 1998*, vol. 10, pp. 515–519 VI.
5. Y. Yao, M. N. Gulari, J. F. Hetke, and K. D. Wise, "A self-testing multiplexed CMOS stimulating probe for a 1024-site neural prosthesis", The 12th International Conference on Solid State Sensors, Actuators and Microsystems, Boston, June 8–12, 2003, Transducer '03, pp. 1213–1216.
6. Shuenn-Yuh Lee, Shyh-Chyang Lee, and Jia-Jin Jason Chen, "VLSI Implementation of Implantable Wireless Power and Data Transmission Micro-Stimulator for Neuromuscular Stimulation", IEICE Transactions on Electronics, vol. E87-C, no.6, June 2004, pp. 1062–1067.

10
Visual Prosthesis Based on Optic Nerve Stimulation with Penetrating Electrode Array

Qiushi Ren[1,2], Xinyu Chai[1,2], Kaijie Wu[1,2], Chuanqing Zhou[1,2] and C-Sight Group[2]

[1]*Institute for Laser Medicine and Bio-Photonics, College of Life Science and Technology, Shanghai Jiao-Tong University*
[2]*C-Sight Group: C-Sight Group is a Research Consortium on Visual Prosthesis Funded by the Chinese Ministry of Science and Technology Under National Key Basic Research Program (973 Program, 2005CB724300). The principal investigators of the C-Sight Group include:*
1. Qiushi Ren, Ph.D. (Program Chief Scientist), Department of Biomedical Engineering, Shanghai Jiao-Tong University
2. Li, Xiao-Xin, M.D., Department of Ophthalmology, Beijing University School of Medicine
3. Liang, Pei-Ji, Ph.D., Department of Biomedical Engineering, Shanghai Jiao-Tong University
4. Zhu, Yi-Sheng, Ph.D., Department of Biomedical Engineering, Shanghai Jiao-Tong University
5. Zhuang, Song-Lin, Ph.D., Shanghai University of Science and Technology
6. Zhao Jian-Long, Ph.D., Shanghai Inst. For Micro-System Fabrication Research of CAS
7. Wang, Jin-Ye, Ph.D., Department of Biomedical Engineering, Shanghai Jiao-Tong University

Introduction

The successful development and application of cochlear implants in the deaf has encouraged several research groups to start projects on visual prosthesis for the blind. The C-Sight (Chinese Project for Sight) is the first multidisciplinary research project on visual prosthesis in China funded by the Chinese Ministry of Science and Technology under the National Key Basic Research Program (973 Program, 2005CB724300) and by the Science and Technology Commission of Shanghai Municipality (STCSM). The goal of the C-Sight Project is to develop an implantable microelectronic medical device that will restore useful vision to blind patients.

The attempts involved electrical stimulation of primary visual cortex [1–3] and retina [4–6] have proved that phosphene can be artificially elicited in blind individuals since the sixties of last century. The optic nerve serves as a compact conduit for all of the information in the visual scene. Because it can be accessed via a minimally invasive procedure and is relatively spared by the most prevalent degenerative eye diseases, it can be chosen as a site for a neural interface or as a stimulating site in the visual prosthesis design. The main challenge is to evaluate the possibility of visual prosthesis based on optic nerve stimulation. In 1998, C. Veraart et al. proved that electrical stimulation of the optic nerve in a 59-year-old volunteer with retinitis pigmentosa can also elicit phosphenes [7–9]. The contact spiral cuff electrode was used in their experiment. However, the spatial resolution of recovered vision was limited because surface stimulating electrodes were used.

With the development of technology, the electrode can be made more and more elaborately. Therefore, the damage of electrode to the optic nerve can be minimized and the spatial resolution can also be greatly enhanced.

In our approach, we proposed the penetrating multi-electrode arrays with specific configurations into the optic nerve as a neural interface to couple the encoded electrical stimuli into the axons of the ganglion cells for vision recovery. Figure 10.1 shows the schematic presentation of our experimental approach. Another important feature of our approach is that the function of photoreceptors of retina is replaced by a self-contained implantable micro-camera.

The full structure of our proposed visual prosthesis is shown in Figure 10.2. The images are captured by a CMOS micro-camera which can be implanted into the lens of the blind eye. Then, the key features of the image can be extracted from the actual scenes with some advanced image-processing algorithm. After

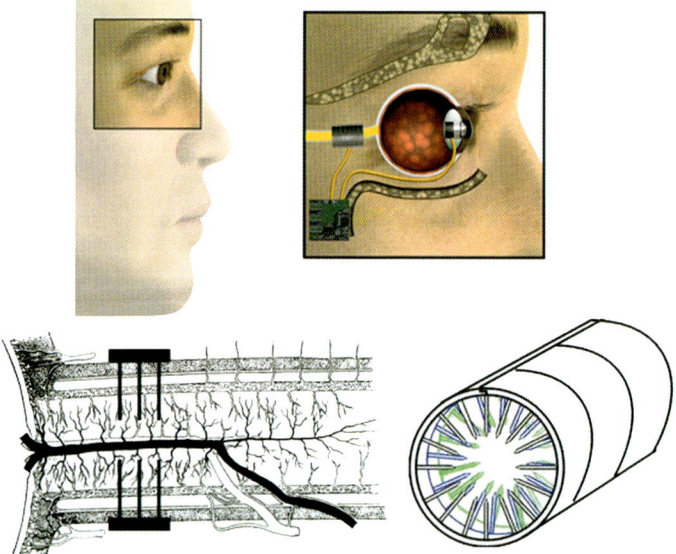

FIGURE 10.1. Schematic drawing of electrical stimulation at optic nerve bundle with penetrating electrode array.

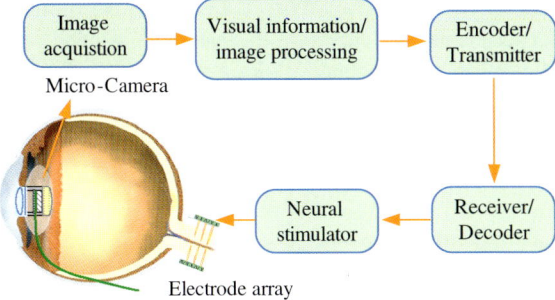

FIGURE 10.2. The structure of visual prosthesis based on penetrating optic nerve electrical stimulation.

that the "retinal encoder" encodes the processed visual information as trains of electrical impulses with a specific spatiotemporal stimulation pattern. Finally, the electrical stimulation pattern is delivered by the penetrating multi-electrode array directly into the optic nerve where action potentials are generated and conducted to the visual cortex.

Animal Experiment

Optic Nerve Electrical Stimulation

In order to evaluate the feasibility of our approach, we carried out a series of animal experiments. Firstly, we chose three different anatomical positions of optic nerve for placing the stimulating electrode. To determine whether the electrodes implanted into the three different places can all elicit cortical potentials, stimulations were applied to the three stimulating positions, respectively. Furthermore, several different electrical stimulus patterns were applied to the optic nerve at one certain position with the purpose of investigating whether different pattern stimulations could elicit different cortical responses.

Materials and Methods

Animal model
Albino rabbits weighting about 2.8 kg (7 ~ 8 months old) were used. They were maintained in a 12-hour light and 12-hour dark diurnal cycle, supplied with water and food. Experimental procedures on rabbits conformed to the ARVO Statement for the Use of Animals in Ophthalmic and Vision Research issued by the National Institutes of Health.

Exposition of optic nerve
The rabbits were anesthetized generally with intraperitoneal injection of pentobarbital sodium (30 mg/kg), and the eyeballs were locally anaesthetized with a few drips of procaine. At the upper site of the right eye's outer canthus, several

stratums of parenchyma were peeled off to expose osteal tissue. Along the bone wall, the upper eyelid was opened and the cartilage of orbit was removed with rongeur. Conjunctivae were opened along eyeball wall until ciliary vasa were touched. Between the two ciliary vasa, the white colored optic nerve (about 3 mm width) was obviously exposed without any pull and push. Throughout the exposing procedure, care should be taken not to damage the blood vessels on the choroid.

Implantation of stimulating electrodes

With the help of surgical microscope, a tungsten electrode was inserted into the exposed optic nerve through the incision of the orbit. There were three positions for placing the stimulating electrode (Figure 10.3). The distances from all of the three sites to the optic nerve head were the same, about 2 mm in front of anterior band for better retina–optic nerve mapping. Site 1 was at the surface of the optic nerve sheath, with the cut-end electrode pressing on the sheath. Site 2 was intra- and para-optical nerve sheath. An incision was cut using a bistoury with tip angle 15°. Then the electrode was inserted into the space between dura and pia of the optic nerve. For site 3, electrode was passed through dura and pia and penetrated into the optic nerve bundles, about 1 mm deep into the optic nerve.

Implantation of recording electrodes

An incision in the skin nearby the occipital bone was cut, which was contra lateral to the stimulated optic nerve, about 8 mm anterior to the lambda and 7 mm lateral to the midline. And a corresponding hole through the site of skull was drilled. A tungsten electrode (Φ 100 μm) was screwed into the skull to contact the dura, used as a recording electrode. A stainless steel electrode inserted into one of the rabbit's earlobe was the ground. The reference electrode, which was the same type as the ground one, was inserted subcutaneously at the forehead.

Electrical Stimulation and Recordings of Electrical Evoked Cortical Potential

For the first experiments three different places of the optic nerve were stimulated respectively. The reference electrodes were all located at the sclera for the three different stimulating positions. A short charge-balanced biphasic pulse was used for the electrical stimulation. The biphasic rectangular pulse consisted of a negative pulse and a second opposing pulse. The pulses duration were 1 ms. The current amplitude was 50 μA, which was generated with an electronic stimulator (RX7 Moray, Tucker-Davis Technologies, USA) and a stimulus isolator (MS16, Tucker-Davis Technologies, USA). While the electrical stimulation was applied to the optic nerve, the electrical evoked cortical potentials (EEPs) were recorded from the recording electrodes on the cortex opposite to the stimulated eye. EEPs were recorded using a data acquisition system (RX7 Moray, Tucker-Davis Technologies, USA) connected to a low noise, battery-operated amplifier (RA16PA Tucker-Davis Technologies, USA) and filtered with a band-pass of 1–300 Hz. Results from 50 trials were averaged and analyzed.

For the experiments where stimuli with different parameters were applied we chose the third stimulating position. Firstly, the amplitude of the pulse was

(c) **Penetrating Optical Nerve Bundle**

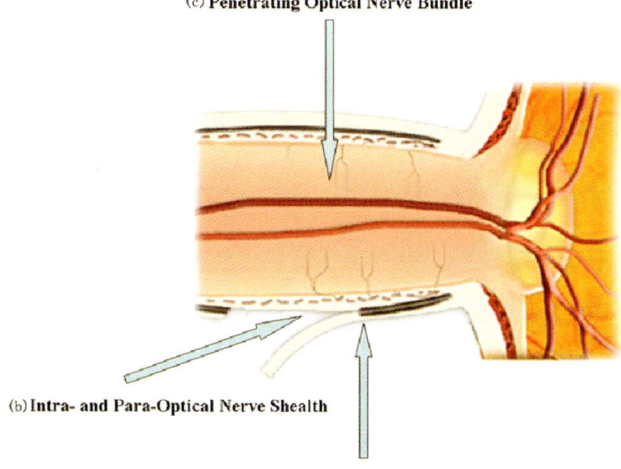

(b) **Intra- and Para-Optical Nerve Shealth**

(a) **At Surface of Optical Nerve Shealth**

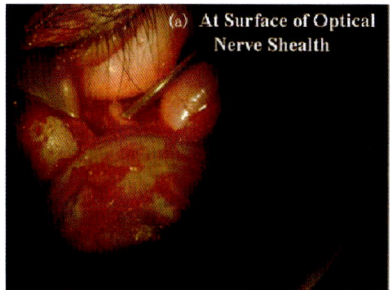

(a) **At Surface of Optical Nerve Shealth**

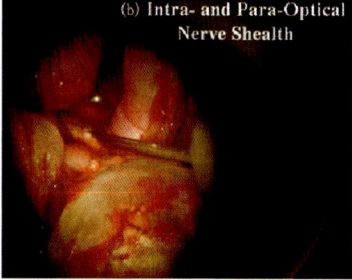

(b) **Intra- and Para-Optical Nerve Shealth**

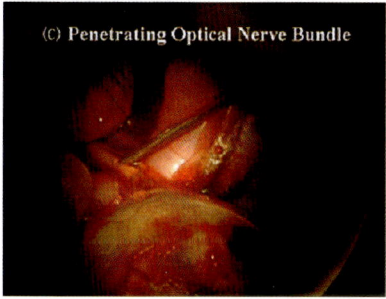

(C) **Penetrating Optical Nerve Bundle**

FIGURE 10.3. Schematic of three anatomical positions for electrical stimulation: (a) the surface of the optical nerve sheath; (b) intra- and para-optical nerve sheath; and (c) intra-optical nerve by the penetrating electrode.

varied (25uA, 50uA, 100uA) while the duration of each pulse was kept constant (0.6 ms). Then we changed the duration of the pulse (0.3ms, 0.6ms, 1.2ms) and the pulse amplitude was kept $25\,\mu A$.

In all the electrophysiological experiments, cortical responses to electrical stimulation were recorded within 1–3 h after electrode implantation.

Result

The EEPs could be recorded with the recording electrode when the electrical stimulation was applied to any of the three stimulating positions, which are shown in Figure 10.4. The implicit times of the first positive peak (P1) and the amplitudes of the EEPs both are significantly different among the three different

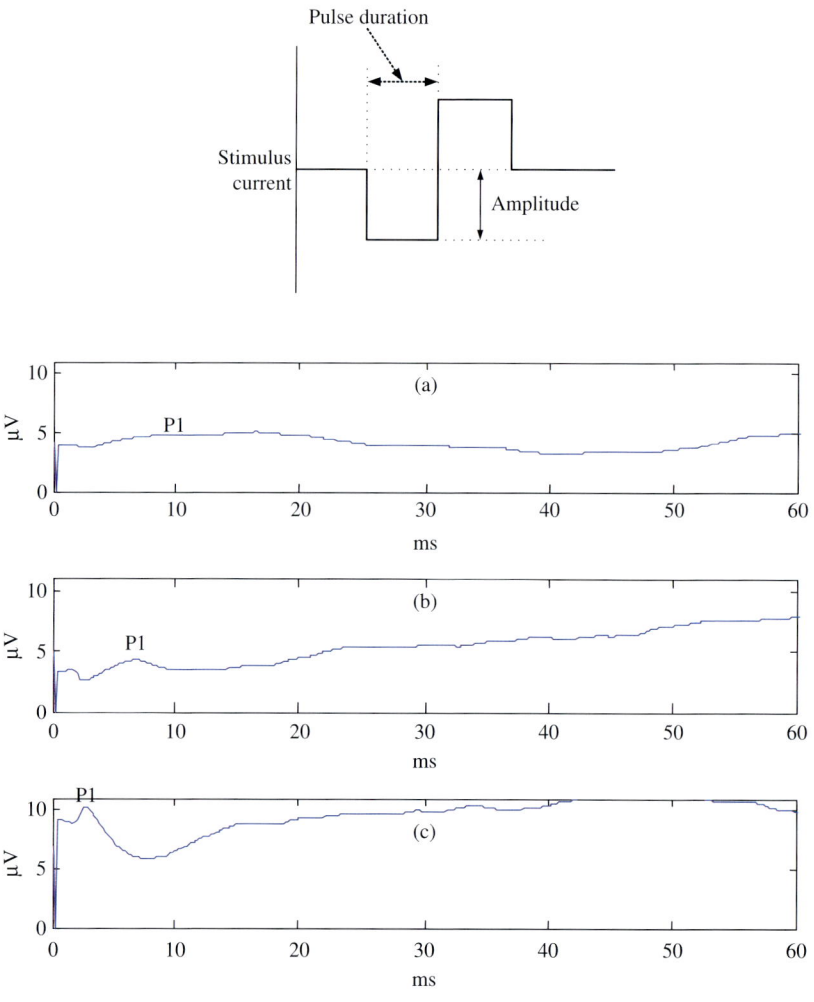

FIGURE 10.4. Diagram of stimulating current pulse (Top). EEPs recorded after biphasic electrical pulse applied to different stimulating places of the optic nerve (Bottom): (a) EEP elicited by stimulating the surface of the optical nerve sheath, (b) EEP elicited by stimulation with the electrode intra- and para-optical nerve sheath, and (c) EEP elicited by stimulation with the electrode intra-optical nerve. For the three stimulating places, the pulses duration were 1 ms and the current amplitude was 50 μA.

stimulation methods. The stimulation of the penetrating electrode implanted into the optic nerve elicits the strongest signal of EEP while that of the electrode placed at the surface of the optic nerve sheath elicits the weakest signal of EEP. And the stimulation with the electrode penetrated in the optic nerve also has the shortest implicit time (Figure 10.4c), while the other two have longer implicit time.

The EEP waveforms elicited by different stimulation parameters are shown in Figures 10.5 and 10.6. Figure 10.5 shows the waveforms when the amplitude of the pulse was changed. As a general rule, the data from the experiments indicates that the higher the stimulation current, the higher the response amplitude of the epidural recording. The stimulation of $100\,\mu A$ elicited a higher response than the other two. The implicit time of P1 elicited by the stimulation of $100\,\mu A$ was similar to that by the stimulation of $50\,\mu A$, which was about 4 ms. The implicit time of P1 elicited by the stimulation of $25\,\mu A$ was longer than the other two, which was about 8 ms. Figure 10.6 shows the waveforms when the duration of the pulse was changed. The result indicates that the longer the pulse duration, the higher the response amplitude of the epidural recording, which can be explained by the different charge densities of the different stimulation.

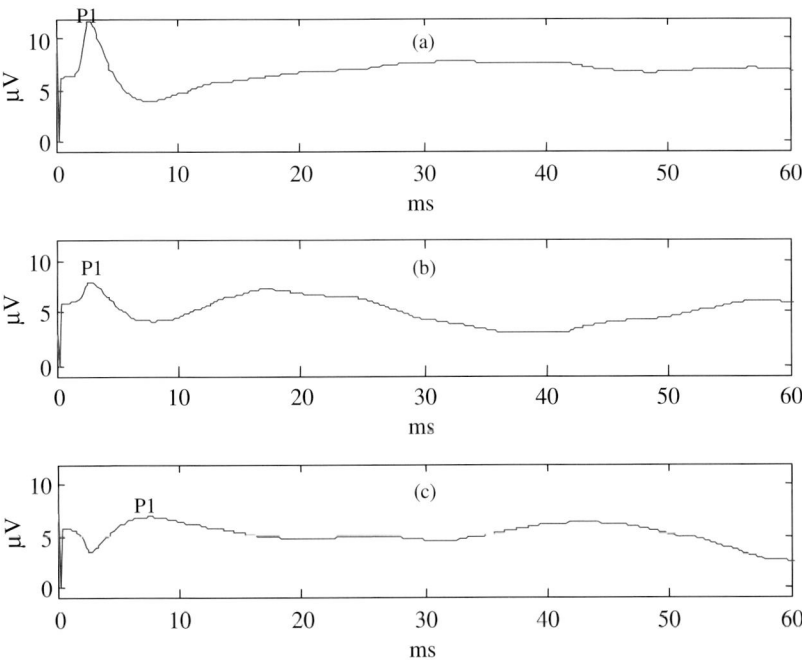

FIGURE 10.5. EEP waveforms elicited by the different pattern stimulations with fixed pulse duration and varied amplitude: (a) pulse duration 0.6 ms, amplitude $100\,\mu A$; (b) pulse duration 0.6 ms, amplitude $50\,\mu A$; and (c) pulse duration 0.6 ms, amplitude $25\,\mu A$.

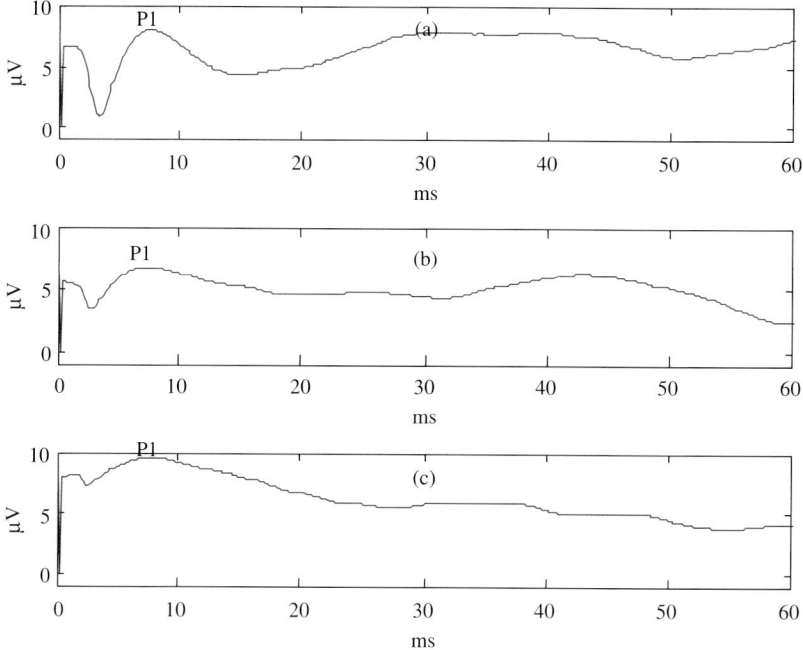

FIGURE 10.6. EEP waveforms elicited by the different stimulations with varied pulse duration and fixed amplitude: (a) pulse duration 1.2 ms, amplitude 25 μA; (b) pulse duration 0.6 ms, amplitude 25 μA; and (c) pulse duration 0.3 ms, amplitude 25 μA.

Stimulation of longer pulse duration has higher charge density than that of shorter pulse duration, so the stimulation whose pulse duration is 1.2 ms elicits the strongest signal of EEP (Figure 10.6a).

The experimental results demonstrated that the visual cortex can be excited by the electrical stimuli to the optical nerve at any of the three anatomical positions i.e., stimulation at the surface of the optical nerve sheath, intra- and para-optical nerve sheath, and intra-optical nerve by the penetrating electrodes. Furthermore our experiments indicate that different stimulation parameters can elicit different responses at the visual cortex, suggesting that this approach may be useful for a visual prosthesis system.

Light Evoked Potential Measured at Optic Nerve Bundle

In order to investigate the mechanism of the response of the retina to the color stimulation and the effect of the optic nerve sheath to the conduction of bioelectrical signal, we conducted another series of experiment. Evoked potentials were recorded from the optic nerve while the eye of the rabbit was stimulated by photic stimuli.

Methods

Tungsten electrodes (Φ100 μm) were used for recording the evoked potential and two positions were chosen for placing the recording electrode. Site 1 was at the surface of the optic nerve sheath, with the cut-end electrode pressing on the sheath. For site 2, electrode was passed through dura and pia and penetrated into the optic nerve bundles, about 1 mm deep into the optic nerve. Three different photic stimuli (white light, blue light, red light) were used to stimulate the eye. In all the experiments the stimulus was a Ganzfeld flash of short duration generated with a visual electrophysiological diagnostic system (TEC, China). For each kind of stimulus, two different frequencies of 1.98 Hz and 7.3 Hz were used here, which were chosen purposely to cancel 50 Hz power-line interference. While the photic stimulation was applied to the eye, the light evoked potentials (LEPs) were recorded from the recording electrodes (Figure 10.7).

Results

Like the recordings of EEP, we recorded and calculated LEP waveforms by a computer using OpenEx software (Tucker-Davis Technologies, USA). The evoked potentials were filtered with a band-pass of 1–300 Hz and results from 50 trials were averaged and analyzed.

The LEP waveforms when the recording electrode was at the surface of the optic nerve sheath are shown in Figure 10.8. There were no distinct waveforms of LEP for any of the stimulation.

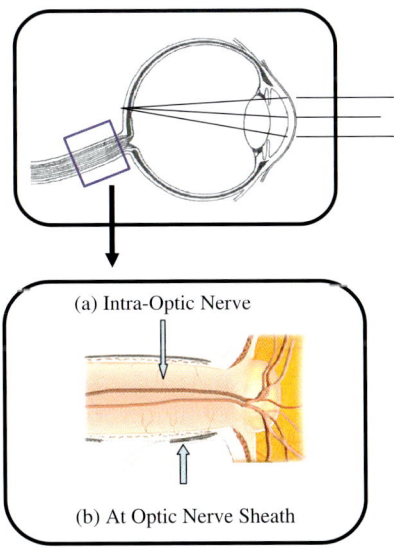

(a) Intra-Optic Nerve

(b) At Optic Nerve Sheath

FIGURE 10.7. Diagram of the photic stimulation and the recording of the light evoked potential.

Vertical: 6.2μV/Div Horizontal: 50ms/Div

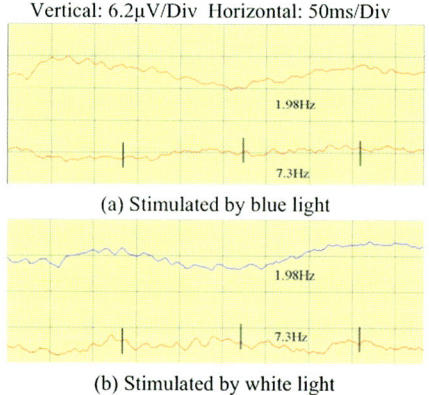

(a) Stimulated by blue light

(b) Stimulated by white light

FIGURE 10.8. LEP recorded from the recording electrode at the optical nerve sheath by different wavelengths of the stimulating light sources.

The result indicates that minimal LEP signals were measured at the surface of the optical nerve sheath position by the recording electrode, which demonstrates that the optic nerve sheath can shield the conduction of bioelectrical signal. So

Vertical: 6.2μV/Div Horizontal: 50ms/Div

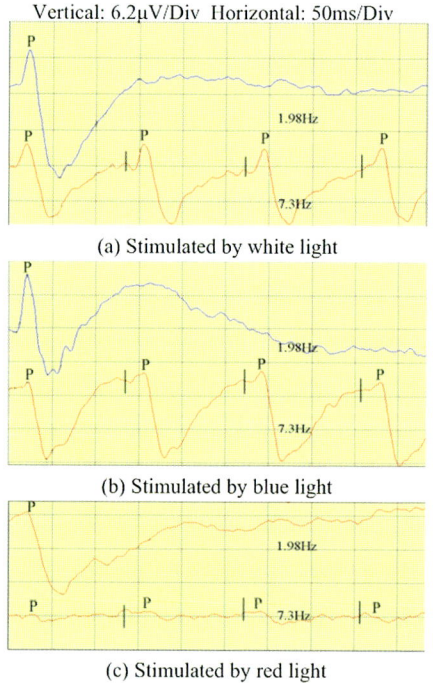

(a) Stimulated by white light

(b) Stimulated by blue light

(c) Stimulated by red light

FIGURE 10.9. LEP recorded from the recording electrode penetrated into the optical nerve by different wavelengths of the stimulating light sources.

when we want to stimulate the optic nerve, it will be more effective to penetrate the stimulating electrode in the optic nerve bundle.

Figure 10.9 shows the waveforms when the recording electrode was penetrated in the optic nerve. Different patterns of the LEP can be recorded when different wavelengths of the stimulating light sources are applied to the rabbit retina, which indicates the different color coding processes can be detected from the recording electrode penetrated into the optical nerve. The blue light stimulation elicited the strongest response and the 7.3 Hz red light stimulation did not elicit distinct response.

The results of our experiment demonstrate that the LEP can be measured only at the position of intra-optical nerve by the penetrating recording electrode. Comparing with the other wavelengths of light stimulation, the LEP was recorded strongest by the blue light stimulation. As shown in Figure 10.9c, there were no LEP signals recorded at the red light stimulation with the stimulation frequency greater than 7.3 Hz.

The Hardware Design of Visual Prosthesis

In the past several years, many approaches have been pursued to provide the neural electrical simulation at various positions of the visual pathway, such as visual cortex and retinal and optic nerve, to restore the vision of the blind patient. Although the most significant difference of these approaches is the interface to the nervous system, all of them share a common set of components, such as micro-camera, visual information processing and extraction system, power and information transmission system, neural electrical simulator and electrode array (Figure 10.2).

Here we present some progress of hardware design in our projects.

Image Acquisition System

As the prosthesis is going to be implanted into the human eye, all the components should be small in size with low weight and low power dissipation. A number of image acquisition devices or mini-cameras are available and suitable for this application. It is possible to integrate both Charge Coupled Device (CCD) and Complementary Metal-Oxide Semiconductor (CMOS) cameras into the visual prosthesis. Although CMOS camera is more susceptible to noise and has lower light sensitivity than CCD, it is more suitable for implantable visual prosthesis application with some crucial features, such as lower power dissipation, single supply operation and camera-on-a-chip integration. As a result of the rapid development of advanced microelectronic fabrication techniques, now A/D conversion can also be integrated into one common CMOS camera microchip with the image sensors. OV6650FS (Omnivision Co.) was chosen as the photo-sensor in our image acquisition system, for it is the minimum CMOS camera on the market and has relatively lower power dissipation. On the other hand, the resolution of CIF (352 × 288) of OV6650FS is enough for visual prosthesis prototype. Noelle R. Stiles et al have found that only 625 pixels are enough for object recognition by blind people [10]. In the first step of visual prosthesis research, the miniature camera will be set in the patient's spectacle. Further it will be implanted into patient's pupil after a more biocompatible subminiature custom camera is developed.

The real-time image acquisition and processing in visual prosthesis consists of the implementation of various image processing algorithms like edge detection, edge enhancement, decimation, etc. [11]. Accordingly it needs high performance CPU to satisfy the high level of computational complexity. Especially when the penetrating electrodes and image are not clear in mapping the relationship, extra algorithms are required to design the appropriate stimulation pattern for the optical nerve prosthesis. For these reasons, one type of high performance DSP (TMS320DM642) was used to fulfill the successive image acquisition and processing. Figure 10.10 shows the basic connection between DSP and CMOS camera.

Image Processing Strategies

The ultimate goal of vision prostheses is to artificially produce visual perception in individuals with profound loss of vision due to disease or injury. With the help of visual perception, these individuals can perform activities, such as reading text, recognizing faces and exploring unfamiliar spaces.

Due to the constraints of fundamental physiology and the complexity of implanting process, none of the visual prosthesis can consist of enough electrodes to make a relationship with image sensors one by one. The image grabbed by the micro-camera must be extracted to form some necessary visual information as simply as possible. Then the neural encoder will transform the visual information to the electrical stimulus pattern. So some advanced strategies should be established to extract effective visual information from the original image. Some useful algorithms [12] which can be applied are given below.

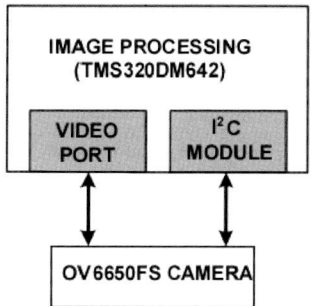

FIGURE 10.10. Functional diagram of the image acquisition system.

Histogram Equalization

Histogram equalization is one of the popular image enhancement transformations. It is used to obtain a uniform histogram and can improve contrast of the image. This algorithm can be used on a whole image or just a part of the image.

Histogram modeling [13] is usually introduced using continuous, rather than discrete, process functions. Therefore, we suppose that the image of interest contains continuous intensity levels (in the interval $[0,1]$) and that the transformation function f which maps an input image $A(x,y)$ into an output image $B(x,y)$ is continuous within the interval. Furthermore, we assume that the transfer law (which may also be written in terms of intensity density levels, e.g. $H_B (s) = f (H_A(r))$) is single-valued and monotonically increasing (as the case in histogram equalization) so that it is possible to define the inverse law $H_A (r) = f^{-1}(H_B (s))$. An example of such a transfer function is illustrated in Figure 10.11.

Edge Detection

The edges of an image hold much information, such as position, texture, shape and size. An edge is where the intensity of an image moves from a low value to a high value or vice versa.

The human visual system is very sensitive to details, especially abrupt changes or edges. These details in the frequency domain are always located in the high frequencies.

One way to detect edges or variations within a region of an image is by using the gradient operator. For instance, the gradient G is a vector with two elements G_x and G_y, where G_x is the gradient in the width direction and G_y is the gradient in the height direction. Since G is a vector, its magnitude G_m and direction angle θ can be given as:

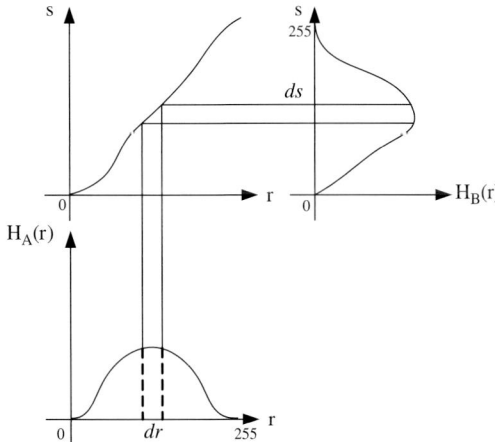

FIGURE 10.11. Example of histogram transformation function.

$$G = [G_x G_y]$$

$$G_m = \sqrt{G_x^2 + G_y^2}$$

$$\theta = \tan^{-1}\left(\frac{G_x}{G_y}\right) \tag{1}$$

There are several well-known gradient filters. In our experiment we use the Sobel gradients, which are obtained by convolving the image with the following kernels:

$$\begin{pmatrix} -1 & 0 & 1 \\ -2 & 0 & 2 \\ -1 & 0 & 1 \end{pmatrix} \begin{pmatrix} -1 & -2 & -1 \\ 0 & 0 & 0 \\ 1 & 2 & 1 \end{pmatrix} \tag{2}$$

Edge Following

If an object in an image has a discrete edge all around it, and also a beginning position on the edge has been found, it is possible to follow the edge of the object and back to the beginning position. Edge following is a very useful operation, particularly as a stepping stone to making decisions by discovering region positions in images. This is effectively the dual of segmentation by region detection.

Edge detectors yield pixels in an image lying on edges. The next step is to collect these pixels together into a set of edges. Thus, the whole object can be present simply by a few edges around it.

Binary Morphological Operations

Binary morphological operations are defined on bi-level images, which consist of either black or white pixels only. Dilation and erosion are popular binary morphological operations. If dilation can be said to add pixels to an object or to make it bigger, then erosion will make an object smaller.

Erosion can be used to eliminate unwanted white noise pixels from another black area. The only condition in which a white pixel will remain white in the output image is that all of its neighbors are white. The effect of erosion on a binary image is to diminish the edges of a white area of pixels and make the object look smaller than ever. The same rules applied to erosion can also be applied to dilation. But the logic must be inverted – using the NAND rather than the AND logical operation. Compared to erosion, dilation will allow a black pixel to remain black only when all of its neighbors are black. This operator is useful for removing isolated black pixels from an image.

Some Preliminary Result of Image Processing Strategies

The goal of our research group is to develop an image-processing system based on DSP for the optic nerve stimulation. In the first step, we test image-processing strategies by computer simulation and then transfer them to DSP.

Figure 10.12 shows the effect of histogram equalization. It is obvious that histogram equalization can improve the contrast of the original image without affecting the structure of the object.

Figure 10.13 shows the process of visual information extraction. It was found that both edge following and edge detection arithmetic operators can gain the border of the object. Based on the edge image, dilation operation can enhance the visual information of the object.

Figure 10.14 shows the unprocessed image captured by the micro-camera OV6650FS, and the processed image with edge detection algorithm by high performance DSP. As shown in the edge image, the detail of source image is not lost, and especially the numbers can still be recognized. It is very easy

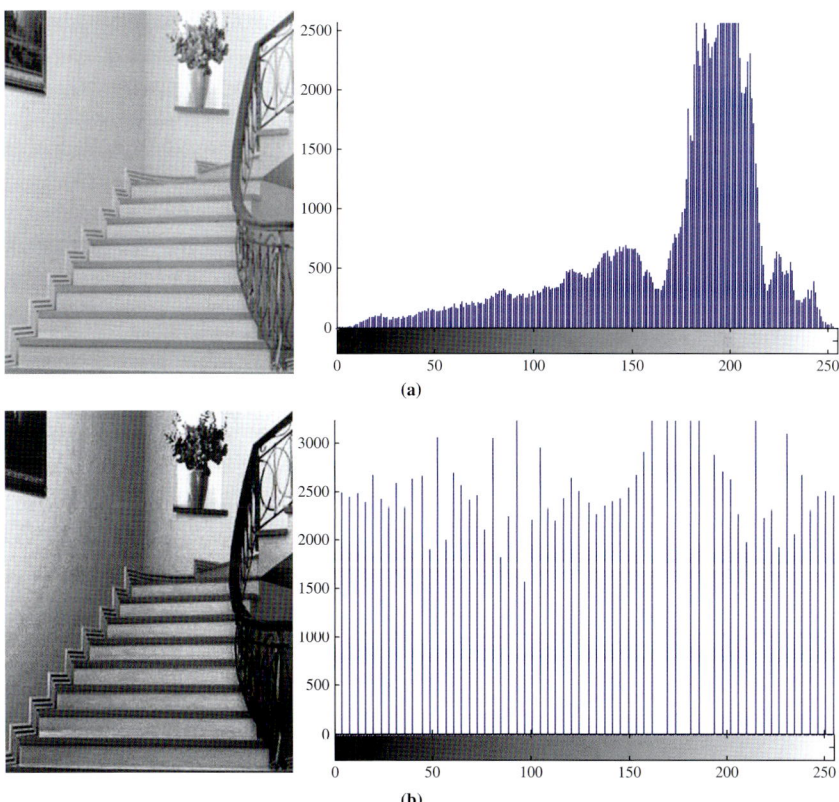

FIGURE 10.12. The effect of histogram equalization: (a) original image; (b) output image.

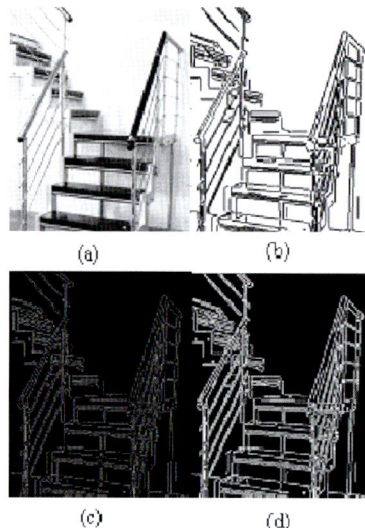

(a)   (b)

(c)   (d)

FIGURE 10.13. The process of visual information extraction: (a) original image, (b) the image after edge following, (c) the image after edge detection, and (d) the image after dilation based on edge image.

to implement other algorithms by our DSP image processing platform. Once more advanced algorithms are fulfilled, the basic function of visual information extraction of retina can be replaced by our optic nerve prosthesis.

Neural Electrical Stimulator

According to the experiment results of Claude Veraart et al., the stimulation based on optic nerve needs biphasic current pulses with variable amplitudes (from several microamperes up to several milliamperes), pulse duration (from several microseconds up to hundreds of microseconds) and pulse frequency. Then multiple biphasic current pulses make up one pulse train.

FIGURE 10.14. The visual information extraction by DSP: (a) original image captured by micro-camera, (b) the image after edge detection by DSP.

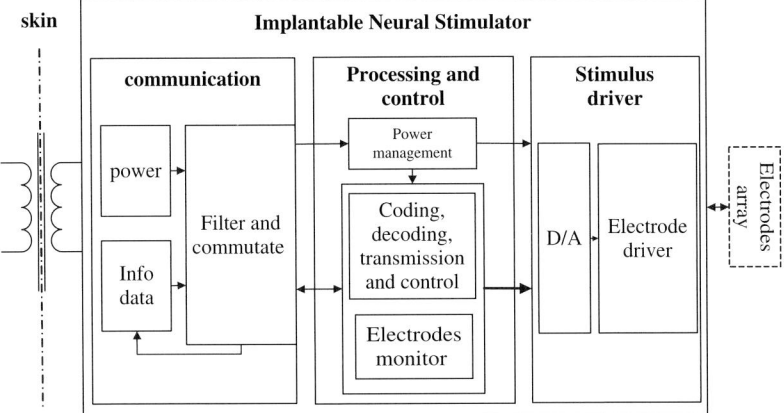

FIGURE 10.15. Schematic of the architecture of the micro-stimulator.

After the important characteristics of a visual object, such as position, dimension, colour, etc., are extracted from the original image, we can encode these information into special stimulation patterns.

The implantable micro-stimulator in our project consists of three parts: communication unit, processing and control unit, and stimulus driver unit. Figure 10.15 reveals the architecture of the micro-stimulator. The external part collects the visual information and encodes the information into electrical stimulation patterns. And then it modulates and transmits the patterns by radio frequency (RF) telemetry with other important control signals. The implantable parts exchange information by RF telemetry too. The transmission of power is also realized by RF telemetry. Once the power and data information are received, these information will be demodulated and decoded by the processing and control unit, which also monitors and records the state of electrode arrays.

The Stimulus Driver Unit converts digital signals of stimulation patterns into analog signals by DAC. The output of DAC is converted to biphasic current pulses by stimulus driver circuitry and delivered to the penetrating electrode array.

Implantable Micro-Camera in Model Eye

Design of Implantable Micro-Camera

This micro-camera which will be implanted in the position of the lens of the blind eye consists of the micro-lens, CMOS and the cable. The raw data of the image acquired by the micro-camera is then transmitted to the next processing unit out of the eye through the cable.

According to the dimension of the normal human eye, the distance between posterior cornea and crystalline lens is about 6 mm, and the pupil size in a normally illuminated room condition is about 3 to 4 mm; we chose the dimension of the micro-camera as 5 mm long and 4 mm wide as demonstrated in Figure 10.16. The micro-lens should be of wide angle and short focal length, such that the visual field could be extended.

The necessary highest resolution of the CMOS could be calculated based on visual discrimination. For example, for a given focal length of 4 mm of the micro-lens, if the best discrimination ability was described with visual angle of 10 minutes of arc or more, the maximum single pixel should not be larger than 5 μm. Much higher resolution could consume more power, hence more heat would be created, which was a considerable problem since it was to be implanted into the eye. In addition, more time was needed for the data processing which also should be taken into account with real time visual purpose.

Evaluation of the Micro-Camera in the Model Eye

In order to evaluate the imaging quality of the micro-camera in different environments, we put the micro-camera in the position of the crystalline lens of a model test eye (Figure 10.17). The model test eye was full of 0.9 % NaCl fluid when the pictures were to be taken in order to simulate the internal circumstance of the anterior chamber of the eye. The images were recorded with the target at different distances from the model test eye. It was demonstrated that the images became only a little blurred when recorded with the micro-camera in 0.9 % NaCl fluid (Figure 10.18). It also could be seen that the depth of field was large enough for the visual prosthesis to "see" clearly in the near and middle distance. In addition, it might have enough resolution for the image information extraction to correctly stimulate the optic nerve or neurons.

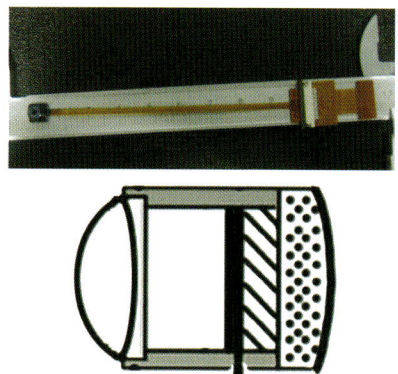

FIGURE 10.16. The implantable micro-camera including micro-lens, CMOS and cable (Top). The sketches of the implantable micro-camera (Bottom).

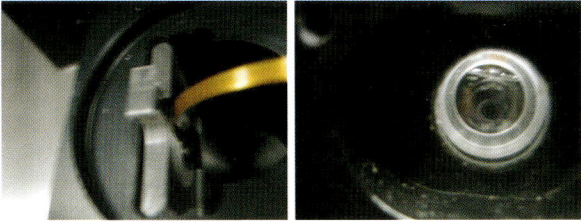

FIGURE 10.17. The setup of the model test eye with the micro-camera position in the pupil. Side view (left) and front view (right).

Discussion

Through the comparison of the images acquired by micro-camera in the 0.9 % NaCl fluid and in the air, it can be seen that this micro-camera seems suitable for behaving as an imaging system for the visual prosthesis if it was packaged with

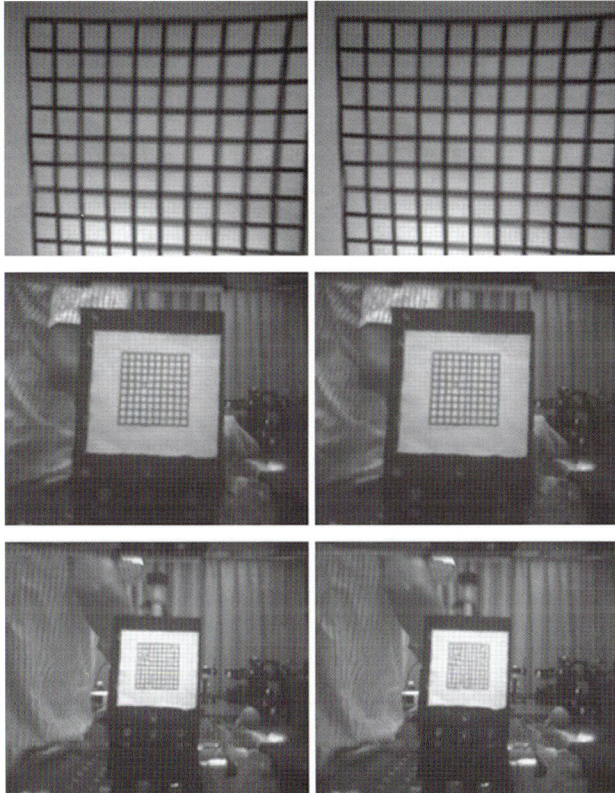

FIGURE 10.18. The comparison of images taken in 40 cm (top), 200 cm (middle) and 300 cm (bottom) from the model test eye with the micro-camera in the air (left column) and in the 0.9 % NaCl fluid (right column).

biocompatible material. However, these are very preliminary results. Further studies need to be carried out to implant the micro-camera into a cat's eye to evaluate the long-term biocompatibility and imaging function.

Conclusion

A type of visual prosthesis with stimulating electrode penetrating into optic nerve is introduced in this chapter. The feasibility of this visual prosthesis has been validated using some animal experiments. Many more animal experiments will be carried out to find out the rules of visual perception elicited by electrical stimulation in our project. To some degree, the hardware design of visual prosthesis depends much on the development of MOEMS technology. But a number of effective image-processing strategies can be explored and evaluated to enhance the performance of visual prosthesis and even decrease the difficulty of complex hardware design. In this chapter, we presented some tests of image-processing strategies by computer simulation and DSP implementation in real time. A high precision neural stimulator is also under study. An implantable CMOS micro-camera with special design is also evaluated. Although much more tests for its long-term biocompatibility must be carried out, the preliminary results showed that the micro-camera was suitable for our visual prosthesis with minimal loss of image quality in the simulation physiological circumstance.

References

1. Brindley GS, Lewin WS, The sensations produced by electrical stimulation of the visual cortex, J. Physiol. (Lond), 1968, 196:479–493.
2. Dobelle WH, Mladejowsky MG, Artificial vision for the blind: electrical stimulation of visual cortex offers hope for a functional prosthesis, Science, 1974, 183:40–44.
3. Schmidt EM, Bak MJ, Hambrecht FT, Kufta CV, O'Rourke NA, Vallabhanath P, Feasibility of a visual prosthesis for the blind based on intracortical microstimulation of the visual cortex, Brain, 1996, 119:507–522.
4. Zrenner E, Miliczek KD, Gabel VP, Graf HG, Guenther E, Haemmerle H, Hoefflinger B, Kohler K, Nisch W, Schubert M, Stett A, Weiss S, The development of subretinal micro-photodiodes for replacement of degenerated photoreceptors, Ophthalmic Res., 1997, 29:269–280.
5. Humayun MS, de Juan Jr E, Weiland JD, Dagnelie G, Katona S, Greenberg R, Suzuki S, Pattern electrical stimulation of the human retina, Vision Res., 1999, 39:2569–2576.
6. Chow AY, Pardue MT, Chow VY, Peyman GA, Liang C, Perlman JI, Peachy NS, Implantation of silicon chip microphotodiode arrays into the cat subretinal space. IEEE Trans. Neural Sys. Rehabilitation Eng., 2001, 9:86–95.
7. Veraart C, Raftopoulos C, Mortimer JT, Delbeke J, Pins D, Michaux G, Vanlierde A, Parrini Sand Wanet-Defalque MC, Visual sensations produced by optic nerve stimulation using an implanted self-sizing spiral cuff electrode, Brain Research, 1998, 813:181–186.

8. Veraart C, Wanet-Defalque MC, Gerard B, Vanlierde A and Delbeke J, Pattern recognition with the optic nerve visual prosthesis, Artif. Organs, 2003, 27:996–1004.
9. Brelén ME, Duret F, Gérard B, Delbeke J and Veraart C, Creating a meaningful visual perception in blind volunteers by optic nerve stimulation, J. Neural Eng., 2005, 2:22–28.
10. Noelle R Stiles, Intraocular camera for retinal prostheses: Restoring vision to the blind, 2004.
11. Parikh NJ, Weiland1 JD, Humayun1 MS, Shah SS, Mohile GS, DSP based image processing for retinal prosthesis, Proceedings of the 26th Annual International Conference of the IEEE EMBS, 2004, pp. 1475–1478.
12. Pratt WK, Digital Image Processing, John Wiley & Sons, 2001.
13. Gonzalez RC and Woods RE, Digital Image Processing, Addison-Wesley Publishing Company, 1992.

11
Dynamic Interactions of Retinal Prosthesis Electrodes with Neural Tissue and Materials Science in Electrode Design

Charlene A. Sanders[1], Evan J. Nagler[1], David M. Zhou[2]
and Elias Greenbaum[1]

[1] *Oak Ridge National Laboratory*
[2] *Second Sight Medical Products, Inc.*

Introduction

Visual sensation communicates greater information about the environment than any other sense. It is a carefully integrated neural interpretation of chemical and electrical signals that are initiated by photons of light and culminate in cerebral processes that create and map a complex range of visual percepts. Useful visual sensation is dependent upon the efficient functioning of all the links in the visual pathway and the transfer of the signal from image to visual cortex without interruption. Artificial sight refers to a number of experimental photochemical and photoelectrical devices that mimic the function of specialized cells in the optical neuronal network and assume their role if they become impaired by injury or degenerative disease. One such device, presently under development, is a microelectrode array retinal prosthesis for the treatment of people who are blind from retinitis pigmentosa (RP) or age-related macular degeneration (AMD). In both diseases, the photoreceptor cells (rods and cones) are gradually destroyed. Patients affected by photoreceptor degeneration slowly lose visual acuity and eventually become blind. Without viable photoreceptor cells, there are few options for regaining vision. Defective cells may be replaced by removing the retina and transplanting a new retina from a compatible donor. Clinical studies of transplant procedures and immunological studies of transplant survival and rejection are presently under way [1]. The only other viable option is an electronic visual prosthesis. The retinal prosthesis (Figure 11.1) is an intraocular electronic device that can be permanently implanted on the inner retinal surface

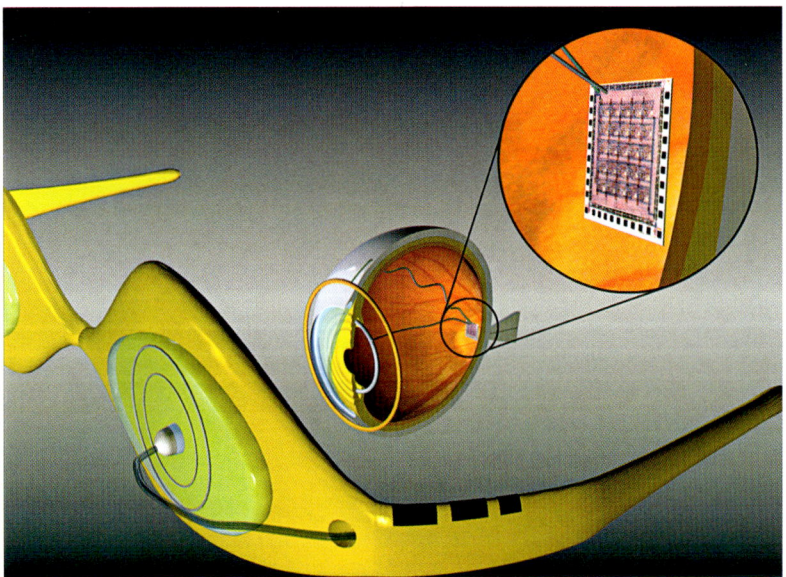

FIGURE 11.1. Illustration of the epiretinal prosthesis.

(epiretinally) and fully integrated with photosensitive imaging and microchip stimulating components. It is capable of sensing and processing environmental visual information and transforming that information into a patterned stimulus on a microelectrode array. Electrical connection to the retinal bipolar and ganglion cell layers is accomplished by positioning the array in close proximity (within $20\,\mu m$) to the retina. The leads and track lines of stimulating electrodes are individually electrically isolated and coated with biocompatible polymer coatings to protect them from intraocular fluids. When energized, the prosthesis can bypass damaged photoreceptor cells and reinitiate threshold membrane potentials in retinal neuronal cells to restore the visual neural pathway.

The rationale for the retinal prosthesis arises from previous experimental analysis of the morphology and electrical response of retinas damaged by degenerative disease. In these eyes, photoreceptor damage is extensive but transsynaptic neuronal degeneration is limited. Initial research with retinal electrical stimulation in patients by Humayun et al. [2] showed that viable neurons could be electrically excited using small extracellular currents delivered by handheld electrodes placed on the retinal surface. Performed in blind human volunteers in the operating room, these short-term tests showed that retinal electrical stimulation resulted in the perception of light in patients who were blind. More recently, a functioning 16-pixel microelectrode array has been chronically implanted on the retina of a patient by Humayun et al. [3]. In the laboratory, external digital imaging and data processing devices were linked with the internal prosthesis and, when the electrodes were current pulsed, the patient perceived light. He could also discern large print letters and everyday objects, such as a plate, cup, or spoon, on a high-contrast background. These results lend credence to the important contribution an implantable intraocular retinal prosthesis can have on the long-term quality of life of blind people.

Along the road to development of the retinal prosthesis, a number of significant questions about electrode design and electrochemical interactions between the electrode and the retinal neural tissue must be formulated and tested. Current pulsed prosthesis electrodes must depolarize retinal neurons without damaging nerve cells or surrounding retinal tissue. They must also be mechanically reliable for the lifetime of the implant recipient. For patients to effectively view their environment, the pixel number of individually addressable electrodes should approach 1000, placed in a macular area no larger than 5 mm × 5 mm. The small electrode area and large electrode number cause problems of electrode arrangement, charge density, long-term stability, and biocompatibility. Here we report how electrode materials research and biomimetic modeling of the electrochemical interface between electrode and intraocular fluids define the operating limits of the retinal visual prosthesis.

Electrochemical Reactions at the Electrode–Vitreous Interface

The Interface

A microelectrode array, implanted in the orb of the eye in close proximity to the retina, is embedded in the vitreous humor, the viscous fluid filling the posterior chamber of the eye behind the lens. The vitreous is a gelatinous electrolyte solution, composed mainly of ascorbate, lactate, collagen, glucose, inorganic salts, and hyaluronate in water [4, 5]. The electrode– tissue interface is the point of contact, and plays a critical role in both the efficacy and the safety of neurostimulatory electrode function. When current is delivered to the electrodes of the prosthesis, a redistribution of charge occurs in the vitreous at the electrode surface. The electrical potential (voltage) across the interface depends on the electrical properties of the electrode material, i.e. conductivity, impedance, and efficiency of electrode charge storage and transfer; and the chemical composition of both phases, i.e. presence of oxide, dissolved metal in solution, etc. To be effective, the electrode must deliver sufficient charge for the cell membrane potential of local neurons to exceed the threshold for neuronal depolarization. For retinal prosthesis, electrode arrays in RP patients, this means injecting 0.05–0.6 μC of charge per phase of a biphasic stimulating pulse (50–300 μA current amplitudes and 1–2 msec phase duration) [6, 7]. The charge density at the electrode–electrolyte interface with a stimulating pulse is the injected charge per phase (amplitude × duration) per unit surface area (expressed as $\mu C/cm^2$) of the electrode. For high charge storage microelectrodes, the total available energy required for threshold excitation of retinal neurons is lower if the energy is clustered on the surface of small diameter electrodes. Because of the unique requirements for high pixel number in a limited active area, the charge density burden of retinal stimulatory electrodes is very high. As a result, the potential exists for charge-related tissue trauma in the area adjacent to implanted neural stimulating electrodes. Interfacial charge density with protracted stimulation of the central nervous system has been implicated in stimulation-induced neural injury. The mechanism for this damage is unknown, but is conjectured to be a result of the passage of current through tissue and neuronal hyperactivity [8]. Mortimer et al. [9] chronically stimulated the cerebral cortex of cats with platinum (Pt) electrodes and biphasic current pulses. They observed a relationship between blood–brain barrier damage in the area surrounding the electrodes and the charge density at the electrode surface. McCreery et al. [10] found that the threshold for damage to the parietal cortex of cats during extended stimulation with activated iridium (IrO_x) microelectrodes was the result of synergistic interaction of charge density and charge per phase beneath the implanted electrode. In later work, McCreery and his colleagues [11] were able to link stimulus-induced temporary depression of electrical excitability of the cat cochlear nucleus with prolonged neuronal activity. Depression was not correlated with histological changes near the electrodes, suggesting that the stimulation protocol was above the threshold for reversible neuronal over-activity but below the threshold for irreversible neuronal damage. Neural damage in the cortex of the cat was observed by Yuen et al. [12] after continuous stimulation at a charge density of $100\,\mu C/cm^2$ for 6 hours. Less damage resulted from the application of $40\,\mu C/cm^2$ for 20 hours. Rose and Robblee [13] found that the safe charge injection limit for cathodic to anodic biphasic $0.2\,msec$ pulsing of Pt electrodes in saline was 100–$150\,\mu C/cm^2$. Based on Pt dissolution data in saline, Brummer et al. [14] suggested limiting charge densities to $< 300\,\mu C/cm^2$ for balanced biphasic pulsing. For iridium oxide microelectrodes in saline, the limits can be extended to $1\,mC/cm^2$ because of the greater charge carrying capacity of IrO_x. [15]. Understanding the electrophysiological interactions between prosthesis electrical components and excitable tissue and the electrochemical reactions at the electrode–vitreous interface are the key elements to developing a safe and practical retinal prosthesis insert.

Interfaction Electrochemical Processes

Electrodes charged at a tissue surface interact with ionic species in the fluid medium bathing the tissue. The voltage, time duration, and conductive properties of stimulating electrodes determine the electrochemical processes that take place at the interface. Charged species in the vitreous can realign spatially around the electrode in an attempt to satisfy electroneutrality at the electrode surface (non-faradaic or capacitative charging), or they can bind to the electrode surface and transfer electrons into the vitreous in the form of electrochemical reactions (Faradaic charging). Faradaic charging is reversible if reaction products remain bound to the electrode surface. It is irreversible if products leave the electrode surface and redistribute in the vitreous. Neurostimulatory electrodes can be divided into two categories: capacitative and faradaic. Capacitative electrodes have a metal core coated with a non-conductive metal oxide surface layer. The oxide insulates the conductive metal, creating a dielectric at the electrolyte interface. Electrons moving through the electrode cannot physically traverse the interface, but can attract or repel ions in close proximity target tissue and initiate an action potential [16]. In contrast, metal electrodes transfer electrons directly across the interface, resulting in faradaic oxidation–reduction reactions in the vitreous and the formation of new chemical species. Guyton and Hambrecht [16] compared the pHs of unbuffered saline solutions after injecting 1 mA monophasic current pulses with 1 mm tantalum oxide (Ta_2O_5) capacitative and Pt electrodes. Pulsing the Pt electrode increased the pH at the cathode and decreased it at the anode. No apparent pH change occurred with similar pulsing of the Ta_2O_5 electrode. It would appear that capacitative electrodes can offer some advantages over metal faradaic electrodes, i.e. they are more resistant to degradation in physiological solutions and are inert to electrical and chemical interactions with extracellular fluids at the interface. However, for the retinal prosthesis, the size constraints on high numbers of densely packed microelectrodes favor the charge transfer properties of faradaic electrodes. In general, capacitative microelectrodes do not have the charge storage ability of noble metal microelectrodes [17], and their charge storage capacity is not high enough to deliver the charge per phase necessary to activate retinal neurons. Attempts to increase charge storage, such as increasing surface roughness, are often accomplished at the expense of electrode strength and durability. Moreover, techniques for making thin oxide films with high dielectric constants that do not leak current are complicated, especially when applied to micron diameter wire. McCreery et al. [8] compared neural damage in the cortex of cats following 7 hours continuous biphasic pulse stimulation ($80–100 \mu C/cm^2$) with Pt faradaic and Ta_2O_5 capacitative electrodes. Both electrodes evoked action potentials in the cortex, and both damaged the cortex equally. The capacitor electrodes did not depolarize neural ganglia more effectively or safely than faradaic electrodes.

Metal microelectrodes, driven at interfacial charge densities above safe potential limits, initiate irreversible electrochemical reactions in the vitreous. A number of byproducts of these reactions can be toxic to the tissue surrounding the electrode. Rosenberg et al. [18] reported inhibition of cell division in bacterial cultures grown in media containing electrolysis products from Pt electrode charging in saline. Toxic byproducts are usually the result of anodic oxidation of vitreous solutes. Free chlorine and hypochlorite (both strong oxidizing agents) are formed when saline-based medium is oxidized at the Pt–s saline interface [19]. Electrode charging in body fluids also raises the probability for in situ water electrolysis. H_2 and O_2 gas evolved during electrolysis [20] can result in local pH changes in the region surrounding the electrode and can be physically damaging to delicate ocular tissue. The electrode itself may be oxidized, resulting in slow dissolution of the metal structure and loss of electrode integrity. Pt oxidizes in saline according to the equation: $Pt + 4Cl \rightarrow Pt\ Cl_4^{2-} + 2e^-$ [20]. Dissolution is related to the aggregate positive or negative phase charge injected and can occur in charged solutions at potentials below the threshold for electrolysis [21]. For safe, long-term prosthesis operation, stimulatory pulses must be of sufficient amplitude and duration to activate neural ganglia without irreversibly generating damaging chemical byproducts in tissue fluid. Microelectrodes for chronic stimulation must have high charge storage capacities and comparatively high thresholds for initiation of interfacial redox reactions.

Pulse Configuration

In considering the thresholds for harmful faradaic reactions at the interface, the nature of the stimulating pulse is a critical factor. DC monopolar (monophasic) electrode charging can supply the minimum pulse amplitude and duration required for neuronal excitation. However, lengthy DC charging can initiate irreversible reactions in the vitreous via faradaic processes [22], resulting in tissue damage. With constant, prolonged monophasic stimulation, the change in vitreous chemical composition with time is unabated; reaction products accumulate and injury results. In 1961, Lilly [23] proposed an electric waveform that did not damage cortical tissue with long-term stimulation. The waveform passed short duration currents first in one direction and then equally in the other (biphasic). The amplitudes of the biphasic currents were equal, resulting in no net charge injection (no direct current component). A fixed duration intraphase delay, inserted between the polar phases of the biphasic pulse, allowed more time for the transfer of charge to the cortex before charge reversal [24]. With this pulse, ionic changes induced in the interstitial fluid during one polar phase were quantitatively reversed during the reverse phase. Faradaic electrochemical reactions produced products that could not discharge from the electrode before the polarity was reversed. The intended result is the injection of no new species in the solution if the stimulating pulse is biphasic, symmetrical, and balanced.

Electrode Test Apparatus

Prospective prosthesis electrodes should be assessed using experimental protocols that test their electrical performance and stability when current pulses are applied in an electrolyte solution. In this way, pulse parameters can be manipulated within the charge density limits for retinal prostheses and toxic products of interfacial reactions can be measured under electrode stimulating conditions expected to be encountered in real-world retinal implants. An electrolysis cell, a model of the eye, approximating the physiological parameters of temperature, photo excitation, and local electrolyte environment found in the human eye, has been designed and integrated with advanced sensing and bioelectrical stimulating instrumentation for the analysis of the performance of potential prosthesis electrodes. A schematic illustration of the apparatus is shown in Figure 11.2. The eye cell is a jacketed, glass chamber filled with 6–8 ml hyaluronic acid (200 µg/ml) in Ames' medium. Ames' medium is a synthetic analogue for

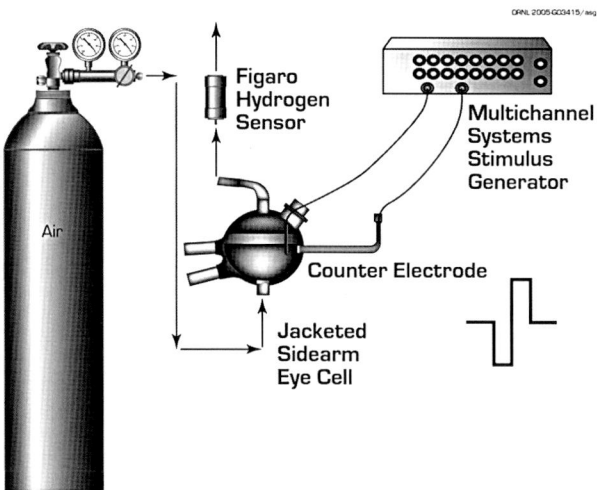

FIGURE 11.2. Electrode test apparatus.

vitreous humor, especially formulated to preserve retinal tissue and photoelectrical response in vitro [25]. During tests, the cell temperature is maintained at 37 °C. Test electrodes immersed in the vitreous solution through a hermetic O-ring assembly are connected to a Multichannel Systems Stimulus generator for application of a pulse stimulus. A Pt plate electrode in the sidearm of the cell, representing the counter electrode placed behind the ear of the transplant recipient, completes the electrical circuit to the generator. The cell is sparged with breathing air (100 ml/min) through an isolated gas flow system, supplying circulating oxygen to the vitreous and allowing for downstream analysis of gaseous products of electrochemical reactions at the electrode–vitreous interface. An in-line, Figaro tin oxide combustible gas sensor, positioned after the cell in the gas flow line, detects hydrogen gas evolved at the cathode above the threshold of water electrolysis. This system can be used for acute or chronic steady-state repetitive testing of the kinetics of interfacial electron transfer reactions at the neural electrode–vitreous interface. For these experiments, the limits of charge injection were between 0.1 and 1 mC/cm^2.

Pt electrodes have been tested for durability and thresholds of electrolysis using the biomimetic eye cell of Figure 11.1. Pt wire (76 μm), fused in insulating glass capillary tubing and polished flush with the glass across the fused end, was immersed in synthetic vitreous and driven with biphasic or monophasic current pulses. Charge densities were kept below 1 mC/cm^2. Repetitive monophasic pulses were applied from the stimulus generator. Current amplitudes were 50–500 μA for 0.1 msec and 1–50 μA for 1 msec pulses. The time between pulses was 16 msec. Steady-state hydrogen production was measured as a function of charge density. The results for hydrogen production for 0.1 and 1 msec pulse durations are shown in Figure 11.3. Thresholds for hydrogen generation correlated with charge density at both pulse durations. The charge density for

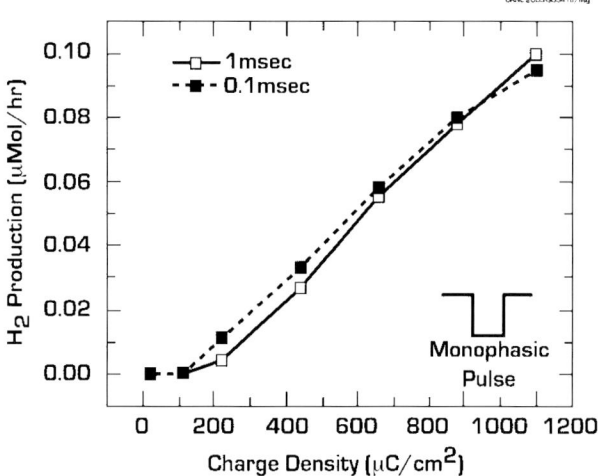

FIGURE 11.3. Electrolytic hydrogen evolution, 0.1 and 1.0 msec monophasic pulses.

threshold electrolysis in both cases was approximately $200\,\mu C/cm^2$. Monophasic pulses applied at charge densities below $200\,\mu C/cm^2$ fall safely within the water window for electrolysis and fit the criterion for safe stimulation of neural tissue.

The same tests were performed on $76\,\mu m$ Pt wire electrodes under the same conditions but with repetitive biphasic pulses. In these experiments, the charge density ceiling was also $1\,mC/cm^2$. Pulse amplitudes were ±1–$50\,\mu A$ and the positive and negative phases of the pulse were symmetrical and balanced. Steady-state hydrogen production was recorded again as a function of charge density (Figure 11.4). With biphasic pulsing, hydrogen was not detected in the air stream at charge densities less than $1\,mC/cm^2$. These results demonstrate proof of

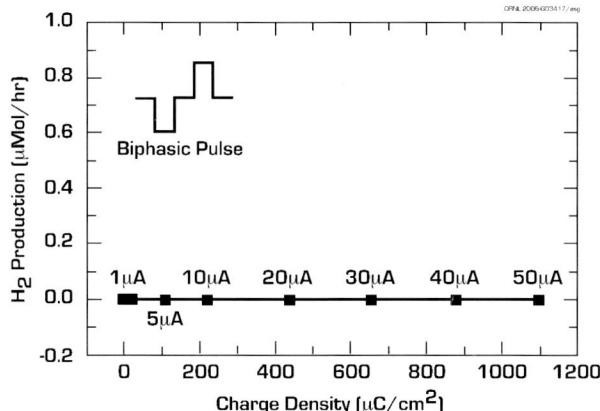

FIGURE 11.4. Hydrogen evolution from Pt electrode with biphasic pulsing, charge density $\leq 1\,mC/cm^2$.

principle and corroborate observed results in in vivo neural stimulation experiments that current reversal and charge balance are the determinants of safe stimulation of neural tissue. The charge density at which breakthrough electrolysis occurs on a 1.05 mm diameter Pt sphere electrode with 1 msec biphasic pulses and 1 msec intraphase delay is just below $1\,mC/cm^2$ (Figure 11.5).

The assumption that a charge density of $1\,mC/cm^2$ will initiate interfacial electrolysis is, however, not based on data derived from tests with small diameter electrodes like those proposed for retinal implants. Benchmark reference points derived from tests with microdiameter Pt electrodes are essential for optimizing electrode design and function. To this point, there are no well-established standard protocols for predicting the onset of unwanted faradaic transformations in the vitreous. The biomimetic eye model and the isolated gas flow apparatus were used to test the limits of stimulation of $150\,\mu m$ Pt electrodes. The electrode disks were electropolished until shiny and smooth to define the surface geometry for calculations of charge density. The electrodes were driven with balanced biphasic pulses with an intraphase delay for 20 min. Pulse amplitude and dwell time were variable and charge densities remained within the prescribed range of 0.1–$1.0\,mC/cm^2$. Table 11.1 shows a compilation of experimental pulse parameters and steady-state hydrogen production. First appearance of H_2 was at $1\,mC/cm^2$, consistent with earlier results from larger electrodes. The onset of electrolysis was reproducible and specific for retinal implant models.

It is conjectured that, if electron transfer initiates electrolysis at $1\,mC/cm^2$, it is likely that other vitreous solutes are also involved in redox reactions at the interface. Table 11.2 shows hydrogen gas/production when a spherical Pt electrode is pulsed with a biphasic $\pm500\,\mu A$ current at variable dwell times (3–500 msec) in saline. In these experiments, total chlorine from charge-induced oxidation of saline was measured using an Orion ion-selective electrode. Saline samples were drawn from the reaction cell before and after 1-hour pulsing. In general, when

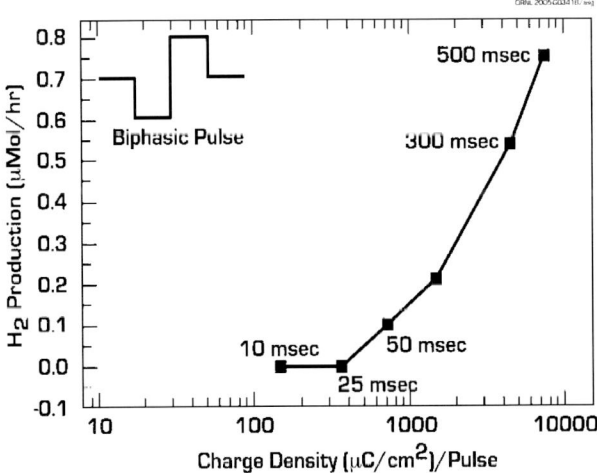

FIGURE 11.5. Threshold electrolysis with biphasic pulsing of a 1 mm Pt electrode.

TABLE 11.1. Threshold gas evolution as a function of charge density.

Charge density (mC/cm^2)	Current amplitude (μA)	Dwell time (msec)	H_2 Production $(\mu Mol/hr)$
0.1	177	0.1	0
0.1	59	0.3	0
0.1	18	1.0	0
0.1	6	3.0	0
0.3	530	0.1	0
0.3	177	0.3	0
0.3	53	1.0	0
0.3	18	3.0	0
1.0	1767	0.1	–
1.0	589	0.3	0.01
1.0	177	1.0	0.01
1.0	59	3.0	0.01

TABLE 11.2. Hydrogen and chlorine production during biphasic stimulation.

Pulse	Dwell time (msec)	$\mu Mol\ H_2/hr$	Cl_2 (ppm)
	1	0.0	0.014
	10	0.0	0.017
	50	0.0	0.025
	100	0.9	0.049
	300	1.8	0.056
	500	2.9	0.063

hydrogen was detected downstream from the Pt electrode, chlorine (HOCl, ClO$^-$, and Cl_2) was also detected. It appeared that these reactions proceeded simultaneously at rates directly related to the charge injected. It can be predicted that monophasic pulses with longer dwell times, i.e. > 1 msec, will also oxidize Cl$^-$ at the anode if the electrolysis threshold is exceeded. Although not relevant for conventional neural stimulation, testing with longer pulse widths for both monophasic and biphasic pulses helps to clarify the relationships between pulse configuration and electrolyte oxidation. The presence of electrochemical byproducts underscores the need for determining the charge injection limits for safe, protracted tissue stimulation without the local accumulation of hazardous byproducts.

Materials Science in Electrode Design

Electrode Materials

Microelectrode core and surface composition are important considerations in the development of chronically implanted retinal neural prostheses. Materials suitable for microelectrodes must meet the minimum criteria of compatibility and capacitance [26]. They must be mechanically and biologically compatible with their implant location and transfer enough charge to depolarize excitable tissue without triggering significant interfacial faradaic oxidation–reduction reactions. Future demands for higher resolution, high pixel, and small diameter electrode arrays will require the consideration of a variety of materials for electrode construction, including Pt and other precious metals, metal composites, and metal alloys. Micron diameter Pt, Ir, Au, and PtIr electrodes can perform well as high density, low impedance cortical stimulatory electrodes. Pt and its more durable alloy PtIr have a long track record as neural stimulatory electrodes, chiefly because of their biocompatibility, relative resistance to corrosion, and high charge carrying capacities [22]. Iridium can be activated by biphasic cycling of the electrochemical potential in electrolyte solutions to form iridium oxide (IrO_x) on the surface of precious metal substrates [27]. Charge is transferred in IrO_x electrodes by stimulus-generated valence changes in the surface oxide layer [15]. IrO_x electrodes can deliver a net safe charge of $1.0\,mC/cm^2$ [28] with 0.2 ms biphasic pulses in bicarbonate-buffered saline, and the charge storage capacity is much greater than that of Pt or PtIr [29]. Moreover, it is hard and durable, resistant to fowling and dissolution in physiological solutions. In vivo comparisons of the performance of IrO_x and PtIr electrodes [30] demonstrated that they could both safely evoke action potentials with extended continuous microstimulation in cat cortex without notable histological damage. Both electrodes performed well for extended periods at charge densities of $800\,\mu C/cm^2$, but poorly at charge densities above $3200\,\mu C/cm^2$. Weiland et al. [31] examined the histology and the electrical properties of the guinea pig cortex implanted with thin-film iridium oxide electrodes. The cortex was energized for two hours with $5–100\,\mu A$, 100 msec biphasic pulses. Electrochemical impedance spectroscopy and cyclic voltammetry measurements showed that the cortex and electrodes were both affected by a temporary impedance shift but, histologically, the tissue was unaffected.

Titanium nitride (TiN) is a material for electrode construction that offers good potential for extending the safe stimulation limits beyond iridium oxide or Pt [32]. The electrochemical performance of titanium nitride deposited on silicon bioelectrodes was compared with similarly constructed iridium oxide electrodes by Weiland et al. [33]. Titanium nitride had similar charge injection capability to Pt ($0.87\,mC/cm^2$), significantly below the capacity of iridium oxide ($4\,mC/cm^2$).

Operational Stability of Pt Retinal Prosthesis Electrodes

The development of a retinal prosthesis for artificial sight includes a study of the factors affecting the structural and functional stability of chronically implanted Pt microelectrode arrays. The maintenance of electrical integrity between the electrode and soft tissue depends, in part, on initial electrode placement in close proximity to the retina and insulation of prosthesis components from intraocular fluids. Only the electrode surface is in direct contact with the retina and the corrosive salt solution bathing the ocular cavity. The 16-pixel microelectrode retinal prosthesis now implanted in the eye of a retinitis pigmentosa patient [3] is constructed in a 4×4 configuration using 520-μm-diameter Pt disc microelectrodes. Durable and dependable cochlear implant electrodes are also made of Pt. However, studies of long-term charging of Pt electrodes in oxidizing solutions indicate evidence of slow oxidation of the Pt surface, dissolution of the metal structure, and loss of electrode integrity. As stated above, Pt dissolution in saline is directly related to the total charge injected but not to rates and thresholds of other redox reactions that may be proceeding at the interface during charging. In vitro dissolution rates of Pt electrodes in saline [34] range from ≤ 10 ng Pt/C for pulses of $\pm 100 \mu C/cm^2$ to 10^3 ng Pt/C for pulses of $\pm 500 \mu C/cm^2$. In vivo tests [35] of cortical stimulation with Pt electrodes for 36 hours with biphasic pulsing at $\pm 100 \mu C/cm^2$ showed an increased concentration of solubilized Pt in the cortical tissue beneath the stimulating electrodes.

Small diameter (76 μm) Pt electrodes, the size projected for the next generation of multipixel retinal prosthesis arrays, were tested for dissolution during chronic charging in synthetic vitreous humor. The electrode was a single Pt wire fused in a glass capillary tube and polished flush with the end of the fused glass tube. Using the biomimetic electrochemical eye cell described above, electrodes were charged continuously for 24 hours with balanced biphasic pulses, 1 msec in duration, and $\pm 50 \mu A$ current amplitude. The time between pulses was 16 msec. Scanning electron microscope (SEM) images taken before and after pulsing are shown in Figs. 11.6 and 11.7. After 24-hour stimulation, the Pt rough electrode surface became visibly smoother and had receded into the capillary glass fused around it. The shortened, flat surface of the charged electrode is indicative of

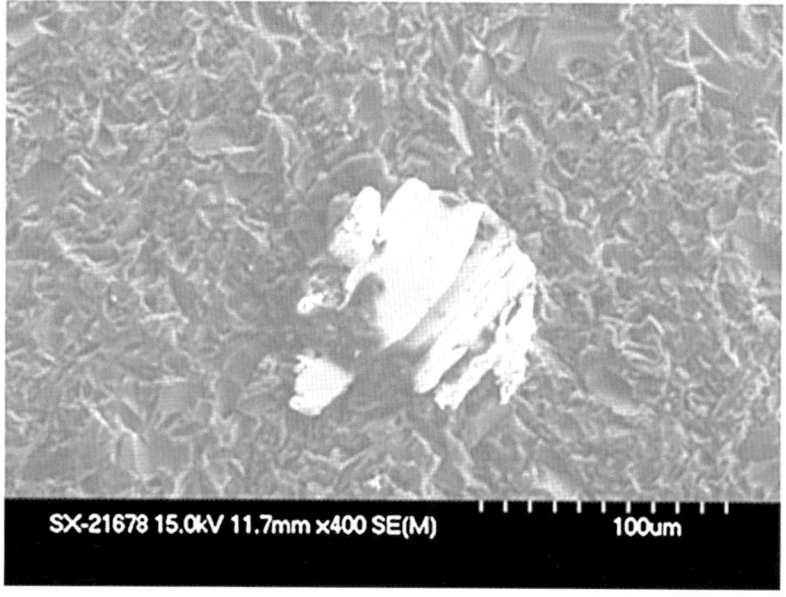

SX-21678 15.0kV 11.7mm ×400 SE(M)   100um

FIGURE 11.6. SEM image of Pt electrode before pulsing.

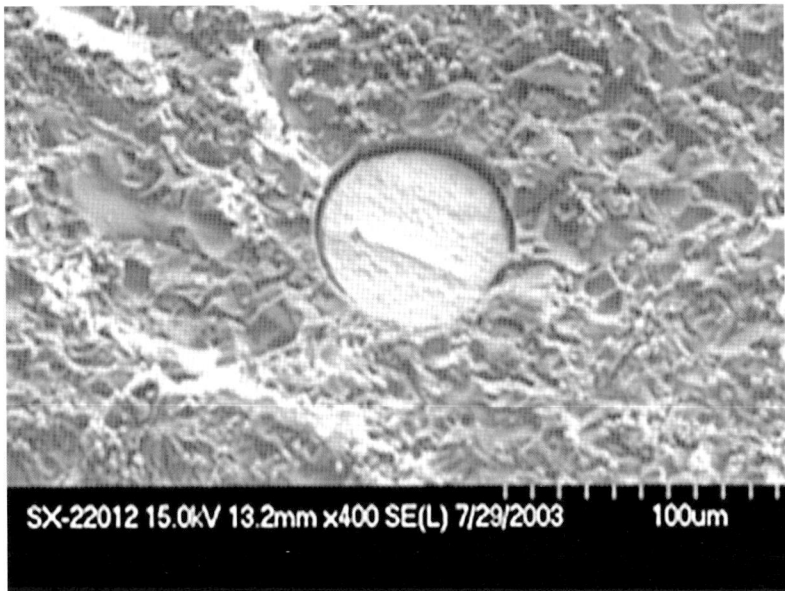

SX-22012 15.0kV 13.2mm x400 SE(L) 7/29/2003   100um

FIGURE 11.7. SEM image of Pt electrode after pulsing for 24 hr (charge density $1 \, mC/cm^2$).

electropolishing of the metal surface by electrochemical oxidation and solubilization in the electrolyte.

Effective Electrode Surface Area and Charge Density

Retinal prosthesis electrode materials that exhibit desirable charge-carrying capacities and low electrical impedances can inject sufficient charge per phase to stimulate retinal neurons. If the charge is constant but electrode diameter decreases, charge density increases. As discussed in Section "**Electrochemical Reactions at the Electrode–Vitreous Interface**," charge density is implicated in the induction of neural injury during chronic stimulation of the mammalian cortex. To mitigate the damage incurred by high charge injection by microelectrodes, the morphology of the microelectrode surface can be modified to increase the effective surface area. Various electrochemical techniques can alter the electrode surface, making it irregular and rough. A rough electrode can transfer the same charge per phase as a smooth electrode with the same geometric dimensions, but the charge on the rough electrode is dissipated over a larger area, thereby decreasing the charge density. PtIr model pacemaker electrodes, deposited with rough coatings of TiN [36], have effective surface areas 500- to 1000-fold higher than their geometric surface areas. Electrochemical impedance spectroscopy (EIS) of TiN-coated electrodes in buffered saline showed that the interfacial capacitance of the coated substrates was several 100-fold higher than planar Pt foil. Pure Pt has been the favored material for retinal prosthesis micro-electrode arrays. The surface area of smooth (planar) Pt wire is approximately equal to its geometric surface area. Other forms of Pt, however, have rougher surfaces and higher effective surface areas that compensate for small diameters of microelectrodes in the charge density equation. Rough Pt (HiQ) made by electrochemical treatment of standard Pt electrodes [37] is being tested in the auditory system as a possible alternative to planar Pt in cochlear implants. HiQ electrodes [38, 39] have lower impedances and 70–75 times the real surface area of conventional cochlear electrodes. High-intensity cochlear implant stimulation can result in decreased auditory brainstem response, due, in part, to DC along the implant during stimulation. Stimulation of mammalian auditory nerves with HiQ Pt electrodes did not change the amplitude and threshold of the brainstem response, as did conventional electrodes. Histology of the electrode implant area showed no significant damage from stimulation with either HiQ or standard Pt electrodes.

Another form of rough Pt, equally interesting for prosthesis implant arrays, is Pt gray. Pt gray has a surface area as high as Pt black, 5 times higher than planar Pt. It is a durable material, not subject to stress cracking like Pt black. Zhou [40] has patented a method for electroplating Pt gray onto the surface of microelectrodes. The Pt gray surface increases the effective surface area, retaining desirable Pt characteristics of low impedance, high conductivity and high threshold for onset of faradaic interfacial redox chemistry. Small diameter (76 μm), planar Pt electrodes were coated with Pt gray by Zhou and tested for dissolution and electrolysis during chronic charging in synthetic vitreous humor.

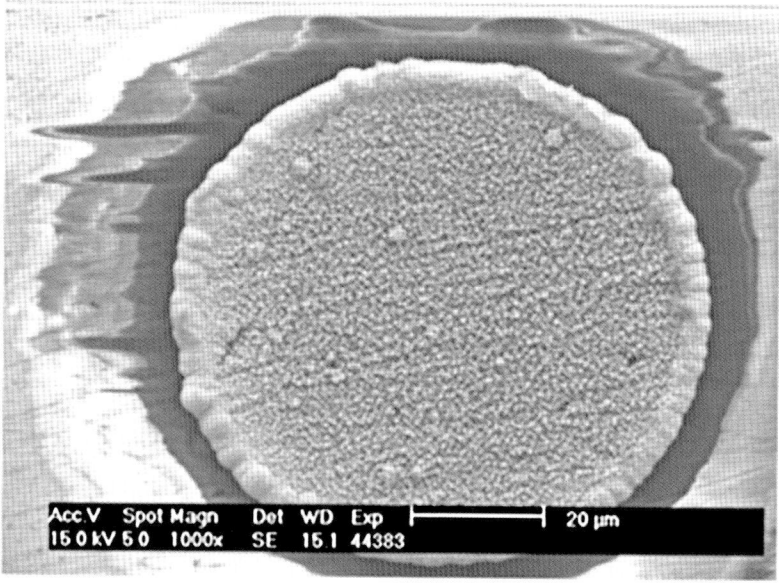

FIGURE 11.8. SEM image of Pt gray before pulsing.

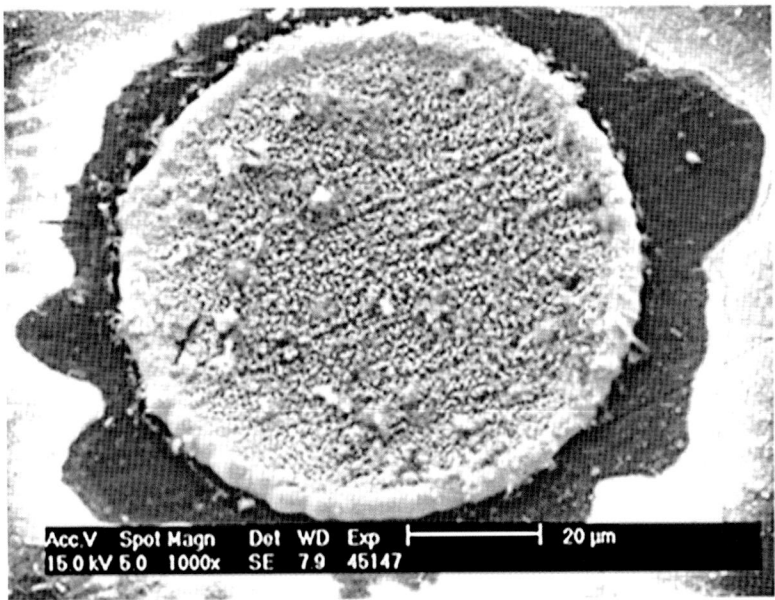

FIGURE 11.9. SEM image of Pt gray after 24 hr biphasic pulsing, charge density 1 mC/cm^2.

The base electrode was a single Pt wire fused in a glass capillary tube and polished flush with the end of the fused glass tube. Pt gray was electroplated on the flat polished surface of the planar Pt. The electrodes were charged continuously in the electrochemical eye cell for 24 hours with balanced biphasic pulses, 1 msec in duration, $\pm 50\,\mu$A current amplitude (charge density 1 mC/cm^2), and 16 msec between pulses. SEM images taken before and after pulsing (Figs. 11.8 and 11.9) show that the rough surface of the Pt gray was very little affected by charging. Planar Pt electrodes, stimulated under the same conditions (see Figure 11.7), solubilized in synthetic vitreous humor. Manipulation of electrode surface morphology is another method for stabilizing and enhancing the performance of chronically implanted retinal prosthesis arrays.

Conclusions

A global effort is presently underway to develop a microelectrode retinal prosthesis for the treatment of people who are blind from retinitis pigmentosa or age-related macular degeneration. Sixteen-pixel arrays already implanted in patients will be replaced by 60-, 256-, and eventually 1000-pixel arrays. The search for a workable design for a 1000-pixel electrode array will require a daunting combination of microsensor technology, materials science, and electronic engineering. Optimal design will encompass studies of electrode structure and core composition, biocompatibility, hermeticity, and shielding of prosthesis electrical components, modeling of neural electrode/vitreous interactions, and telemetric external electronic imaging and signal-processing technology. To avoid damage when the prosthesis is driven, there must be good understanding of the chemical and electrical properties of the microenvironment where the electrodes contact living ocular tissue. Reaching electrical threshold potentials for axonal depolarization in retinal tissue requires that electrodes be close to target neurons and have sufficient charge-carrying capacity to induce depolarization of targeted neuronal cell membranes. The charge per phase transferred and the electrode surface area determine the tissue focal area stimulated and the concentration of the charge. In microelectrode arrays mimicking the function of myriads of defective retinal photoreceptor cells, charge density at the surface is an increasing hazard to safe, efficacious retinal stimulation. Charge injection above thresholds for purely capacitive processes at the interface initiates irreversible electrochemical reactions in tissue fluids, resulting in localized electrolysis of water, oxidation of vitreous solutes, pH shifts, and dissolution of the electrode material. The safe window for stimulation, i.e. the interval where only reversible reactions occur in the vitreous, can be extended by choosing waveform stimulation patterns which are reversible. Reversible pulses are biphasic (bipolar), symmetrical and balanced, passing charge in one direction and then in the other, with no net charge injection. Charge injection limits and thresholds for redox reactions in the vitreous can be determined in vitro by modeling of the eye in a biomimetic eye-cell electrolysis apparatus. From the results of those studies, standard protocols for safe retinal stimulation can be established and applied to the next generation of epiretinal visual prostheses.

References

1. R. D. Lund, S. J. Ono, D. J. Keegan, and J. M. Lawrence, "Retinal Transplantation: Progress and Problems in Clinical Application," *Journal of Leukocyte Biology* **74**, 151–160 (2003).
2. M. S. Humayun, E. de Juan, Jr., G. Dagnelie, R. J. Greenberg, R. H. Propst, and D. H. Phillips, "Visual Perception Elicited by Electrical Stimulation of Retina in Blind Humans," *Arch Ophthalmol.* **114**, 40–46 (1996).
3. M. S. Humayun, J. D. Weiland, G. Y. Fujii, R. Greenberg, R. Williamson, J. Little, B. Mech, V. Cimmarusti, G. Van Bieme, G. Dagnelie, and E. de Juan Jr., "Visual Perception in a Blind Subject with a Chronic Microelectronic Retinal Prosthesis," *Vision Research* **43**, 2573–2581 (2003).
4. H. Davson and L. T. Graham, Jr. (Eds.), *The Eye*, Academic Press: New York and London, 136 (1974).
5. C. Long (Ed.), *Biochemists' Handbook*, D. Van Nostrand Co., Inc.: Toronto, New York, and London, 711 (1961).
6. M. S. Humayun, E. de Juan, Jr., J. D. Weiland, G. Dagnelie, S. Katona, R. Greenberg, and S. Suzuki, "Pattern Electrical Stimulation of the Human Retina," *Vision Research* **39**, 2569–2576 (1999).
7. M. Mahadevappa, J. D. Weiland, R. J. Greenberg, and M. S. Humayun, "Perceptual Thresholds and Electrode Impedance in Three Retinal Prosthesis Subjects," *IEEE Transactions on Neural Systems and Rehabilitation Engineering* **13**, 201–206 (2005).

8. D. B. McCreery, W. F. Agnew, T. G. H. Yuen, and L. A. Bullara, "Comparison of Neural Damage Induced by Electrical Stimulation with Faradaic and Capacitor Electrodes," *Annals of Biomedical Engineering* **16**, 463–481 (1988).

9. J. T. Mortimer, C. N. Shealy, and C. Wheeler, "Experimental Non-Destructive Electrical Stimulation of the Brain and Spinal Cord," *J. Neurosurg.* **32**, 553–559 (1970).

10. D. B. McCreery, W. F. Agnew, T. G. H. Yuen, and L. Bullara, "Charge Density and Charge Per Phase as Cofactors in Neural Injury Induced by Electrical Stimulation," *IEEE Transactions on Biomedical Engineering* **37**, 996–1000 (1990).

11. D. B. McCreery, T. G. H. Yuen, W. F. Agnew, and L. A. Bullara, "Stimulation with Chronically Implanted Microelectrodes in the Cochlear Nucleus of the Cat," *Hearing Research* **62**, 42–56 (1992).

12. T. G. H. Yuen, W. F. Agnew, L. A. Bullara, S. Jacques, and D. B. McCreery, "Histological Evaluation of Neural Damage from Electrical Stimulation: Considerations for the Selection of Parameters for Clinical Application," *Neurosurg* **9**, 292–299 (1981).

13. T. L. Rose and L. S. Robblee, "Electrical Stimulation with Pt Electrodes. VIII. Electrochemically Safe Charge Injection Limits with 0.2 ms Pulses," *IEEE Transactions on Biomedical Engineering* **37**, 1118–1120 (1990).

14. S. B. Brummer, J. McHardy, and M. J. Turner, "Electrical Stimulation with Pt Electrodes: Trace Analysis for Dissolved Platinum and Other Dissolved Electrochemical Products," *Brain Behav. Evol.* **14**, 10–22 (1977).

15. S. B. Brummer and L. S. Robblee, "Criteria for Selecting Electrodes for Electrical Stimulation: Theoretical and Practical Considerations," *Annals New York Academy of Sciences* **405**, 159–171 (1983).

16. D. L. Guyton and F. T. Hambrecht, "Capacitor Electrode Stimulates Nerve or Muscle Without Oxidation-Reduction Reactions," *Science* **181**, 74–76 (1973).

17. T. L. Rose, E. M. Kelliher, and L. S. Robblee, "Assessment of Capacitor Electrodes for Intracortical Neural Stimulation," *Journal of Neuroscience Methods* **12**, 181–193 (1985).

18. B. Rosenberg, L. Van Camp, and T. Krigas, "Inhibition of Cell Division in *Escherichia coli* by Electrolysis Products from a Platinum Electrode," *Nature* **205**, 698 (1965).

19. S. B. Brummer and M. J. Turner, "Electrochemical Considerations for Safe Electrical Stimulation of the Nervous System with Platinum Electrodes," *IEEE Transactions on Biomedical Engineering* **24**, 59–63 (1977).

20. S. B. Brummer and M. J. Turner, "Electrical Stimulation of the Nervous System: The Principle of Safe Charge Injection with Noble Metal Electrodes," *Bioelectrochemistry and Bioenergetics* **2**, 13–25 (1975).

21. J. McHardy, L.S. Robblee, J.M. Marston, and S. B. Brummer, "Electrical Stimulation with Pt Electrodes. IV. Factors Influencing Pt Dissolution in Inorganic Saline," *Biomaterials* **1**, 129–134 (1980).

22. E. Margalit, R. J. Greenberg, D. V. Piyathaisere, G. Lazzi, and E. de Juan Jr., "Retinal Prosthesis for the Blind," *Survey of Ophthalmology* **47**, 335–356 (2002).

23. J. C. Lilly, "Injury and Excitation by Electric Currents: The Balanced Pulse-Pair Waveform," in D. E. Sheer (ed.): *Electrical Stimulation of the Brain*, Austin, University of Texas Press 60–64 (1961).

24. C. van den Honert and J. T. Mortimer, "The Response of the Myelinated Nerve Fiber to Short Duration Biphasic Stimulating Currents," *Annals of Biomedical Engineering* **7**, 117–125 (1979).

25. Ames and F. Nesbett, "In Vitro Retina as an Experimental Model of the Central Nervous System," *J. Neurochem.* **37**, 867–877 (1981).

26. R. Plonsey and R.C. Barr (Eds.), "Functional Neuromuscular Stimulation," *Bioelectricity: A Quantitative Approach*, Plenum Press: New York, 271–299 (1988).

27. R. D. Meyer, S. F. Cogan, T. H. Nguyen, and R. D. Rauh, "Electrodeposited Iridium Oxide for Neural Stimulation and Recording Electrodes," *IEEE Transactions on Neural Systems and Rehabilitation Engineering* **9**, 2–11 (2001).

28. X. Beebe and T. L. Rose, "Charge Injection Limits of Activated Iridium Oxide Electrodes with 0.2 ms Pulses in Bicarbonate Buffered Saline," *IEEE Transactions on Biomedical Engineering* **35**, 494–495 (1988).

29. D. R. Merill, M. Bikson, and J. G. R. Jefferys, "Electrical Stimulation of Excitable Tissue: Design of Efficacious and Safe Protocols," *Journal of Neuroscience Methods* **141**, 171–198 (2005).

30. D. B. McCreery, L. A. Bullara, and W. F. Agnew, "Neuronal Activity Evoked by Chronically Implanted Intracortical Microelectrodes," *Experimental Neurology* **92**, 147–161 (1986).

31. J. D. Weiland and D. J. Anderson, "Chronic Neural Stimulation with Thin-Film, Iridium Oxide Electrodes," *IEEE Transactions on Biomedical Engineering* **47**, 911–918 (2000).

32. M. Janders, U. Egert, M. Stelze, and W. Nisch, "Novel Thin-Film Titanium Nitride Micro-Electrodes with Excellent Charge Transfer Capability for Cell Stimulation and Sensing Applications," in *Proc. 19th Int. Conf. IEEE/EMBS* 1191–1193 (1996).

33. J. D. Weiland, D. J. Anderson, and M. S. Humayun, "*In Vitro* Electrical Properties for Iridium Oxide Versus Titanium Nitride Stimulating Electrodes," *IEEE Transactions on Biomedical Engineering* **49**, 1574–1579 (2002).

34. J. McHardy, L. S. Robblee, J. M. Marston, and S. B. Brummer, "In Vitro Dissolution of Pt Stimulation Electrodes," *J. Electrochem. Soc.* **126**, C152 (1979).

35. L. S. Robblee, W. F. Agnew, L. A. Bullara, J. McHardy, and S. B. Brummer, " *In Vivo* Dissolution of Pt Stimulation Electrodes," *J. Electrochem. Soc.***126**, C152 (1979).

36. Norlin, J. Pan, and C. Leygraf, "Investigation of Interfacial Capacitance of Pt, Ti, and TiN Coated Electrodes by Electrochemical Impedance Spectroscopy," *Biomolecular Engineering* **19**, 67–71 (2002).

37. J. M. Otten, "Method for Reducing the Polarization of Bioelectrical Stimulation Leads Using Surface Enhancement, and Product Made Thereby," *US Patent* **5326448** (1994).

38. M. Tykocinski, Y. Duan, B. Tabor, and R. S. Cowan, "Chronic Electrical Stimulation of the Auditory Nerve Using High Surface Area (HiQ) Platinum Electrodes," *Hearing Research* **159**, 53–68 (2001).

39. C. Q. Huang and R. K. Shepherd, "Reduction in Excitability of the Auditory Nerve Following Electrical Stimulation at High Stimulus Rates: V. Effects of Electrode Surface Area," *Hearing Research* **146**, 57–71 (2000).

40. D. M. Zhou, "Platinum Electrode and Method for Manufacturing the Same," *US Patent* **US 2003/0192784** (2003).

12
In Vitro Determination of Stimulus-Induced pH Changes in Visual Prostheses

A. Chu, K. Morris, A. Agazaryan, A. Istomin, J. Little, R. Greenberg and D. Zhou

Second Sight Medical Products, Inc.

Introduction

Inspired by the success of Cochlear implants, which restores the hearing for the deaf, research efforts worldwide are developing visual prostheses aimed at restoring vision for the blind [1–4]. Several recent developments from research teams and industrial developers working on visual prostheses have raised hopes for a retinal implant and provided other strategies in restoring vision to blind individuals. Intraocular retinal implants developed by Second Sight Medical Products have been chronically implanted in six patients over the past 3 years in an FDA-approved IDE study. Figure 12.1a illustrates part of the design for an intraocular retinal prosthesis [5]. In this model, a small camera that would be housed in the patient's glasses captures visual information, such as the letter "E." This information is then relayed to a microprocessor called Visual Processing Unit (VPU) located externally. After processing this information, the VPU wirelessly sends the information to a microelectronic receiver, implanted behind the ear of the patient, underneath the skin of the scalp. This information is then converted into tiny electric impulses and transmitted through a cable across the eye wall to a microelectrode array to stimulate the remaining retinal neurons of the patient. An example of such implanted microelectrode array, developed by Second Sight, is shown in Figure 12.1b. The electrode array is composed of 16 platinum (Pt) disks arranged in a 4×4 square array. The array is kept tightly against the retinal surface by a medical tack.

Neural prostheses require microelectrodes and stimulation devices that minimize electrochemical damage to surrounding tissue or nerve from chronic use. Electrical stimulation using metallic electrodes in an aqueous electrolyte introduces charges into the environment via electrochemical reactions. At low intensities, charge injection is dominated by capacitive mechanisms [6, 7]. With increasing current intensities, reversible and irreversible Faradaic reactions may occur (Table 12.1) [8]. Almost all Faradic reactions produce or consume

A

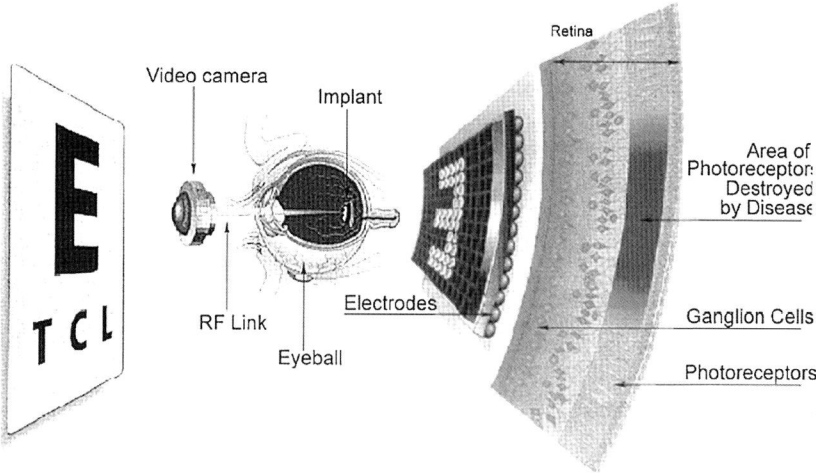

B

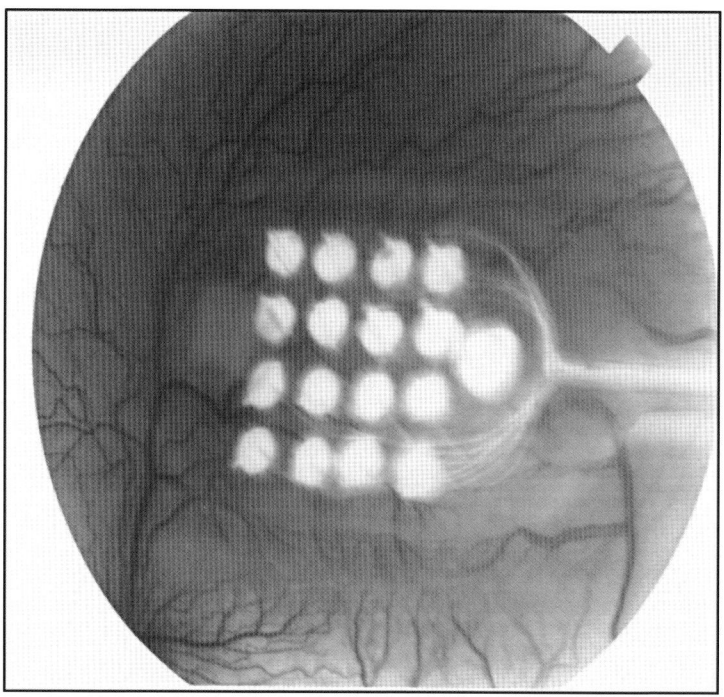

FIGURE 12.1. (a) An early model of an intraocular retinal prosthesis. A small video camera captures the visual information which is processed in an external unit (not shown) and transmitted to an implanted electrode array that stimulates remaining retinal neurons of a patient (Reproduced from DeMarco et al. [5], with permission from Institute of Electrical and Electronics Engineers, Inc.). (b) A photograph of a 4 × 4 electrode array made by Second Sight Medical Products, implanted in dog eye.

TABLE 12.1. Examples of reversible and irreversible electrochemical reactions associated with platinum electrode stimulation [Reproduced from Huang et al. 2001, with permission from Springer-Netherlands.].

Oxidation and reduction	$Pt + H_2O \Leftrightarrow PtO + 2H^+ + 2e^-$
Corrosion of electrode metal	$Pt + 4Cl^- \Rightarrow [PtCl_4]^{2-} + 2e^-$
Hydrogen generation	$2H_2O + 2e^- \Rightarrow H_2(g) + 2OH^-$
Oxygen generation	$2H_2O \Rightarrow O_2(g) + 4H^+ + 4e^-$
H-atom absorption	$Pt + H_2O + e^- \Leftrightarrow Pt - H + OH^-$
Oxygen reduction	$O_2(g) + 2H_2O + 4e^- \Rightarrow 4OH^-$

hydrogen or hydroxyl ions. Since the presence of these ions at the electrode surface alters hydrogen ion concentration, one can expect stimulus-induced pH shift. When translated into a biological environment, these pH shifts could potentially have detrimental effects on the surrounding neural tissue and implant function [14, 17]. Stimulation parameters for neural prosthesis must be controlled to ensure minimal pH changes.

The importance of monitoring pH changes during neural stimulation has been demonstrated in a study for cochlear implants [8]. The studies from cochlear research show that the extent of pH changes is related to stimulus rate and intensity. Such findings provide great groundwork for further investigation in retinal implants. Differences between cochlear implant and retinal implant in terms of stimulation parameters, electrode size and material, and electrolyte may provide further insight into the mechanisms of safe charge injection in a neural prosthesis. A similar study has been carried out to establish a test system to monitor pH changes in retinal implants under pulse stimulation in our group [9].

The pH electrodes used in above studies are commercial microneedle-type electrodes. These pH electrodes would not be suitable to be integrated with existing implantable stimulating electrode arrays. A solid state, planar pH electrode has been considered for such an implant. A considerable amount of research has focused on the fabrication and characterization of oxide-based pH electrodes [10–13]. A comparison of some typical iridium oxide (IrOx)–based pH electrodes reported in the literature with respect to fabrication method and pH-sensing characteristics has been made by Madou's group [10]. The IrOx-based solid-state micro-pH sensors have been used to measure extracellular pH in ischemic heart [11, 12] and esophageal and gastric pH in vivo [13]. A potentiometric pH electrode based on melt-oxidized IrOx film reported by Yao et al shows high chemical stability [10]. The oxide film produced in a lithium carbonate melt has the composition of $Li_xIrO_y \cdot nH_2O$. However, the concerns about its biocompatibility due to the content of lithium salt may limit such electrode as an implantable sensor for in vivo pH sensing. Most oxide-based pH sensors reported are wire type and employed a very high temperature–based processes to form stable oxide layers. It posed limitations to adopt such processes for the MEMS-based electrodes and electrode arrays.

The aim of this study was to further investigate various stimulation conditions on the pH changes in the electrode–electrolyte interface. These results provide an insight to the electrochemical mechanisms at the interface of the electrolyte medium and retinal stimulation electrodes. It also provides information for the safety margin of stimulation parameters. In addition to using a needle-type pH electrode, an IrOx-based planar micro-pH electrode array was constructed. The array consists of both stimulating and sensing electrodes. The pH changes due to electric stimulation were recorded successfully by the planar micro-pH electrode array. A 2D distribution of pH change can be established by using such combined microelectrode arrays.

Experimental

Working Electrodes

Two stimulating electrode array designs were used to generate pH changes by stimulation currents. Most results were achieved using a silicone-substrate Pt 16-electrode array. Subsequent testing using a polyimide thin film Pt electrodes array verified results from the silicone array. Electrode arrays were placed in small Petri dish test cell with aqueous electrolyte. Stimulation was conducted in monopolar configuration with large Pt wire as common electrode.

Test Solutions

Tests were conducted in an unbuffered salt solution of $0.14\,mM$ NaCl with $2.7\,mM$ KCl in purified DI water. Buffered pH standard solutions from VWR were used to calibrate pH electrodes. All experiments were carried out at room temperature ($\sim 22\,^{\circ}C$), and the test cell was exposed to ambient air.

Electrode Stimulation

Charge-balanced square-wave biphasic current pulses were generated from a computer-controlled current stimulator. The stimulator was capacitively coupled to minimize dc current. Monophasic negative (cathodic) and positive (anodic) current pulses were generated using a stimulator with no capacitive coupling. Unless otherwise stated, all experiments were conducted with current pulses of 60 Hz frequency and 1 ms pulse width.

pH Measurements

The pH level was detected using a needle-type commercial micro-pH electrode made by WPI with a tip diameter of $100\,\mu m$. In some earlier experiments, a combination micro-pH electrode with Ag/AgCl reference electrode from Thermo-Orion was used. A Thermo-Orion A420 pH meter was used for pH measurements. Prior to each day's testing, the pH electrode was calibrated using standard pH solutions. The pH microelectrode was positioned close to the stimulating electrode, using a micromanipulator. Figure 12.2 shows a typical test cell

A

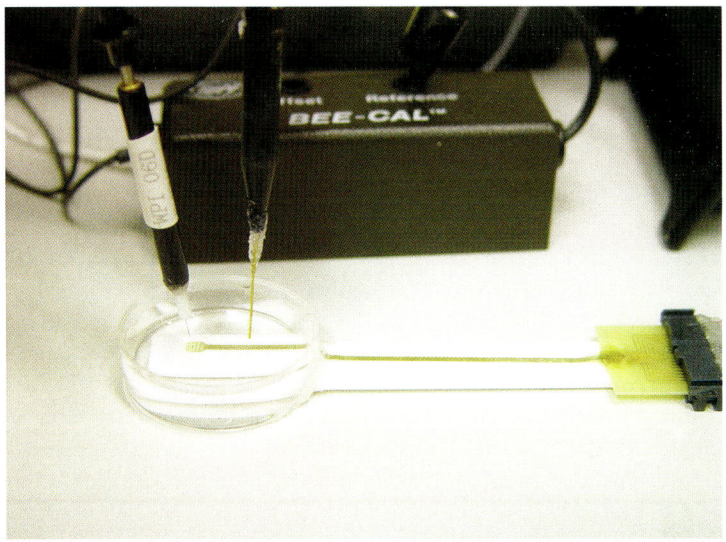

B

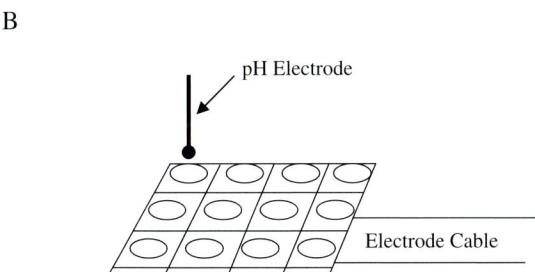

FIGURE 12.2. (a) A photograph of a test cell setup for both stimulation and pH measurements with a stimulating electrode array and a needle pH electrode. (b) A schematic diagram of test cell set-up with position of pH electrode relative to stimulating electrode site in an electrode array.

set up. For pH measurements, the needle pH electrode was positioned ~1 mm above a 200-um Pt disk electrode. Initial readings were taken seconds before start of stimulation; and throughout the duration of stimulation, measurements were recorded at about 5-second intervals. The pH change results were verified with a pH indicator, phenol red (Fisher). Phenol Red is a color pH indicator that changes from yellow to red over a pH range of 6.8–8.4. For planar micro-pH electrode arrays, a miniature single junction Ag/AgCl electrode was used as a reference electrode. The oxide-based pH electrode's potentials in mV were measurements using the pH meter.

Results

From previous publications, reactions of charge-injecting Pt electrodes are shown to involve pH changes at the electrode–electrolyte interface (Table 12.1) [8]. For several possible cathodic reactions during cathodic stimulation, hydroxyl ions are produced, which will lead to increase in alkalinity and the pH value. In contrast, possible anodic reactions will produce hydrogen ions, leading to an increased acidity and a lowered pH value. As reported in our earlier work [9], no pH changes were detected, if a buffered solution was used. The buffering agents in the solution may neutralize any pH changes induced by electrochemical reactions. The buffering effect of the saline solutions may mask any possible electrochemically induced pH changes. To avoid such effect in order to highlight the electrochemically induced pH changes, an unbuffered saline solution was employed in this experiment.

The stimulus-induced pH shifts were seen from stimulation with a single stimulation electrode pulsed at monophasic pulses of either cathodic-only or anodic-only pulses at $60\,\mu C/cm^2$. The pH shift for cathodic-only stimulation increased nearly linearly with stimulation time at a rate of about 2 pH/minutes. When stimulation was stopped after about 5 minutes, when the pH significantly exceeded the accuracy range of the pH meter, rapid pH decay was noted. The decay of pH after stimulation could indicate the diffusion of byproducts away from pH electrode surface. For anodic-only stimulation, pH was noted to decrease at about 3 pH/minute initially, then the rate of pH decrease slowed gradually. After 5 minutes, stimulation was stopped and the pH shifted toward initial baseline (Figure 12.3).

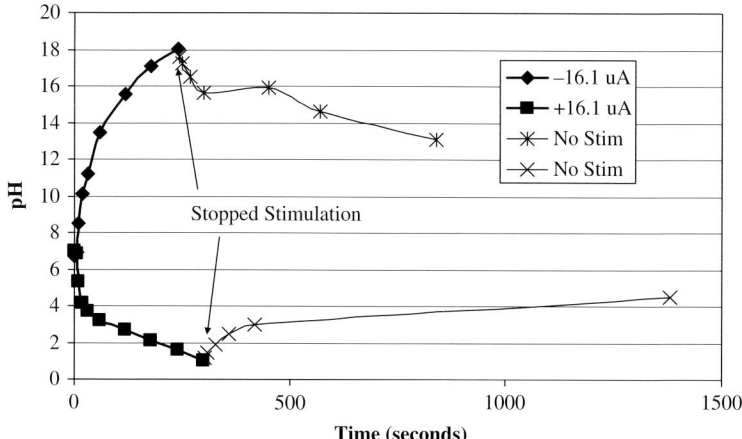

FIGURE 12.3. A typical H shift from monophasic cathodic or anodic stimulation. A typical pH shift around stimulation electrode was increased by cathodic-only stimulation and decays shortly after stimulation was stopped. pH decrease was noted for electrode stimulated with anodic-only pulses. After current pulses were stopped, pH shifted toward baseline.

Effects from intensity and frequency of stimulation pulses were also examined. The pH changes with stimulation time were monitored at different pulse currents and frequencies. To avoid gas evolution reactions, very low charge densities (30 to 120 μC/cm^2) were used for monophasic stimulations. It was noticed that pH changes increased with the increased pulse current and pulse frequencies. Figure 12.4 shows both cathodic-induced pH increase and anodic-induced pH decrease under different pulse current amplitudes. It is clear that the higher the pulse current, the more the pH changes are detected at any given time. This indicates that an electrode with a lower threshold in clinical applications, which requires less pulse current, will minimize possible damage due to the pH shift. The frequency effects on the pH changes during pulse stimulation were measured at 60 μC/cm^2 and the results showed similar effects (data not shown). Higher frequency pulse (60–120 Hz) induced more pH shift for the electrodes pulsed with lower frequency (10–30 Hz) at the same current amplitudes. For neural stimulation, pulsing should use low frequencies to improve the pH balance at the electrode–electrolyte interface.

The pH changes in the test solution were confirmed using a pH indicator, phenol red. Time for color change of phenol red indicator was recorded. For cathodic-only pulses, phenol red color change was noted around the stimulating electrode to change from yellow to pink in about 10 minutes, which indicates an increased pH, shown in Figure 12.5a. For anodic-only stimulation, phenol red color change could not be easily detected due to small contrast of yellow to light yellow color change; however, color change could be seen around the Pt counter

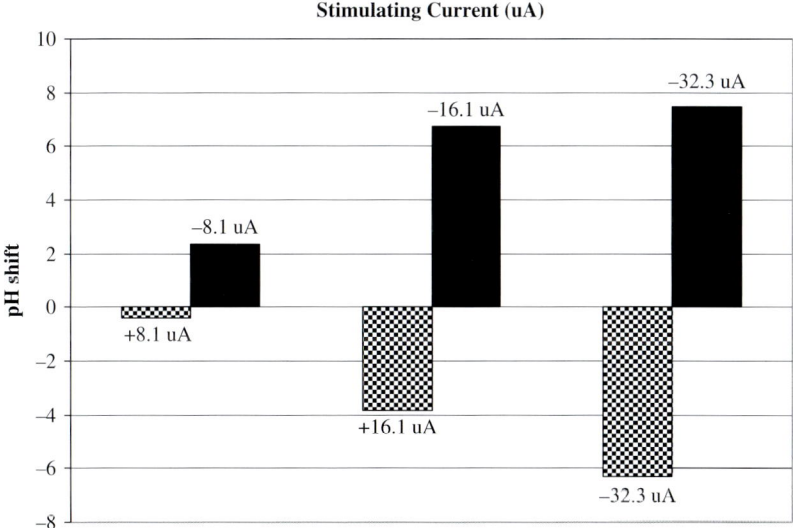

FIGURE 12.4. Induced pH shifts with various stimulation current intensities. Cathodic pulse current is shown to cause an increased pH while anodic pulse current caused a decreased pH, correlating with amplitude of current.

A

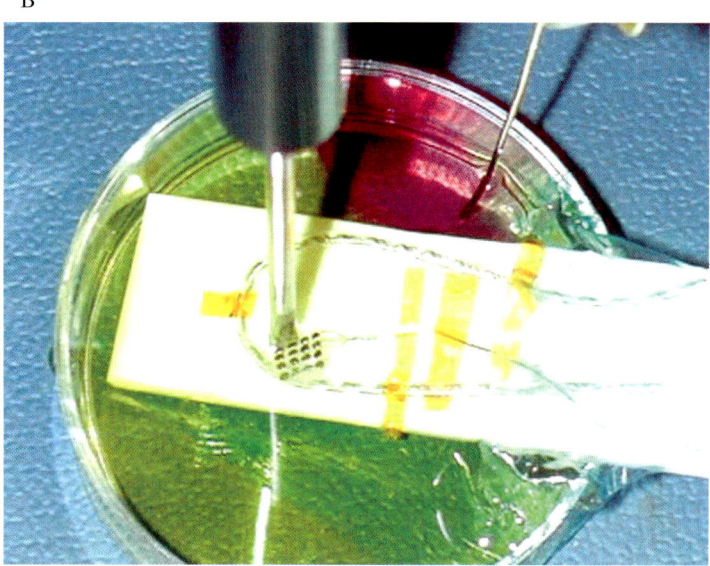

B

FIGURE 12.5. (a) pH indicator color change around cathodically stimulating electrodes. Phenol red pH indicator changed color from yellow to pink, due to increase in pH, around working electrode. (b) pH indicator color change around counter electrode of anodically stimulating electrodes. pH decrease due to anodically stimulating electrodes could not be verified by phenol red pH indicator; however, phenol red around return electrode of test cell showed yellow to pink color change, indicating the increase in pH.

electrode, shown in Figure 12.5b. A different pH indicator which gives a sharp color change in lower pH range (< 6) should be employed in future work to detect pH changes around anodic stimulating electrodes.

Typical neural prosthesis stimulation use charge-balanced, biphasic pulses with the intention that by-products produced in the first phase of the current pulse are reversed by the second phase [8, 14]. Figure 12.6 shows the minimal pH shift over 15 minutes of biphasic pulsing at $0.35 \, mC/cm^2$ with a cathodic-first pulse followed by anodic pulse stimulation in unbuffered saline solution. No phenol red color change was noted and the decrease in pH was less than 1 unit. Anodic-first, followed by cathodic counter pulse biphasic stimulation, showed similar results. pH shift was minimal over 15 minutes stimulation. No phenol red color change was observed and the change in pH was less than 1 unit. These results demonstrate the importance to use biphasic pulses for neural stimulations to minimize pH changes.

While the single needle-type pH electrode can detect the pH change around the stimulation electrode, it cannot be used to detect the pH change in the electrode–electrolyte interface when the stimulation electrode array is tightly against the tissue or retinal surface. Another limitation of such needle-type pH electrodes is that it cannot detect the pH change very close to the stimulation electrode surface without disturbing current and field distributions. During our preliminary studies, we found that pH changes were higher when the sensor tip is closer to the stimulation electrode (Data not shown). Additional factors influencing measurements by the needle electrode included the size of the pH senor, which contributed a blocking effect, and the movement (vibration) of solution around the stimulation electrode, due to the stirring effect. For a sensitive

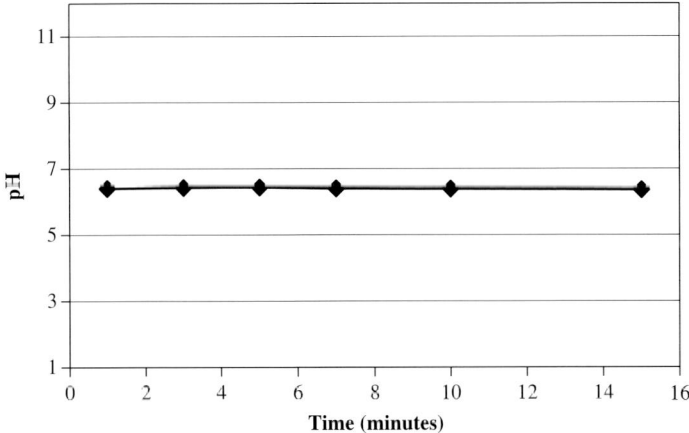

FIGURE 12.6. A typical pH shift from biphasic, charge-balanced, cathodic-first stimulation. Minimal pH shift was detected from stimulation using biphasic pulse at $0.35 \, mC/cm^2$ in unbuffered saline solution.

pH measurement, the pH-sensing electrode closely arranged around the stimulating electrode is important. Planar microelectrode arrays were constructed in
our group to monitor pH changes during pulse stimulation. The microelectrode
arrays consist of both stimulating electrodes and pH-sensing electrodes. Stimulating electrodes on these arrays could be made from Pt or other noble metals
and their alloys, while the pH-sensing electrodes are IrOx based. The array can
be in both thick-film and thin-film forms. The IrOx is reactively sputtered on
the metal seed layer. The electrode sizes range from 50 to 350 μm disks in
diameter.

The IrOx pH-sensing electrodes were characterized by cyclic voltammetry.
Cyclic Voltammetry (CV) was measured using an EG&G M273 potentiostat
and M270 Electrochemical analysis software. The Cyclic Voltammogram for a
typical IrOx electrode is shown in Figure 12.7. The potential scan range was
from −0.6 to +0.9V against an Ag/AgCl reference electrode at 50 mV/s. The
electrolyte was 100 mM phosphate buffered saline (PBS). The thin-film IrOx
electrode exhibited characteristic redox peaks in the voltammogram. Two pairs
of anodic and cathodic peaks within −0.4 to +0.8 V vs. Ag/AgCl are believed
to be from redox reactions of IrOx.

Although the redox reaction mechanisms of IrOx are still not clear, most
researchers believe that during electrochemical reactions, oxidation states
changes in the hydrated IrOx layer is accompanied by the injection/ejection of

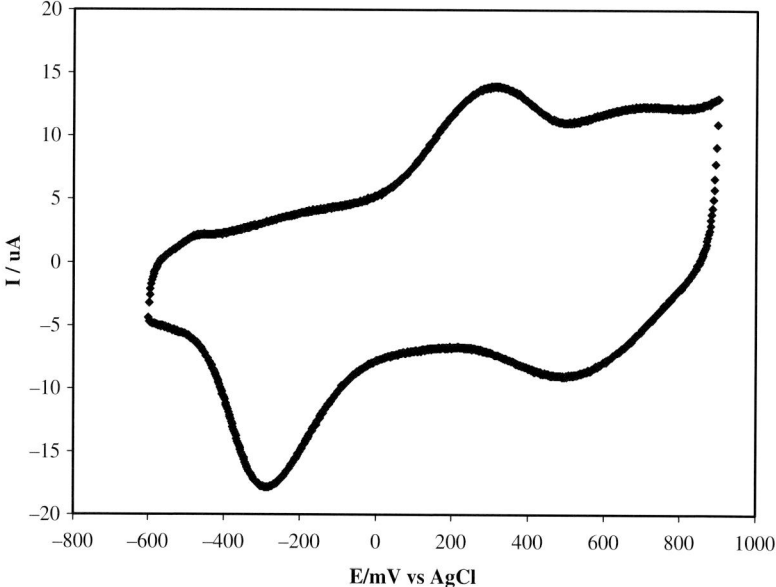

FIGURE 12.7. A typical cyclic voltammogram of Electroplated IrOx. The voltammogram
is recorded with a voltage scan rate of 50 mV/sec in 0.1 M phosphate-buffered saline,
pH 7.4.

H^+ [10, 15]. The electrode potential changes with the hydrogen ion concentration and is expressed by Nernstian equation:

$$E = E° + 2.303RT/F \,(\text{pH}) = E° - 59.16\,\text{pH} \tag{1}$$

where $E°$ is the standard electrode potential with the value of 926 mV *vs.* standard hydrogen electrode (SHE), or 729 mV vs. Ag/AgCl reference electrode, R is the gas constant, T is the absolute temperature (K) and F is the Faraday constant. The potential/pH slope is expected to be $\sim 58\,\text{mV/pH}$ at 22°C.

An example of such micro-pH electrode array in thin-film construction is shown in Figure 12.8. The array is made of flexible polymer with imbedded thin-film metal traces. There are 16 electrodes with a 4 x 4 arrangement on the array. The pH response of the IrOx electrodes were calibrated in three standard solutions. The buffer solutions are air-saturated without stirring.

Using a series of calibration solutions, the response curve or calibration curve of a pH electrode as per the Nernstian Eq. (1) can be experimentally determined by plotting the electrode voltage versus the pH of the calibration solution. The linear range of the calibration curve is applied to determine the pH in any unknown solution. The slope of the calibration curve within the linear range is used to determine the response (Nernstian) slope or electrode sensitivity in mV/pH. Figure 12.9 shows the open circuit electrode voltages change with pH and an average Nernstian slope calculated for 16 electrodes in a micro-pH electrode array. The electrode based on this oxide film exhibits very promising pH-sensing performance, with a nearly ideal Nernstian response $57 +/- 1.1\,\text{mV/pH}$ in the tested pH range of 4–10. The intercept at pH 0 gives an $E°'$ value of $756 +/- 15\,\text{mV}$ against an Ag/AgCl reference electrode, or 953 mV *vs.* SHE. The experimentally determined $E°'$ of 953 mV is slightly higher than the theoretical value (926 mV). The $E°'$ value is related in the stoichiometry of

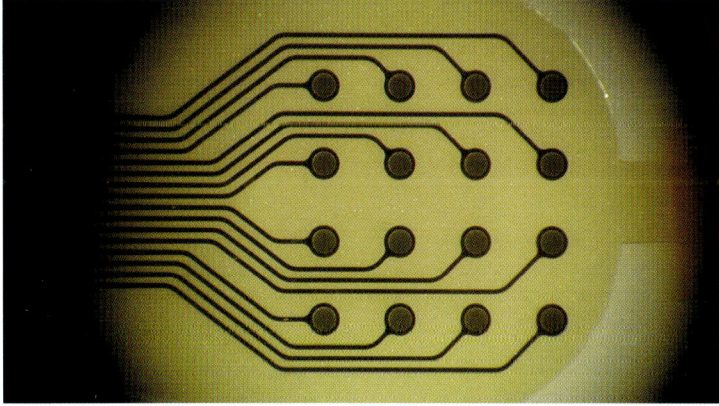

FIGURE 12.8. An example of a micro-pH electrode array in thin-film construction. There are 16 sputtered IrOx electrodes with a 4×4 arrangement on the array.

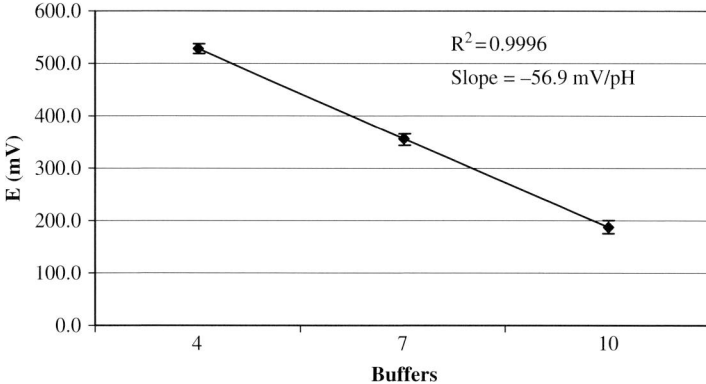

FIGURE 12.9. A calibration curve of an IrOx-based pH electrode array measured in standard pH solutions vs. an Ag/AgCl reference electrode at room temperature. The slope of the calibration curve within the linear range is used to determine the response (Nernstian) slope or electrode sensitivity in mV/pH.

the oxide film and the difference of its oxidation state. The higher $E^{o\prime}$ value measured in the standard pH buffers suggests the higher oxidation states in the stoichiometry of the oxide films [10].

The potential response of the planar micro-pH electrodes is fast, with a 90% response time obtained in less than 1 second for all pH changes. The open-circuit potential of the electrode is stable during calibrations. It is evident from Figure 12.9 that the potential/pH slopes and the electrode potentials show excellent agreement among electrodes from the same array.

In order to measure pH profiles in 2D distributions, all, except one, of the electrodes in the array were coated with pH-sensing materials. The stimulus-induced pH shifts were measured with this type of multi-pH-sensing electrode array. The stimulating electrode was pulsed at monophasic pulses of either cathodic-only or anodic-only pulses. The pH changes on 15 electrodes were recorded throughout a duration of stimulation of about 10 minutes. Similar to the needle pH electrode experiments, for cathodic-only pulses, pH increase was noted around the stimulating electrode while for anodic-only stimulation, pH decrease is detected.

Figure 12.10 shows a 2D pH changes with the electrode positions after stimulation for only one minute [16]). The electrode site in dark color is the Pt-stimulating electrode. All other 15 electrodes are pH-sensing electrodes. The charge density of stimulation pulses applied on the Pt electrode was $0.15\,\mathrm{mC/cm^2}$. It is interesting to note that the pH changes on electrodes presented a clear 2D distribution. The four electrodes closest to the Pt-stimulation electrode had most pH changes $(2.9+/-0.3\,\mathrm{pH})$, while the four corner ones had slightly less pH changes $(2.0+/-0.4\,\mathrm{pH})$. The electrodes far away from the stimulating site showed relatively small pH changes $(0.2+/-0.15\,\mathrm{pH})$.

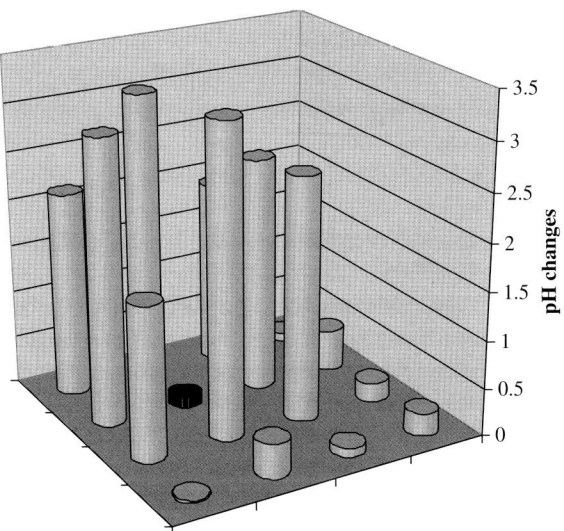

FIGURE 12.10. Two-dimensional pH distribution measured by an IrOx-based planar micro-pH electrode array after monophasic stimulation for one minute. The electrode site in dark color is the Pt-stimulating electrode. All other 15 electrodes are pH-sensing electrodes (Reproduced from Zhou et al., 2004 [16], with permission from The Electrochemical Society, Inc.).

The fact that pH-sensing electrode sites, which have equal distance from the stimulating electrode, reported slightly different pH level in Figure 12.10 might be attributed to distribution of asymmetrical byproducts, affected by diffusion and convection processes, and to differences in pH-sensing electrodes' response time.

Examination of the pH changes at longer stimulation time from this reveals that pH changes slowed down after about two minutes stimulation. The pH distribution due to the difference in distance from stimulation site still exists, although the pH difference from close sites and far sites are decreased. This indicates that the byproducts of electrochemical reactions build up around the electrode surface and diffuse away from the stimulation site with time. During neural stimulation, a biphasic pulse (a counter pulse immediately follows the stimulation pulse) should minimize such diffusion.

The solid-state pH electrodes developed have several advantages in comparison to commonly used glass pH electrodes. Unlike the glass pH electrodes, which require a high input impedance pH meter, the oxide-based pH sensor has low electrode impedance. In contrast to the sluggish response of glass pH electrodes, the solid-state pH sensor also presented a faster pH response. This method to prepare oxide-based pH electrode makes miniaturization of a pH sensor possible and is compatible with thin-film and MEMS microfabrication technologies.

Conclusions

Experiments in unbufferd saline solution showed pH shifts significantly when electrode is pulsed monophasically even at very low charge density ($30 \, \mu C/cm^2$). An increase in pH is detected in surrounding solution as an electrode is pulsed cathodically, and conversely, a decrease in pH is detected when an anodic pulse is used. The results support what is previously known about electrochemical reactions occurring at the electrode–electrolyte interface. Stimulation using biphasic charge-balanced pulses did not introduce a notable pH change, regardless of which leading pulse phase for the charge density up to $0.35 \, mC/cm^2$ for smooth solid Pt electrodes. To minimize such possible pH effects, a biphasic charge-balanced pulse should be used for neural stimulation. Body fluid has high pH buffering capacity and it will neutralize pH changes during in vivo stimulation.

The in vitro pH changes due to electric stimulation were recorded successfully by the planar micro-pH electrode array in this experiment. A 2D distribution of pH change was established by using such combined microelectrode arrays. This in vitro experiment was conducted in a solution when all electrode sites faced up without any blockage and physical constrain. It is possible that the local pulse-induced pH change near the electrode site might not completely agree with what is found under open-space condition, especially when the electrode array is not followed by the curvature of the retina surface and the distance of electrode to tissue is different from that of electrode to electrode in the array. Continued study in vivo may help in better understanding the differences between open space in vitro testing and actual implanted conditions. The planar thin-film micro-pH electrode arrays reported in this work hold promise for the detection of possible stimulus induced pH changes at the electrode–tissue interface.

Acknowledgments. This work was supported by *the National Institute of Health/National Eye Institute*, under grant #1R24EY12893-01. The authors wish to thank Chase Byers, Teresa Swan, and Kjersti Morris for their contributions to this work.

References

1. Rizzo J, Wyatt J, Humayun M, de Juan E, Liu W, Chow A, Eckmiller R, Zrenner E, Yagi T, Abrams G (2001) Retinal prosthesis: An encouraging first decade with major challenges ahead. Ophthalmology 108(1):13–14.
2. Weiland JD, Humayun (2003) Past, present, and future of artificial vision. Artificial Organs 27(11):961–962.
3. Greenberg R (2000) Visual prosthesis: a review. Neuromodulation 3(3):161–163.
4. Humayun M, Weiland J, Fujii, G, Greenberg R, Williamson R, Little J, Mcch B, Cimmarusti V, Van Boemel G, de Juan E (2003) Visual perception in a blind subject with a chronic microelectronic retinal prosthesis. Vision Research (43)24:2573–2581.
5. DeMarco SC, Lazzi G, Liu WT, Weiland JD, Humayun MS (2003) Computed SAR and thermal elevation in a 0.25mm 2-D model of the human eye and head in response

to an implanted retinal stimulator Part I: models and methods. IEEE Transactions on Antennas and Propagation 51(9):2274–2285.

6. Zhou D, Greenberg R (2001) Tantalum capacitive microeletrode array for a neural prosthesis. In: Butler M, Vanysek P, and Yamazoe N (eds) Chemical and Biological Sensors and Analytical methods II. The Electrochemical Society, San Francisco, pp 622–629.

7. Robblee LS, Rose TL (1990) Electrochemical guidelines for selection of protocols and electrode materials for neural stimulation. In: Agnew WF and McCreery DB (eds) Neural Prosthesis. New York, Prentice-Hall, pp 26–66.

8. Huang CQ, Carter P, Shepherd RK (2001) Stimulus induced pH changes in cochlear implants: an in vitro and in vivo study. Annals of Biomedical Engineering 29:791–802.

9. Chu AP, Morris K, Greenberg R, Zhou D (2004) Stimulus induced pH change in retinal implants. In: Proceedings of the 26th Annual International Conference of the IEEE Engineering in Medicine and Biological Society. EMBS, San Francisco, pp 4160–4162.

10. Yao S, Wang M, Madow M (2001) A pH electrode based on melt-oxidized iridium oxide. Journal of Electrochemical Society 148(4):H29–H36.

11. Marzouk SAM, Ufer S, Buck RP, Johnson T, Dunlap L, Cascio W (1998) Electrode-posited iridium oxide pH electrode for measurement of extracellular myocardial acidosis during acute ischemia. Analytical Chemistry 70(23):5054–5061.

12. Marzouk SAM, Buck RP, Dunlap L, Johnson T, Cascio W (2002) Measurement of extracellular pH, K^+, and lactate in ischemic heart. Analytical Biochemistry 308(1):52–60.

13. Wipf DO, Fuyun G, Spaine T, Baur J (2000) Microscopic measurement of pH with iridium oxide microelectodes. Analytical Chemistry 72(20):4921–4927.

14. Donaldson N, Donaldson PEK (1986) When are actively balanced biphasic ("Lilly") stimulating pulses necessary in a neurological prosthesis? II pH changes; noxious products; electrode corrosion discussion. Medical and Biological Engineering and Computing 24:50–56.

15. Zhou D, Greenberg R (2005) Microsensors and Microbiosensors for Retinal Implants. Frontiers in Bioscience 10:166–179.

16. Zhou D, Chu A, Agazaryan A, Istomin A, Greenberg R (2004) Towards an implantable micro pH electrode array for visual prosthesis. In: Cahay M, et al (eds) Nanoscale Devices, Materials, and Biological Systems: Fundamental and Applications. The Electrochemical Society, Honolulu, pp 563–576.

17. Mortimer JT, Kaufman D, Roessman U, (1980) Intramuscular electrical stimulation: tissue damage. Annals of Biomedical Engineering 8:235–244.

13

Electrochemical Characterization of Implantable High Aspect Ratio Nanoparticle Platinum Electrodes for Neural Stimulations

Zhiyu Hu[1], Dao Min Zhou[2], Robert Greenberg[2] and Thomas Thundat[1]

[1]*Oak Ridge National Laboratory*
[2]*Second Sight Medical Products, Inc.*

Abstract: Two-dimensional (2D) and three-dimensional (3D) patternable conductive structures such as entire neural stimulation circuit are created by using metal nanoparticles and a nanopowder molding process. Fabricated structures retain a height-to-width ratio of up to 10:1. The described process is able to fuse the stimulating electrode, connection trace, and contact pad into one continuous, integrated structure where different sections can have different heights, widths, and shapes. The batch process is suitable for mass production, and the fabricated electrode is robust and very flexible. Additionally, the completed structure can be packed onto a biocompatible flexible substrate, such as poly-dimethylsiloxane, parylene, and polyimide as well as other temperature-sensitive or vacuum-sensitive materials at room temperature which make it more suitable for biomedical applications. Experimental data show that the electrodes and wires have about the same electrical resistivities as their bulk materials and desirable electrochemical properties, including low impedance. The nanoscale feature on the electrode surface enhanced the interface and contact quality between electrode and bio-substrate that led to better electrochemical performance.

Introduction

Implantable microelectronic devices for neural prostheses require that stimulation electrodes cause minimal electrochemical damage to tissue or nerve from chronic stimulation. Neurological stimulation requires high-quality electrochemically stable electrodes which should have low electrode impedance, high charge storage capacity, and low voltage excursion. A stimulating electrode array must meet several requirements that include a high electrode density, high charge injection capability, corrosion resistance, and could be packaged with biocompatible materials [1]. The most commonly used electrodes are made by four methods based on their fabrication processes: direct-wiring of conductive wires or fibers, electroplating, vacuum-deposited thin films, and micromachining. While more and more prosthetic devices are in development, there are increasing demands for high-performance and high-density electrodes. The electrode array in contact with the living tissue forms an interface between the electronic device and the biological tissues. The size of charge-injection electrodes becomes smaller in order to increase selectivity and accommodate more electrodes on the arrays to achieve high resolutions for neural recording and stimulation. For such high-density microelectrode arrays, the design of electrode arrays and choice of electrode materials become increasingly important. An electrode must be able to deliver higher charge density without generating irreversible electrochemical reactions such as metal corrosion or dissolution, gas evolution, or introduction of toxic chemical reaction products.

Using metal nanoparticles and microfabrication in an innovative nanopowder molding process, 2D and 3D patternable structures were produced with a height to width ratio greater than 10:1. The unique fabrication process of this molding method is the ability to fuse the entire nerual stimulation circuit, including stimulating electrode, connection trace, and contact pad, into one integrated continuous structure in which the different sections of structure might have a different height, width, and shape. The fabricated electrode is robust, flexible and is also suitable for mass production with high reproducibility. More importantly for biomedical applications, the entire fabricated structure can be packed at room temperature into bio-compatible flexible substrates, such as poly-dimethylsiloxane (PDMS), parylene and polyimide, and other temperature-sensitive or vacuum-sensitive materials. The molded electrodes and wires not only have the same electrical resistivities as their bulk materials, but also have desirable electrochemical properties, including low electrochemical impedance.

For many biomedical applications, it is necessary for any electrode array to be a patternable, flexible, and biocompatible package. Traditionally, noble metals such as gold or platinum (Pt) are the preferred conductive materials used for coating electrodes and forming conducting circuits. Vacuum-deposited thin films are the most widely used method for making various neural stimulation devices [2, 3]. There are many materials that could be used to make thin conductive films on both hard surfaces (i.e. silicon, glass) and flexible substrates [i.e. PDMS, parylene, polyimide or other polymers]. The most frequently used conductive materials are Ti, Ti alloys, TiN, Au, Pt, Pt alloys, Pt/W, Pt/Ir, Ir, IrOx, etc. Biocompatibility is one of the main concerns for material selection [4, 5].

Vacuum deposition is a good technique for making patternable conducting electrodes and conductive lines on various 2D surfaces, but microscopic observation reveals that the deposited metal thin film is actually a collection of nanometer scale metal particles stacked on top of each other. Generally, the bonding forces among those particles are not strong enough to support the structure itself. As a direct result, the vacuum-deposited metal films retain insufficient physical strength to stand alone and are unable to preserve any structural integrity by themselves without a supporting substrate. In addition, the most commonly used metals such as gold and Pt have poor adhesive capability to most substrates (such as silicon, glass, or PDMS) by themselves. In order to improve the bonding quality between the metal deposition layer and the substrate, one or more layers of adhesive materials (such as Cr, Ti) are often required to be deposited prior to the gold or Pt. Such composite structure complicates the electrochemical effects. Thin film thickness results in higher resistance and impedance. The height-to-width ratio is normally low, in a range of 1:10, indicating a thin and wide structure with high electrical resistance and high impendence that are undesirable as they limit the charge injection and cause thermal heating to surrounding tissues.

Direct-wiring of conductive wires or fibers is often used in situations when there are fewer constraints in space, structural complexity, position precision, and electrode density [6]. But these mostly handmade devices are very difficult to produce with high precision and in high density when low cost and high production volume are required. Direct-wiring is a very labor-intensive task. The hands-on process and the physical size of the metal wires limit the packing density and possible pattern structure. These properties are often associated with low productivity, high fabrication cost, and low yield.

Micromachining methods can be successfully used to fabricate 2D or 3D electrode arrays on silicon-based wafers with a process that involves multi-step photolithography and multi-layer vacuum thin film depositions [7]. Unfortunately, lithography-based processes often use silicon, quartz and other semiconductor materials as substrates which are stiff and fragile. It is also not so easy to create a structure that would fit with the specific contour of bio-substrate.

Electroplating methods have been reported to create electrode and other conductive structures with reported aspect ratios of 1:1 [8]. However, the thickness of the trace wire is still limited to a range of a few micrometers due to the high stress of a plated metal layer. The adhesion of the plated layer to the seed layer is also a concern for any plated layer thicker than two micrometers [9]. Modifying the electrode surface by electroplating is also limited to very thin layers of a few micrometers or a soft structure such as Pt black. Platinum black is produced during the rapid electroplating of Pt to give a very rough and porous surface [10]. Platinum black has a very weak structural and physical strength and is therefore not suitable for applications where the electrode is subject to even minimal physical stress.

Here we present another electrode fabrication method – nanopowder molding: sintering metal nanoparticles near their melting temperature to form high aspect ratio electrode arrays with the desirable dimensions in a 2D/3D pattern and layout. The fabricated structures exhibit desirable mechanical properties and electrochemical performance such as low electrical resistivity and low electrochemical impedance. This process yields high-quality electrode and trace wires in one process that is suitable for low-cost, high-volume production. The molded Pt microelectrodes has been characterized by some electrochemical techniques, such as cyclic voltammetry (CV), electrochemical impedance spectroscopy (EIS), and current pulse stimulation (PS).

Experimental

The Pt nanoparticles (diameter $300 \sim 700$ A) powders (99.99% pure) from Technic Inc were used in the preparation microelectrode structures. In order to fit into our tube furnace (2 in. diameter.), quartz substrate was diced to a 4 in. $\times$ 1 in. from 4 in. $\times$ 4 in. wafer which has a thickness of 0.09 in. PDMS substrate was prepared from Dow Corning Sygard 184 silicon elastomer base/curing agent. The high temperature sintering of nanoparticles was carried out with a Thermolyne 21100 Tube Furnace with a quartz tube chamber and manual programmable temperature controller. Electrical resistance of the molded conductive wires was measured by a Signatone Checkmate Probe Station with HP 34401A millimeter with 4-wire resistance measurement setting.

The nanopowder molding fabrication process is illustrated in Figure 13.1. This method utilizes a planar substrate (such as glass, silicon, or quartz) which can withstand high temperature ($> 1000\,^{\circ}C$). The desirable electrode pattern and channels are fabricated on the substrate using different techniques. Additional deep holes (for the tall electrode pad) were drilled using an ultrasonic drilling machine. All these channels may have different width and depth at different sections. With a fabrication precision of better than one micrometer, it is possible to fabricate a structure with feature size of a few micrometers. Once fabricated, the channels are cleaned and dried. The channels are then filled with Pt nanoparticle (Figure 13.2a) slurry after mixing nanaparticles with DI water. The electrode is formed when the substrate is heated to a very high temperature ($> 800\,^{\circ}C$) as shown in Figure 13.3. Precisely controlled melting processes could be used to create nanoscale features (Figure 13.2b) on fabricated Pt electrode surfaces that were formed by fusing nanoparticles at near melting temperature. The rough electrode surface could offer better interface and contact between the electrode and the bio-substrate that might reduce the contact resistance variation and lead to stable retina stimulation. Since the melting point of nanoparticles is lower than the melting point of bulk material, the particles may fuse together even before reaching its bulk melting point. The heating of the substrate is accomplished through a programmable temperature furnace to control the heating and then cooling behavior of the melt. The melt is then slowly cooled down to ensure good metallic quality of the electrodes. A thin layer of PDMS or other polymers is deposited on the electrode surface of the substrate. Prior to deposition of the PDMS pad, the electrode surface is pre-wetted with DI water to enhance the adhesion of the metal to the PDMS and reduce the adhesion of the substrate to PDMS. After curing of PDMS, the PDMS pad is then removed from the substrate to separate the electrodes from the substrate. Therefore the patterned conductive structure could be faithfully transferred to a soft substrate, while maintaining the integrity of conductive structure without the need to expose low-temperature

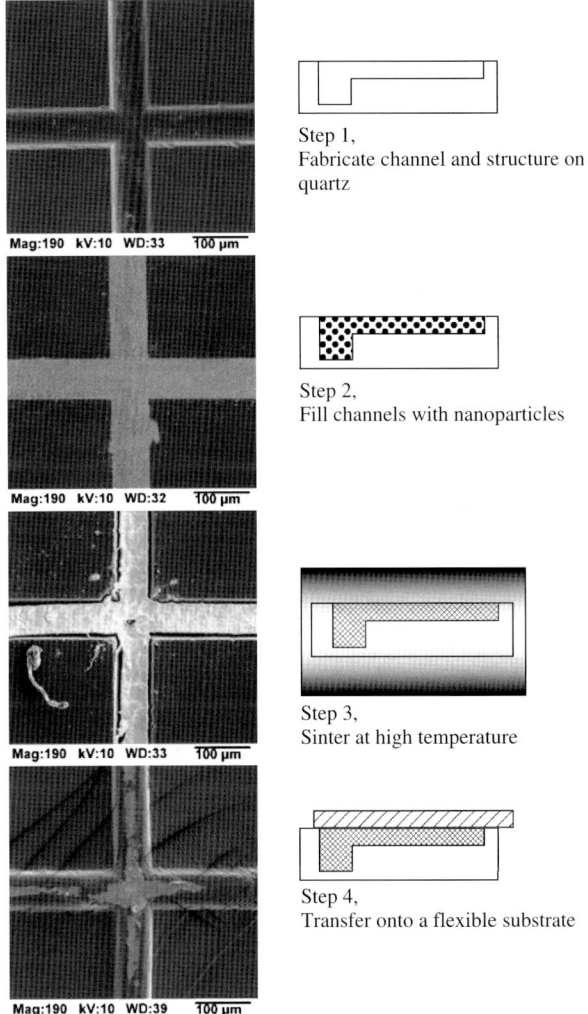

Step 1,
Fabricate channel and structure on quartz

Step 2,
Fill channels with nanoparticles

Step 3,
Sinter at high temperature

Step 4,
Transfer onto a flexible substrate

FIGURE 13.1. Nanopowder molding fabrication process flow.

soft materials to high temperature. The fabricated structure is highly flexible and robust with a width of 50–250 µm and thickness of ~ 10 µm. There is no sign of Pt and silicone fusing; the electrode can be easily released from quartz substrate.

All electrochemical experiments were carried out in 10 mM phosphate buffered saline (PBS) at room temperature (~ 22°C), and the test cell was exposed to ambient air. CV measurements were made by using EG&G M273 potentiostat, and EIS measurements were carried out by a Gamry EIS system. Charge-balanced square-wave biphasic current pulses were generated from a computer-controlled current stimulator from Multichannel System. Stimulation was conducted in monopolar configuration with large Pt wire as common electrode. The Pt

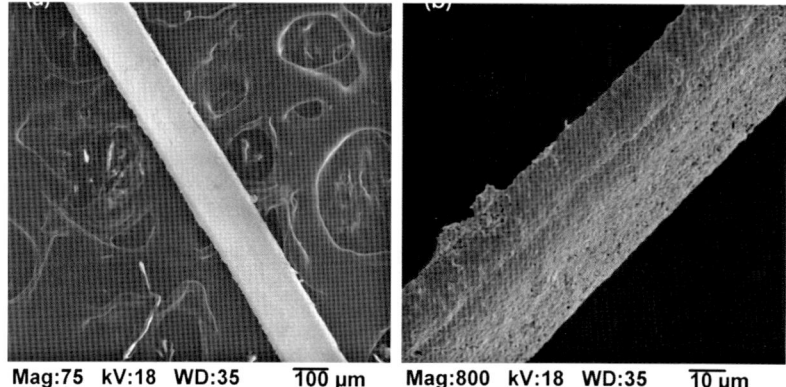

Mag:75 kV:18 WD:35   100 µm   Mag:800 kV:18 WD:35   10 µm

FIGURE 13.2. (a) Scanning electron micrograph of Pt nanopowder. (b) Nano-scale feature on molded Pt wire surfaces which could enhance interface and contact quality between electrode and bio-substrate and also result better electrochemical performance.

electrode samples with diameters of 50–150 µm were mounted on a ceramic strip were used as a supporting substrate for the electrode. The electrode surfaces were cleaned thoroughly before tests.

Results and Discussions

Although as a physical property of material, the electrical resistivity of a given material in its bulk material normally is a constant. Depending upon the fabrication process, the actual effective resistivity of created conductive trace wires and electrodes may be much higher than when it is in its bulk material. For example, vacuum-sputtered thin film might exhibit an effective resistivity value that could be several times higher than the bulk value. Using a probe station and four-wire precision resistance measurement method, we measured the electrical resistivities of molded Pt electrodes. The measurement indicates that molded Pt electrodes have an electrical resistivity of 11.78 µΩ-cm which is very close to the value of the bulk materials. Since electrical resistivity is a physical property of material, having bulk material resistivity means that we could not expect any less resistivity from the same material. In comparison, vacuum-deposited thin film gold trace with the same width might have over a few hundreds ohms resistance [8]. In our experiment we did not observe any noticeable resistivity change after extended mechanical bending.

Electrochemical Impedance Spectroscopy (EIS) is a non-destructive method to characterize biomaterials. For electrode materials, solution resistance, electrode–electrolyte interface impedance, charge transfer resistance and surface roughness/double layer capacitance can be measured and their frequency response can be determined in a fast frequency scan. To evaluate the impedance of the molded electrodes, the EIS measurements were carried out in PBS with

(a)

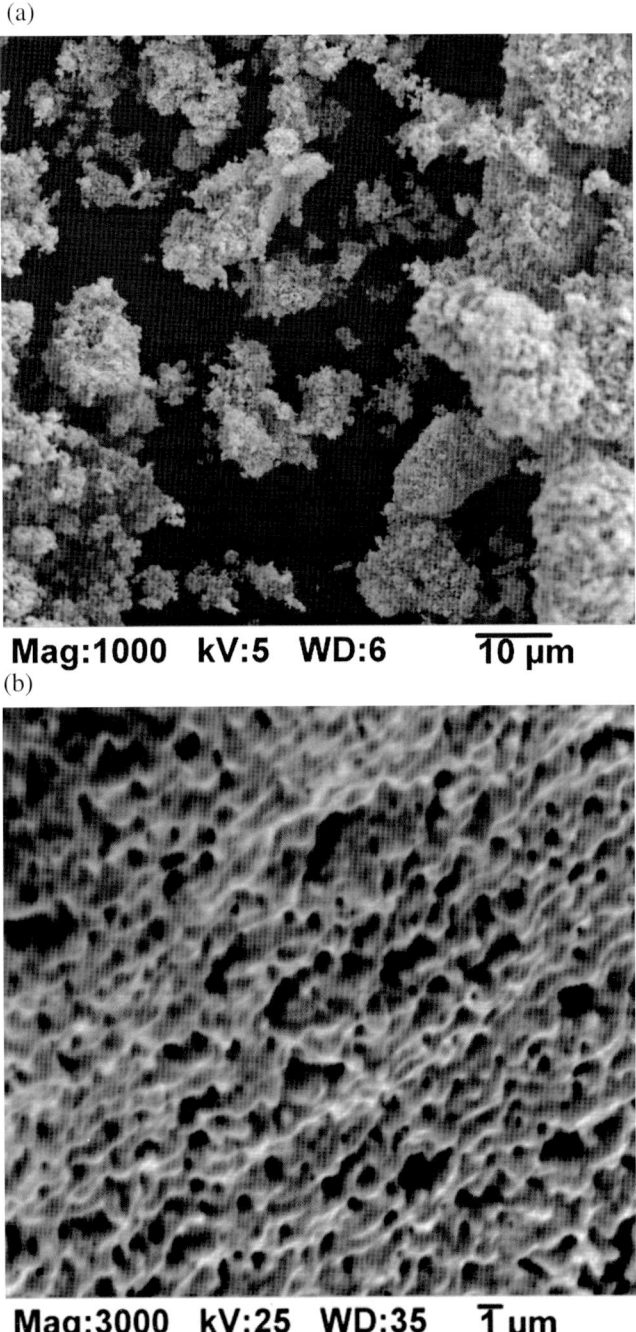

Mag:1000   kV:5   WD:6   10 µm

(b)

Mag:3000   kV:25   WD:35   1 µm

FIGURE 13.3. Molded Pt electrode trace wires after high temperature sintering.

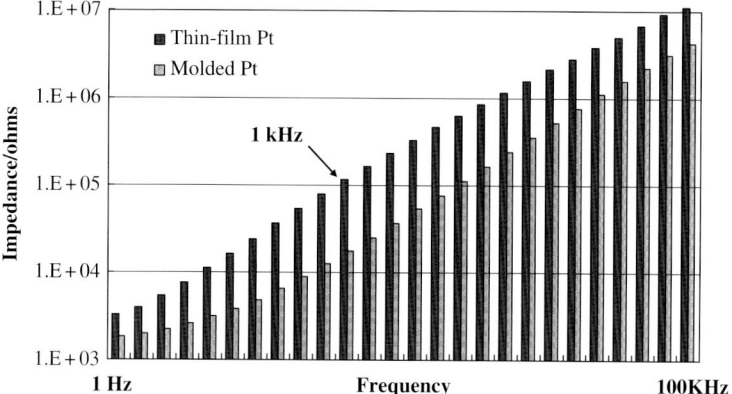

FIGURE 13.4. Electrochemical impedance spectroscopy (EIS) measurement indicates that molded Pt electrodes have lower electrode impedances than those of sputtered thin films.

a frequency scan from 1 Hz to 100 kHz. The EIS results indicate that molded Pt electrodes have lower electrode impedance than those of sputtered Pt thin films as shown in Figure 13.4. Lower electrode impedance is desirable for neural stimulation as it will reduce voltage excursion and lower the power consumption during stimulation.

There is a tendency for charged species to be attracted to or repelled from the metal–solution interface. This gives rise to a separation of charge, and the layer of solution with different composition from the bulk solution is known as the *electrochemical double layer*. As a result of the variation of the charge separation with the applied potential, the electrochemical double layer has an apparent capacitance (known as the *double layer capacitance*) [11]. It can be seen that in order to achieve large charge storage capacity, for a given material and cell setup, the real surface area of the electrode is a key factor: the higher the surface area, the higher the capacitance. For high-density microelectrode arrays used in neural stimulation, the geometric surface area is limited by the application, and an effective way to increase the electrochemical surface area is to increase the roughness of the electrode without changing the array size.

The electrode capacitance results along with electrode surface areas are listed in Table 13.1. The electrode capacitance is determined by EIS measurements at

TABLE 13.1. Electrode Capacitance determined from EIS measurement at 1 kHz of molded Pt Samples.

Electrode	Width (um)	Length (um)	Area ($\times 10^{-4}$ cm^2)	Capacitance (uF/cm^2)
A	50	150	0.75	81
B	100	150	1.5	52
C	150	150	2.3	63

1 kHz at a very small ac voltage of 10 mV. The results indicate that the molded electrodes have better surface roughness than what are found in thin-film metals. For sputtered thin-film Pt surface on polymer substrate, a capacitance is typically in the range of $10–25\,uF//cm^2$.

Cyclic voltammetry (CV) has been employed to determine the operational potential window (the water window) limited by the H_2 and O_2 evolution potentials due to electrolysis of water on the cyclic voltammogram. The area estimated by integrating the cyclic voltommogram within the water window indicates the charge delivery capacity of an electrode. Figure 13.5 shows the comparison of cyclic voltammograms of sputtered and molded Pt electrodes. The CV measurements were carried out in PBS solution with a potential scan rate of 50 mV/s. The Pt electrodes presented a well-defined voltammogram with enlarged reduction and oxidation peaks within the water window. Such increased cyclic voltammograms for molded Pt over sputtered Pt thin film suggest increased charge delivery capacities for neural stimulation for both materials. Another advantage of these electrodes is that no visible active peaks were detected in cyclic voltammetry measurements. This implies that no measurable active metals, which may cause corrosion or tissue damage, are present in Pt samples. This is a very important consideration point for all neural stimulation applications.

Most neuro-stimulation applications use a biphasic, cathodic first current pulse. Under such current pulse, the electrode's voltage response (voltage excursion) is a direct indicator of charge injection capability. For a given electrode at a given pulse current, the lower the electrode voltage, the higher the charge

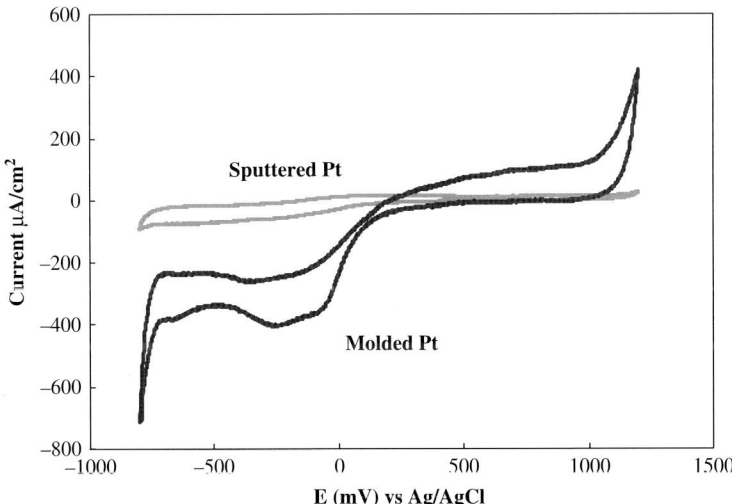

FIGURE 13.5. In cyclic voltammetric (CV) test, molded Pt electrode presented a well-defined voltammogram with enlarged reduction and oxidation peaks and had a higher charge storage capacity. The CV measurements were carried out in PBS solution with a potential scan rate of 50 mV/s.

injection capability. The voltage excursions were measured under pulse stimulation currents to evaluate the molded electrode charge injection capacity. A programmable multichannel stimulator made by Multichannel Systems was used to generate stimulus. A TDS 3014B oscilloscope along with the WaveStar program was used to measure and record the voltage excursion. The electrodes were stimulated in PBS under a charge-balanced, cathodic first, biphasic pulse current at $100\,\mu C/cm^2$ for Pt at 50 Hz. The voltage excursion curves are shown in Figure 13.6. Under the same stimulation conditions and same geometric surface area, molded Pt electrodes presented much lower voltage excursion for the charge density levels tested than what are found in their thin films. A low voltage with linear voltage excursion is favorable for neural stimulation as it will have less effect from irreversible electrochemical reactions which may produce harmful byproducts and/or cause electrode corrosion.

Electrical stimulation of biological tissue with metal electrodes requires the flow of ionic charge in the biological tissue. This flow of charge can be induced by two mechanisms: capacitive and Faradaic. The Faradaic mechanism of charge injection involves electrode transfer across the electrode–tissue interface. This may induce harmful electrochemical reactions and can cause tissue or nerve damage [12]. Hydrogen/oxygen evolution due to water hydrolysis induced by stimulus will alter pH, causing metal corrosion and possible tissue damage in the electrode/tissue interface [13]. In the case of biphasic pulse, no hydrogen

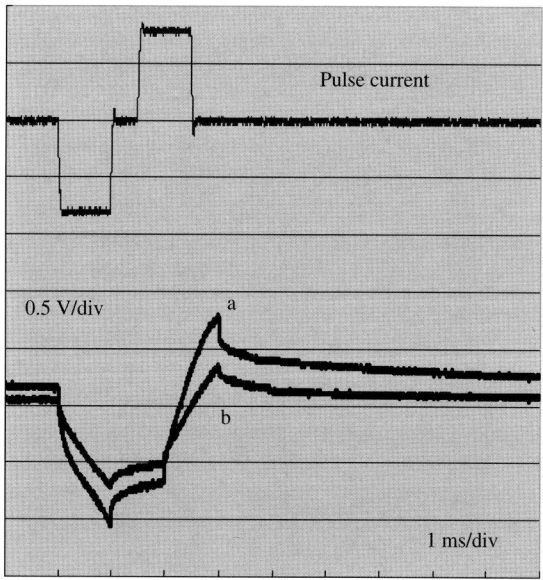

FIGURE 13.6. The electrodes were stimulated in PBS under a charge balanced, cathodic first biphasic pulse at $100\,\mu C/cm^2$ for Pt at 50 Hz. Under the same stimulation conditions, nanomolded Pt electrodes (b) presented lower voltage excursion than that of their sputtered thin-films (a).

production is observed up to the maximum deliverable current of 800 mA which is equivalent to a charge density of $200\,\mu C/cm^2$.

Based on the results from tested samples, the molded Pt has improved qualities in all aspects of electrochemical performance which suggests they could be the preferred candidates in neural stimulation applications versus widely used thin film electrodes. The experiments also suggest that the molded electrode surface with preserved nanoparticle morphology has increased the nanoscale surface roughness and gained more surface area than that of a smooth surface.

Conclusions

The described nanopowder molding fabrication technology could create high-aspect-ratio solid metal electrodes and wires. This rather simple process provides a possible alternative fabrication method other than existing technologies. Use of this molding technology to make 3D electrode surfaces (protruding bump-like structures) will increase the electrochemical surface area and make it possible to position the stimulating electrode closer to the neurons. The rough surface due to the nanoparticles incorporated on the electrode surface and increased surface area. Initial experiment results suggest that molded structure increases the charge injection capacity therefore minimizing stimulus threshold. For micro-electrodes with high DC resistance, the potential exists for unacceptable heat generation. Lower electrical resistance is also desirable for reducing the heat generation.

Molded Pt electrodes outperform their sputtered thin film counterparts in all aspects of electrochemical properties with lower electrode impedance, higher charge storage capacity, and lower voltage excursion. Batch process is suitable for low-cost mass production of high-density 2D/3D conductive structure. In the foreseeable future, there will be a constant need for high-quality neural stimulation electrodes and trace wires, especially when more and more prosthestic devices are in development. The maturation of this electrode fabrication technology could bring broad benefits and impact to all prospective applications.

Acknowledgments. This research was supported by the DOE Lab 01–04 and the Oak Ridge National Laboratory, managed by UT-Battelle, LLC, for the US Department of Energy under contract DE-AC05-00OR22725. Support from the DOE Office of Biological and Environmental Research is gratefully acknowledged. Authors also thank Anat Burger, Christen Smith, Karolyn Hansen, Charlene Sanders, Elias Greenbaum, and Chase Byers for their generous help and assistance during the completion of this work.

Reference

1. Margalit, E. et al. Retinal prosthesis for the blind, *Survey of Ophthalmology* **47**(4), 335–356, 2002.

2. Weiland, J.D. and Anderson, D.A. Chronic neural stimulation with thin-film, iridium oxide electrodes, IEEE Trans. Biomed. Eng. **47**(7), 911–918, 2000.
3. Slavecheva, E. et al. Sputtered iridium oxide films as charge injection material for functional electrostimulation, *J Electrochem Soc.* **151**(7), E226–237.
4. Seo, J.M. et al. Biocompatibility of polyimide microelectrode array for retinal stimulation, *Materials Sci. Eng.* **C24**, 185–189, 2004.
5. Meyer, J.U. Retina implant – a bioMEMS challenge, *Sensors and Actuators* **A97–98**, 1–9, 2002.
6. Humayun, M.S. et al. *Vision Rev.* **39**, 2569–2576, 1999.
7. Bell, T.E., Wise, K.D. and Anderson, D.J. A flexible micromachined electrode array for a cochlear prosthesis, *Sensors and Actuators* **A66**, 63–69, 1998.
8. Gross, M., Altpeter, D., Stieglitz, T., Schuettler, M. and Meyer, J.U. Micromachining of flexible neural implants with low-ohmic wire traces using electroplating, *Sensors and Actuators* A-Physical **96**(2–3), 105–110, 2002.
9. de Haro, C., Mas, R., Abadal, G., Muñoz, J., Perez-Murano, F. and Domínguez, C. Electrochemical platinum coatings for improving performance of implantable microelectrode arrays, *Biomaterials* **23**(23), 4515–4521, 2002.
10. Marrese, C. Preparation of strongly adherent platinum black coatings, *Anal. Chem.* **59**, 217–218, 1987.
11. Bard, A. and Faulkner, L. in *Electrochemical Methods*, Chapter 1, John Wiley & Sons, 1980.
12. Mortimer, J.T., Kaufman, D. and Roessman, U. Intramuscular electrical stimulation, tissue damage, *Annals of Biomedical Engineering* **8**, 235–244, 1980.
13. Huang, C.Q., Carter, P.M. and Shepherd, P.K. Stimulus induced pH changes in cochlear implants, An *In Vitro* and *In Vivo* Study, *Annals of Biomedical Engineering* **29**, 791–802, 2001.

14
High-Resolution Opto-Electronic Retinal Prosthesis: Physical Limitations and Design

D. Palanker, A. Vankov, P. Huie, A. Butterwick, I. Chan, M.F. Marmor and M.S. Blumenkranz

Department of Ophthalmology and Hansen Experimental Physics Laboratory, Stanford University

Abstract: Electrical stimulation of the retina can produce visual percepts in blind patients suffering from macular degeneration and retinitis pigmentosa (RP). However, current retinal implants provide very low resolution (just a few electrodes), whereas many more pixels would be required for a functional restoration of sight.

This article presents a design of an optoelectronic retinal prosthetic system with a stimulating pixel density of up to $2500\,pix/mm^2$ (corresponding geometrically to a maximum visual acuity of 20/80). Requirements on proximity of neural cells to the stimulation electrodes are described as a function of the desired resolution. Two basic geometries of subretinal implants providing required proximity are presented: perforated membranes and protruding electrode arrays.

To provide for natural eye scanning of the scene, rather than scanning with a head-mounted camera, the system operates similarly to "virtual reality" devices. An image from a video camera is projected by a goggle-mounted pulsed infrared LCD display onto the retina, activating an array of powered photodiodes in the retinal implant. The goggles are transparent to visible light, thus allowing for the simultaneous use of remaining natural vision along with prosthetic stimulation. Optical delivery of visual information to the implant allows for real-time image processing adjustable to retinal architecture, as well as flexible control of image-processing algorithms and stimulation parameters.

Introduction

As the population ages, age-related vision loss from retinal diseases is becoming a critical issue. Two retinal diseases are the current focus of retinal prosthetic work: retinitis pigmentosa (RP) and age-related macular degeneration (AMD).

In these diseases, the "imaging" photoreceptor layer of the retina degenerates, yet the "processing circuitry" and "wiring" subsequent to photoreceptors are at least to some degree preserved. The RP occurs in about 1 out of 4000 live births, corresponding to 1.5 million people worldwide. This disease is the leading cause of inherited blindness. The AMD is the major cause of vision loss in people over 65 in the Western world. Each year 700,000 people are diagnosed with AMD, and 10% of these people become legally blind. Currently, there is no effective treatment for most patients with AMD and RP. However, if one could bypass the photoreceptors and directly stimulate the inner retina with information relaying the visual scene, one might be able to restore some degree of sight.

One important factor affecting this strategy is that the absence of normal signaling from photoreceptors can lead to some progressive degeneration and miswiring of retinal circuitry [1, 2]. This type of degeneration is a general property of neural circuits. Thus, for an electronic implant to properly transmit visual signals to the inner retina, any degeneration of circuitry must not drastically change how these signals are interpreted by the higher brain. This is true in the case of cochlear implants, which bypass degenerated primary auditory sensory neurons; both the nerve and the downstream neural circuitry retain the ability to transmit interpretable auditory information.

Indeed, some first steps have been taken toward the development of an electronic retinal implant. It has been demonstrated that degenerated retina can respond to patterned electrical stimulation in a manner consistent with vision [3–6]. Human patients implanted with an array of 16 (4×4) electrodes of 0.4 mm in size can recognize reproducible visual percepts with patterned stimulation of the retina [3–6]. The patterns perceived by the patients did not always geometrically match the stimulation pattern, which is not surprising knowing the complexity of the retinal spatial organization. However, the one-to-one correspondence between the perceived and the stimulation patterns gives hope that with some learning and image processing the patients might be able to perceive useful visual information from this type of stimulation [7].

A large percentage of patients with AMD preserve visual acuity in the range of 20/400 and retain good peripheral vision. Implantation would be worth its risk for such patients only if it provided substantial improvement in visual acuity. In contrast, patients with advanced RP would benefit little unless the enlargement of the central visual field was enough to allow reasonable ambulation. Normal visual acuity (20/20) corresponds to an angular separation of lines by 1 min [8], which corresponds to spatial separation on the retina of about 10 μm, or in other words, spatial frequency $F = 100$ lines/mm on the retina. To provide such spatial frequency the stimulus pixels should have a linear pixel density at least twice higher: $P \geq 2F$, i.e. two pixels per line (Nyquist sampling theorem). In other words, to resolve two white lines at least one black line should be located in between. Thus the maximal spacing between pixels that will allow for resolving two lines separated by 10 μm is 5 μm. Similarly, spatial resolution corresponding to visual acuity of 20/400 corresponds to a pixel spacing of about 100 μm, while acuity of 20/80 (enough for reading with some visual aids) requires pixels smaller than 20 μm. For these estimates, it is understood that retinal stimulation by one electronic pixel may not produce a perceptual pixel-like "phosphene," and may generate more complex perceptions dependent on the precise number and connections of stimulated cells. What is essential in this analysis is the fact that pixel density determines maximal amount of information or maximal spatial resolution that can be provided by the stimulating array, and thus the best possible visual acuity, if the brain will be able to utilize all this information. Encoding the information, i.e. conversion of the image from the video camera into the map of stimulating signals, is a separate issue.

It has been previously estimated that 625 pixels can suffice for minimally resolving images in a tiny (1.7° or less) central field [9]. For functional restoration of sight a retinal implant should ideally cover a larger field of view – at least 10° (3 mm in diameter), and support a visual acuity of at least 20/80 (corresponding to a pixel size of 20 μm and density of 2500 pix/mm^2) in the central 2–3° of stimulating area. We must emphasize that these numbers are just rough estimates since functional acuity might be increased due to the eye scanning (hyperacuity) or require less pixels for some tasks through pattern recognition. On the other hand, retinal disorganization and the need for learning the new type of stimuli could necessitate more pixels.

Electrical stimulation of neural cells in the retina has been achieved with an array of electrodes positioned on either the inner[3, 9, 10] or the outer side of the retina [11–13]. Setting the electrodes into the subretinal space so as to stimulate bipolar cells, although surgically more complicated, has the potential advantage that signal processing in the retina is partially preserved. Full utilization of this advantage will probably require intervention at relatively early stages of retinal degeneration, before significant remodeling of the retinal neural network takes place[2]. Exciting the ganglion cells with electrodes positioned on the epiretinal side abandons the visual processing by the inner retinal network directly stimulating the output of the retinal circuitry.

One concern with either technique, pertaining to the goal of high-resolution stimulation, is that the electrodes will always be some distance from the target cells. This occurs because the inner limiting membrane and nerve fiber layer intervene in the case of epiretinal placement, or because of photoreceptor remnants in the case of subretinal implantation. In addition, a diseased retina may have an uneven thickness or wavy structure. Large distances between the cells and closely spaced electrodes result in cross-talk between neighboring electrodes, and the need for a higher charge density and power for cell stimulation. This, in turn, can lead to erosion of electrodes and excessive heating of the tissue. Furthermore, any variability in the distance between electrodes and cells in different parts of the implant will result in variations of the stimulation threshold, making it necessary to adjust the signal intensity in each pixel.

As shown below [14] for chronic stimulation with pixel density of 400 pix/mm^2, which geometrically corresponds to visual acuity of 20/200, the electrodes need to be within 15–20 μm of the target neurons. For visual acuity of 20/80, the separation between electrodes and target cells should not exceed 7 μm [14]. *Thus, ensuring a close proximity of cells to the electrodes is one of the most important unresolved issues in the design of a high-resolution retinal prosthesis.* In this article we describe several techniques that may assure proximity of electrodes to the target cells. One of these techniques prompts migration of retinal cells into proximity of stimulating electrodes positioned in the sub-retinal space [15]. During migration the cells preserve axonal connections to the rest of the retina thus maintaining the signal transduction path. Another technique is based on an array of electrodes protruding from the sub-retinal chip [14–16].

A very significant problem with many designs of visual prosthetic systems is that they include head-mounted cameras linked (wirelessly) to the pixels on the patient's retina, so that eye movements are dissociated from vision. This dissociation greatly compromises the process of natural viewing. When the eye scans a scene, each movement is coupled to a strong expectation that the image will change accordingly. In addition, small eye movements during fixation are actually required for image perception: if an image is stabilized on the retina, it fades from perception within 100 ms [17]. In this article we describe the design of a system with a microcomputer-assisted interface and *direct optical projection of the processed image onto photosensitive pixels in the retinal implant using near-infrared light. This system should allow for natural eye scanning and enable the simultaneous use of implant-stimulated vision and any remaining normal vision at any level of luminance.*

Another important aspect of a macular chip design is adjustable image processing. Synaptic connections from foveal photoreceptors radiate out to bipolar and ganglion cells at some distance from the visual center. Thus, an image centered on the foveola will be processed by bipolar and ganglion cells in a circular zone outside foveola. Prosthetic chips will need to have stimulus signals that match this neural anatomy. The system described below includes *location-dependent image processing based on a precise tracking system that monitors the location of the implant in real time.* Stimulation of neurons by the retinal implant differs from natural retinal signal processing. Therefore, to enable the translation of stimulus patterns into the conscious recognition of objects, visual chips may require some form of image processing and neural "learning", much as is required by modern cochlear implants. Tracking the implant in real time allows for the position-dependent image processing that may be required to translate visual information into electrical signals that can be properly interpreted by the higher brain.

In the article below, we describe a system that addresses all three issues raised above: (1) proximity of electrodes to the target cells, (2) delivery of information to the retinal implant linked to the natural eye movements, and (3) location-dependent image processing.

Proximity between Electrodes and Cells as a Resolution-limiting Factor

Delivery of high spatial frequencies via thousands of small electrodes is possible only if each of them will be affecting cells only in a close proximity – a zone similar to the size of the electrode. However, simultaneous injection of current from a dense array of small monopolar electrodes having a remote return will result in interference of the local electric fields producing a very far-extending field equivalent to an effect of one electrode similar in size to the whole array, i.e. the field will extend to a distance of about 1.5 mm into the tissue if an array of 3 mm in diameter is used. Such interference can be avoided using sequential activation of the electrodes; however, it does not seem to be a practical solution for the high resolution implants. Pulses of 1 ms in duration applied at repetition rate of 25 Hz (video rate) provide a duty cycle of 1/40. Thus not more than 40 electrodes can be activated without a temporal overlap. Thus, for an implant containing thousands of electrodes, sequential activation is not possible. More practical solution of the cross-talk problem is having the coaxial return electrodes around each pixel in the array. These return electrodes will localize the fields produced by different electrodes and thus isolate their effects from one another, and allow for simultaneous activation. For this reason, in the discussion below, we consider an array where each microelectrode is surrounded by the return electrode.

Voltage and Current Required for Cell Stimulation

We will use a simple passive model of extracellular stimulation [18] of the neural cell, which allows for analytical solutions, and thus markedly simplifies assessment of various geometrical and electrochemical effects.

The typical resting potential of a neural cell in the range from -60 to -70 mV and depolarization of the cell membrane by a few mV are sufficient to affect ion channels and thus elicit a physiological response, including graded potential neurons such as bipolar retinal cells [19, 20]. With extracellular stimulation a change in a cross-membrane potential is induced by application of an electric field to the surrounding medium. Since the impedance of the cellular membrane is much higher than that of the cellular cytoplasm, the interior of a cell quickly polarizes in the external electric field and its cytoplasm equipotentializes [18], as shown in Figure 14.1. Thus the cross-cellular potential actually charges the cellular membrane on two poles of the cell: polarizing the anode side and depolarizing the cathode size. In a uniform electric field these potential steps are equal, i.e. $\Delta U_{membr} = \Delta U_{cell}/2$, but in a non uniform field, which is a case with small electrode in front of a large cell, only a small area of the cell membrane is affected thus producing much larger cross-membrane voltage drop on a proximal end than on a remote end of the cell, so at the proximal end $\Delta U_{membr} \approx \Delta U_{cell}$. The polarization time constant $t_p = 0.5L \cdot c_{mem} \, (g_{cyt} + g/2)$, where $c_{mem} \approx 1 \, \mu F/cm^2$ is the membrane capacitance per unit area [21], g_{cyt} is the cytoplasm resistivity

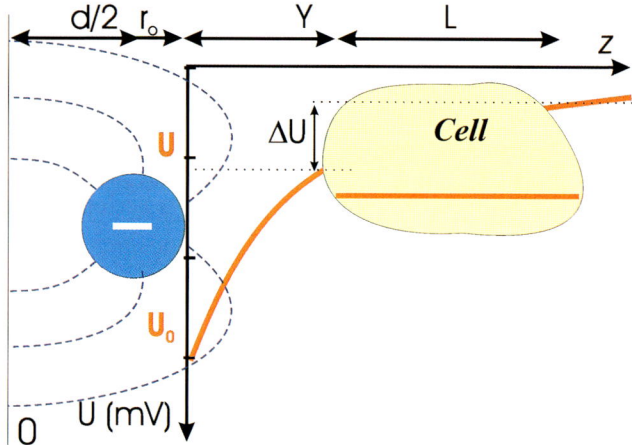

FIGURE 14.1. Schematic representation of a cell in a dipole electric field. Cell of the width L positioned at distance Y from the surface of an electrode with radius r_o. The cross-cell voltage is ΔU. Surface at 0 potential is the return electrode located at the center of the dipole – at distance $d/2$ from the center of the electrode.

(220 Ohm cm [22]), and g. is the resistivity of the extracellular medium (100 Ohm cm [18]). For a cell soma of $L = 10\,\mu$m in width the polarization time is very short: $t_p \approx 130$ ns. However, it is significantly higher in the distributed structure of narrow neural processes having significant internal resistance [23].

Typical retinal neural cells extend their processes over several tens and sometimes even hundreds of microns, thus greatly exceeding the dimensions of the electrodes in a high-density implant ($20\,\mu$m pixel spacing is required for maximal visual acuity of 20/80). The dipole electric potential is reciprocal to the square of distance and thus is negligible at distances much larger than the size of the dipole. So the cross-cellular potential of large cells in the dipole field is practically determined by the potential on their proximal membrane. Assuming the dipole having a length d and a potential of U_0 on its electrode of radius r_0 (in the liquid), the potential U at distance Y in front of it will be

$$\Delta U_{\text{mem}} = U = U_0 r_0 \left(\frac{1}{Y + r_0} - \frac{1}{Y + r_0 + d} \right) \tag{1}$$

With a resistance of the solution

$$R_{\text{s}} = \frac{\gamma}{4k \cdot r_0} \tag{2}$$

where $k = p$ for a sphere, $p/2$ for a hemisphere and 1 for a disk electrode, the current will be

$$I_{\text{th}} = \frac{U_0}{R} = \Delta U_{\text{mem}} \frac{(Y + r_0)(Y + r_0 + d)}{r_0 \cdot d \cdot R} = \Delta U_{\text{mem}} \frac{4k(Y + r_0)(Y + r_0 + d)}{\gamma \cdot d} \tag{3}$$

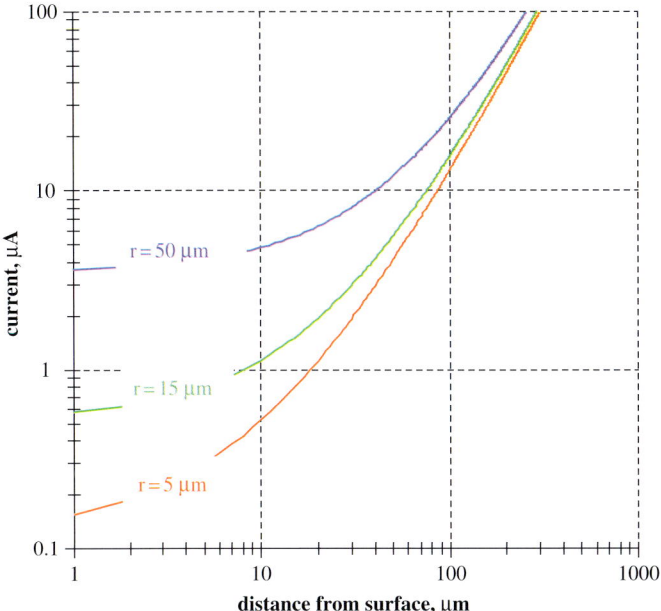

FIGURE 14.2. Threshold current (in μA) required for generation of a 5 mV cross-membrane voltage drop on a proximal surface of a large cell, plotted as a function of distance between the cell and the electrode surface. Calculated for electrodes of 5, 15, and 50 μm in radius.

From the experimental values of the threshold current required for retinal neural cell stimulation – $I_{th} = 0.4$–$3\,\mu A$ (at pulse duration 1 ms)[24] measured with a disk electrode of $r_0 = 62.5\,\mu m$ in diameter in a very close proximity to cells ($Y \approx 0$) and having a remote return electrode ($d \gg r_0$) – we can deduce the $\Delta U_{mem} = I_{th} \cdot g/4r_0 = 1.6$–$12\,mV$. The current required for depolarization of the cell membrane by ΔU_{mem} rapidly increases with the separation of a cell from the electrodes (Y), asymptotically reaching quadratic dependence, as shown in Figure 14.2. If this separation varies in different parts of the array, the stimulation threshold (and the visual outcome) will vary accordingly. *Close and stable proximity of cells to the electrodes is thus an important issue in the design of a high-resolution retinal implant.*

Tissue Heating

Electric current in a conductive medium generates Joule heat. Since the threshold voltage and current strongly depend on distance between electrodes and cells so will the heating. For single pulses at the threshold of neural stimulation this temperature rise is negligible – below $10^{-2}\,°C$. However, with application of repetitive pulses to a large array of electrodes heat might accumulate, and thus a regime of chronic stimulation requires careful consideration. Average power dissipation can be estimated as $P = \Delta U \cdot I_e \cdot N \cdot \tau \cdot \nu$, where ΔU is the voltage applied to the implant electrodes, I_e is a current from one electrode, N is number of electrodes on the implant, τ is pulse duration, and ν is a pulse repetition rate. The total voltage step ΔU is composed of three principal parts: (1) resistive loss in the solution R_s, described above, (2) pseudocapacitive voltage step in the solution, and (3) "serial resistance" of the electrode composed of resistance of charge transfer and of the metal oxide layer.

"Pseudocapacitive" behavior of the electrochemical reduction–oxidation reactions on IrOx and Pt electrodes [25, 26] is due to electrochemical storage of charge at the surface of the electrode during each phase of the pulse. An upper estimate of the energy losses can be given assuming that there is no energy "recycling", i.e. all the energy stored during each phase of the bi-phasic pulse is converted into heat during the opposite phase. In reality, some part of the stored energy is "recycled", but it is quite difficult to accurately estimate the reversible component of these rather complex electrochemical processes. Pseudo-capacitive voltage step $\Delta U = q/c_p$, where pseudocapacitance of the interface $c_p = j/(\mathrm{d}V/\mathrm{d}t)$ is about 5 mF/cm^2 for IrOx, as measured with 0.5 ms pulses [26]. This value is close to the ratio of the maximal charge density q_{max} to the corresponding maximal voltage U allowed for reversible electrochemical reactions on IrOx: $q_{max} = 4\,\mathrm{mC/cm}^2$, $U = 0.8\,\mathrm{V}$ [26], thus $c_p = 5\,\mathrm{mF/cm}^2$. For Platinum (Pt), $q_{max} = 0.4\,\mathrm{mC/cm}^2$ [27], and at similar voltage $c_p = 0.5\,\mathrm{mF/cm}^2$.

The charge transfer resistance and the resistance of the oxide layer for uniformly accessible hemispherical electrode are reciprocal to the electrode surface area [28], i.e. R_{ser} scales with the electrode radius r_o as $R_{ser} \sim 1/r_o^2$, but for the disk electrode, where current density might be distributed non-uniformly, the scaling law might be more complex [29]. For IrOx electrodes 36 μm in radius, $R_{ser} = 3.1\,\mathrm{kOhm}$ (estimated from the data in [26]).

The number of electrodes, N, on a 3 mm diameter implant can be estimated assuming that a pixel width is four times the radius of electrode. In this case, $N \approx 17, 500, 1900$, and 170 for electrodes of 5, 15, and 50 μm in radius, respectively. Assuming an average current on each electrode corresponding to the results of the appropriate electrode size as plotted in Figure 14.3, the pulse duration being $\tau = 1\,\mathrm{ms}$ (0.5 ms per phase), and repetition rate $\nu = 25\,\mathrm{Hz}$, the average power dissipation from the implant with IrOx electrodes is plotted in Figure 14.4.

When a continuous train of pulses is applied, the heat diffusion into the surrounding liquid creates a steady distribution of temperature. Under the steady-state conditions in the spherically symmetric case the temperature decreases reciprocal with the distance, i.e. it drops by 50% at the distance similar to the radius of the array. Thus for practical purposes the cells within a few hundred microns of an implant will be subject to temperatures similar to those at its surface. The power dissipation from the disk-shaped array of electrodes of diameter D surrounded by liquid with thermal conductivity λ having the ambient temperature at infinity is [30]:

$$P_{heat} = 4\lambda \cdot \Delta T \cdot D \tag{4}$$

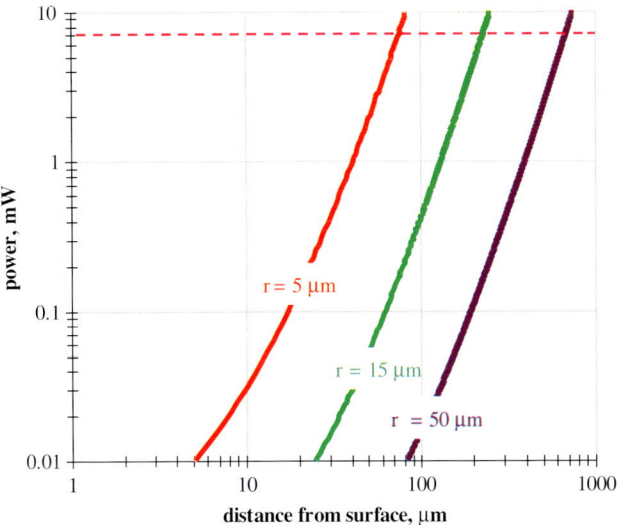

FIGURE 14.3. Average power dissipating in liquid and on the IrOx electrodes of 5, 15, and 50 μm in radius with stimulating pulses at the level plotted in Figure 14.2. Implant diameter is 3 mm, pulse duration 1 ms and rep. rate of 25 Hz are assumed. Maximal tolerable power dissipation of 7 mW, corresponding to a maximal temperature rise of 1 °C, is shown as a dashed horizontal line.

(This solution assumes equal thermal conductivity on both sides of the chip.) Assuming the size of an electrode array being $D = 3$ mm, λ being a thermal conductivity of water (0.58 W/m K), and maximal temperature elevation ΔT being 1 °C results in $P_{heat} \approx 7$ mW. This limit on power dissipation by the implant translates into a limitation on the number of electrodes as a function of distance between them and the target cells. As shown in Figure 14.3, for electrodes of 5 μm in radius this limit is reached when cells are separated from the implant by about 75 μm. With larger electrodes this effect is even less restrictive: 230 and 700 μm separation is tolerable for electrodes of 15 and 50 μm in radius.

Electrochemical Limitations

Current across an electrode–electrolyte interface can be produced by two mechanisms: (i) charging/discharging of the electrical double layer, known as capacitive coupling, and/or (ii) electron transfer due to electrochemical reactions at the electrode surface, known as faradaic process. The capacitive coupling between the electrode and the medium, i.e. transmission of the current by repetitive charging and discharging of the electrical double layer, has an advantage that it does not involve any electrochemical reactions that could affect the tissue. However, typically the capacitance of the double layer on metal electrodes (on the order of 0.01 mF/cm^2) is much lower than what is needed for cellular stimulation. Materials with high surface area, such as carbon black [31] or TiN [26, 32] can, in principle, strongly increase capacitance of the double layer—up to $0.95\,mF/cm^2$ [26], but may have a slow dynamic response.

Faradaic process can provide higher charge values. However, for chronic stimulation the electrochemical reactions should (a) not cause noticeable decomposition of the electrode material, and (b) its products should not be cytotoxic. Oxidation/reduction of Iridium (Ir) oxide on the surface of an Ir electrode satisfies both criteria, and such electrodes have been extensively discussed in literature [26, 33–35]. The Pt electrodes have also been used for cell stimulation [3, 27]. The iridium oxide injects charge by a fast, reversible faradaic reaction involving reduction and oxidation between the Ir3+ and Ir4+ states of the oxide, with the exchange of charge balancing counter-ions with the electrolyte [33]. Unlike Pt, these redox reactions are confined to the oxide film, with no generation of soluble species required to transfer charge. Accordingly the maximum amount of charge per unit area that can be passed without corrosion and gas formation for IrOx electrodes is $4\,mC/cm^2$ [26, 35], while for Pt it is much lower: $0.4\,mC/cm^2$ [27, 36]. Some studies suggest even more conservative limit of $0.15\,mC/cm^2$ [37].

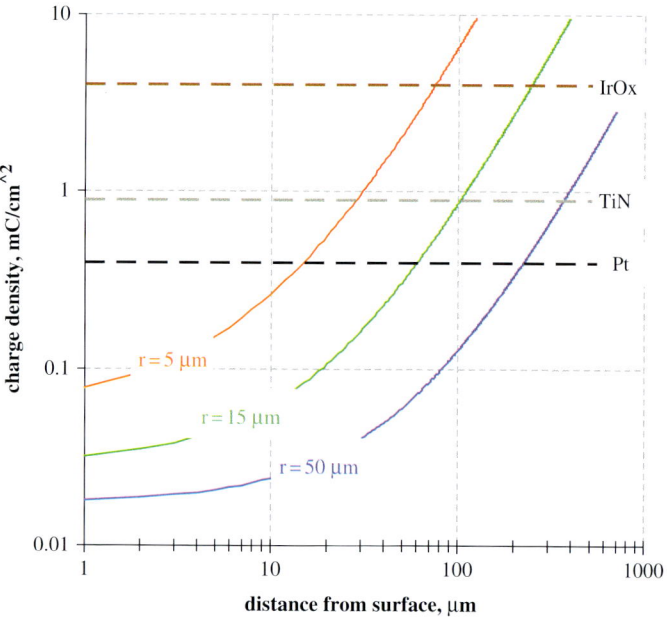

FIGURE 14.4. Threshold charge density on electrode with a pulse duration of 0.5 ms/phase, plotted as a function of distance between the cell and the electrode surface. Calculated for electrodes of 5, 15, and 50 μm in radius. Maximal safe charge densities for Pt, TiN, and IrOx electrodes are shown as horizontal lines.

Current density on the surface of the electrode is

$$j_{th} = \frac{I_{th}}{S} = \Delta U_{mem} \frac{(Y + r_0)(Y + r_0 + d)}{\gamma \cdot d \cdot r_0^2} \tag{5}$$

It is limited by the electrochemical tolerance of the electrode material. As shown in Figure 14.4 (for a $\Delta U_{mem} = 5\,\text{mV}$), this limits the distance between electrodes and cells as a function of the electrode size and material. For example, the Pt electrodes of $10\,\mu\text{m}$ in diameter can only be used at distances up to about $15\,\mu\text{m}$ from the cells. IrOx electrodes of the same size can operate safely at distances up to $70\,\mu\text{m}$. Larger electrodes are less restricted by electrochemical safety.

Divergence of Electric Field

Electric fields (and the associated electric currents) emanating from the stimulating electrodes diverge, i.e. they spread with the distance. Thus the dynamic range of targeted stimulation, i.e. the ratio of the electric field in front on an electrode to electric field produced by the neighboring electrodes, decreases with the distance from the array, as illustrated in Figure 14.5a. Electric potential of a spherical electric field, characteristic for a monopolar electrode having a return electrode at infinity, can be described as following:

$$\varphi \propto \frac{1}{r} = \frac{1}{\sqrt{z^2 + x^2}} = \frac{1}{z\sqrt{1 + (x/z)^2}} \tag{6}$$

The width of its normalized lateral distribution (along the x axis) increases linearly with the distance z from the plane of the array. For example, the full width at half maximum of this distribution: $X = 2\sqrt{3} \cdot z \approx 3.46z$, i.e. the electric potential at a distance of $X = 3.46z$ is half of the maximum value. A dipole electric field, approximating a coaxial pair of electrodes located in close proximity, is more directional: $\varphi \propto \frac{\cos \alpha}{r^2} = \frac{z}{r^3} = \frac{1}{z^2(1+(x/z)^2)^{3/2}}$, but it still diverges linearly with the distance z. The full width at half maximum of this distribution, $X - 2\sqrt{2^{2/3} - 1} \cdot z \approx 1.53z$, while the full width at 1/10th of the maximum, $X = 2\sqrt{10^{2/3} - 1} \cdot z \approx 3.81z$; i.e. cells located at distance z in front of one electrode will be affected by the neighboring electrode by 1/10th of its maximal value if the electrodes are separated by the distance $X/2$. Divergence of the electric field increases the area affected by each electrode and thus reduces the maximal achievable resolution. This effect, analogous to out-of-focus imaging, filters out the higher spatial frequencies in the stimulation pattern thus limiting the spatial resolution, as shown in Figure 14.5b, c. The lower the level of tolerable interference from neighboring electrodes (determining the dynamic range of targeted stimulation), the wider should be the pixel spacing, effectively lowering the resolution. For example, to resolve spatial frequency of 25 lines/mm (geometrically corresponding to maximal visual

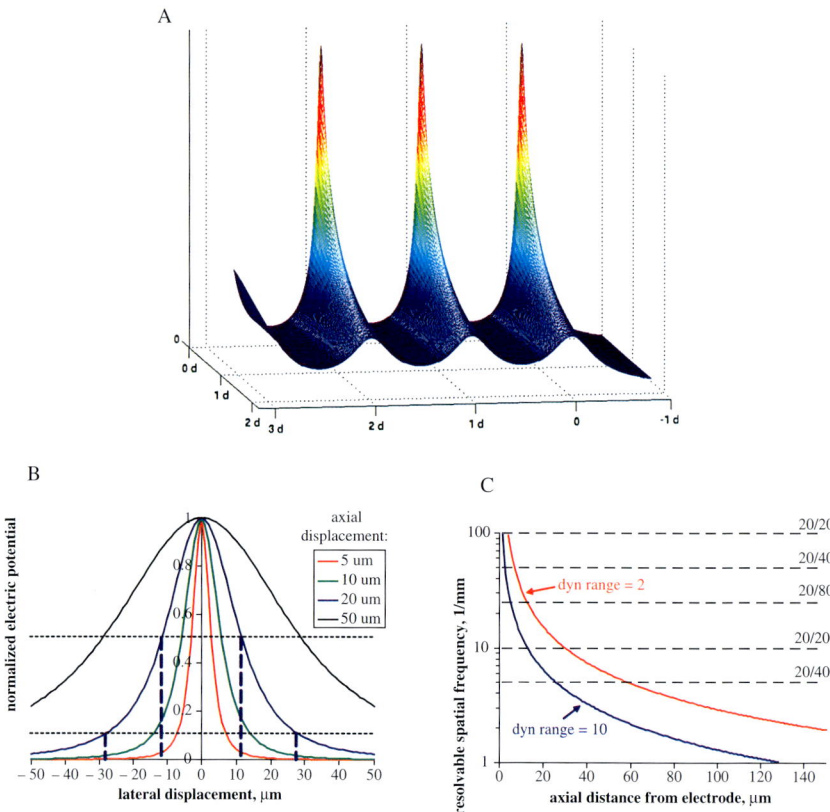

FIGURE 14.5. (a) Distribution of the electric potential produced by an array of dipoles. Dynamic range of targeted stimulation (ratio of peaks to troughs) decreases with the distance from the array. (b) Broadening of the dipole electric field with the distance from the array: normalized electric potential of a dipole at 5, 10, 20, and 50 μm from the electrode. (c) Maximal resolvable spatial frequency as a function of the distance from the stimulating array. Plotted for the dynamic range of 2 (threshold at half-maximum) and 10 (threshold at 1/10th of a maximum). For comparison, the spatial frequencies corresponding to various levels of the visual acuity are plotted on the same scale.

acuity of 20/80) with a dynamic range of 10, a proximity of 5 μm is required, while with the dynamic range of 2, the cells can be withdrawn up to 14 μm from the array. For a spatial frequency of 10 lines/mm (geometrically corresponding to maximal visual acuity of 20/200) the requirements for proximity are a little more relaxed: 13 and 33 μm, respectively. Achicving such a close proximity between electrodes and cells along the whole surface of the stimulating array is one of the major challenges for a high-resolution retinal stimulation system.

Attracting Retinal Cells to Electrodes

Migration of Retinal Cells into Perforated Membrane

We have discovered a robust property of retinal tissue that promises to reliably provide an extremely close proximity of the retinal cells to the implant, thus allowing high-resolution electrical stimulation. In experiments in vitro, in which retinas were placed photoreceptor-side down upon the membrane, a robust migration of retinal tissue into small apertures was observed in all samples of the rat, chicken, and rabbit retina [15]. Migration of the outer nuclear layer, outer plexiform layer, and inner nuclear layer occurred through apertures larger than $5\,\mu m$ (Figure 14.6). The cellular invasion of the aperture appeared to include both glial and neural cellular elements, and the rate of tissue migration increased with aperture size. A transmission electron micrograph of a section through an aperture demonstrated the presence of neuronal processes and synaptic structures connecting the migrating cells. These findings indicate the possibility of signal transduction from the stimulated cells to rest of the retina [14–16]. Culturing of the retina upside down (tested on the P7 rat retina), i.e. nerve fiber layer toward the membrane, did not result in cellular migration, as shown in Figure 14.6.

The RCS rats were used as a model for in vivo experiments, since their photoreceptors degenerate as in RP. Experiments with subretinal Mylar films perforated with apertures of $15–40\,\mu m$ in diameter showed robust migration of the inner nuclear layer after 5 and 9 days [14]. Since unlimited tissue migration through a membrane could be problematic (draining retinal cells and proliferating

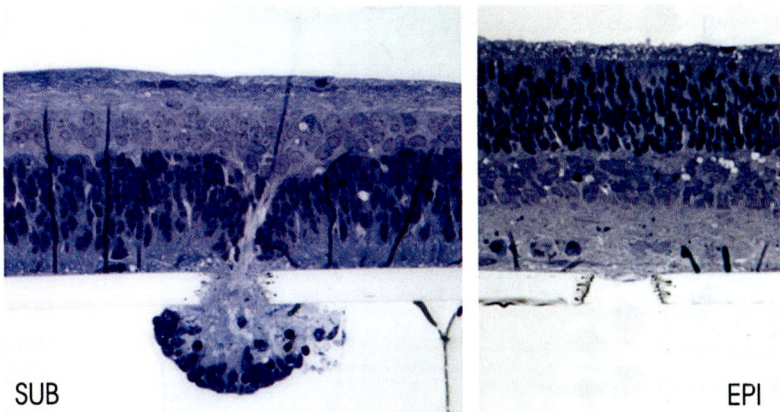

FIGURE 14.6. Histological sections of the RCS rat retina 3 days after explantation onto a perforated 13-μm-thick Mylar film. (Left) Retinal tissue placed photoreceptors down onto the membrane migrates through the aperture. (Right) Retina placed with its nerve fiber layer onto the perforated membrane does not migrate even when the ILM was removed.

under the prosthesis), we explored the placement of perforated membranes with a basal seal to prevent growth out the bottom. These experiments were performed in vitro with cultured rat retinas and in vivo with the RCS rat retinas. In the experiment illustrated in Figure 14.7, the 10 and 20 μm perforations lay atop the 40 μm chambers, with their bottoms sealed with a membrane. When retinas

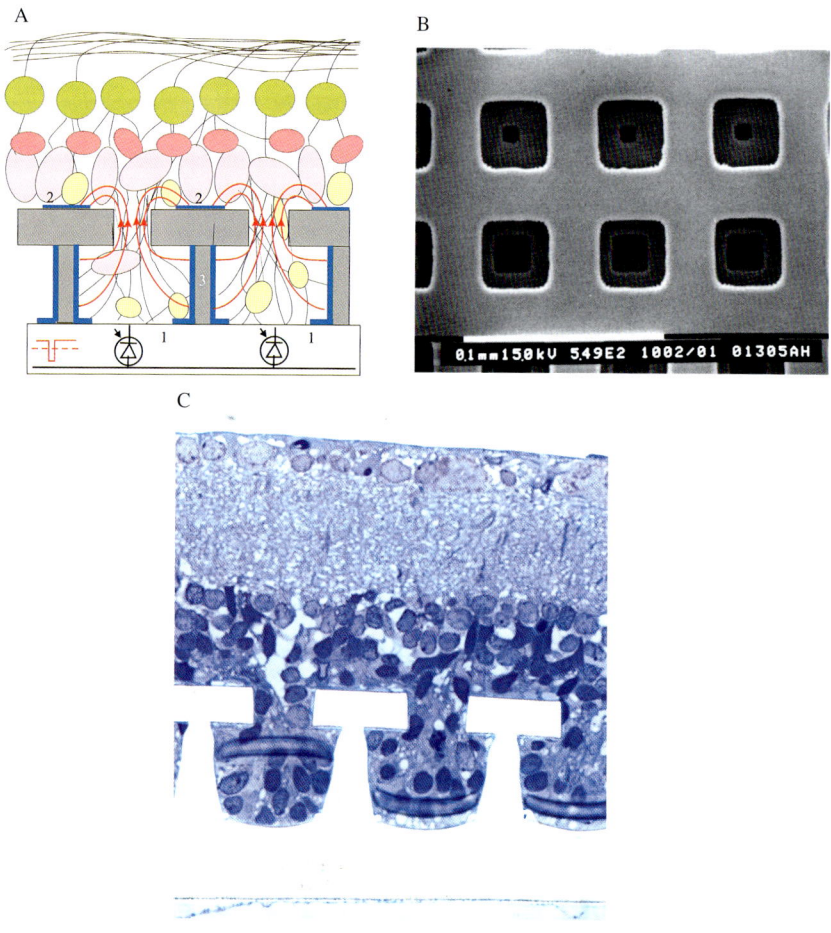

FIGURE 14.7. (a) Schematic of a 3-layered membrane with entry channels on top, wider inner chambers, and a closed at the bottom. Voltage can be applied between the inner electrode (1) and the common return electrode (2). (b) View from below at the lithographically fabricated chamber array with the entrance apertures of 10 and 20 μm in width and chambers of 40 μm in width. (c) RCS rat retina grown into the chamber array 15 days after the implantation. Retinal tissue migrated through the holes into the chambers of 40 μm in width. This way the apertures achieve very close proximity to the cells in the Inner Nuclear Layer. The amount of tissue migrating into the chambers can be adjusted by controlling the depth of the chambers.

werc cultured over this 3-layer structure for 7–14 in vitro, as well as implanted into subretinal space in vivo, tissue was observed to migrate into the chambers in both cases (Figure 14.7c).

Major concerns are whether the neural cells that migrate into the pores will survive for an extended period of time, whether the neural circuitry will be disrupted and whether the migrated tissue will change through glial overgrowth or cell death. The long-term behavior of retinal cells migrating into perforated membranes is currently under detailed investigation to optimize the membrane structure for preserving neural connections and assuring efficacy of an electric interface. It is important to emphasize though that for efficient stimulation the target cells should be in proximity of the apertures, where the current density is highest, but could be above the apertures rather than inside the chambers.

Migration Around Protruding Electrodes

Another promising technique for providing close proximity between the neural cells inside the retina and the stimulating sites of the implant involves protruding electrodes. As diagrammatically shown in Figure 14.8a, stimulating electrodes (1) would extend by several tens of micrometers above the surface of photodiodes and be exposed only at the top of the "pillars," with a common return electrode (2) on the surface of the wafer. This array would be positioned in the sub-retinal space, so that cells could migrate into the empty space between the pillars, similarly to the migration we observed with the perforated membrane. This way the electrodes will penetrate into retina without mechanical stress and associated injury. The depth of penetration is determined by the length of the pillars and their density. The pillars can be manufactured using conventional photolithographic technology. An array with pillars of $70\,\mu m$ in height and $10\,\mu m$ in diameter lithographically fabricated from SU-8 photoresist is shown in Figure 14.8b. Such arrays having various pillar density have been implanted into the subretinal space in adult RCS rats. Histology performed on eyes enucleated 15 days after the implantation is shown in Figure 14.8c. As one can see in this figure, the retina is well preserved with the pillars that penetrated into the inner nuclear layer. On the left end, where the pillar spacing is wider, there is more volume for the tissue migration, the retina rests lower and the pillars penetrate deeper into the tissue.

Delivery of Information and Power to the Implant

Projection System

The projection system is designed to allow natural eye scanning for image perception, flexibility of image processing between the camera and the implant, and utilize any remaining natural vision and neural function. The system controls

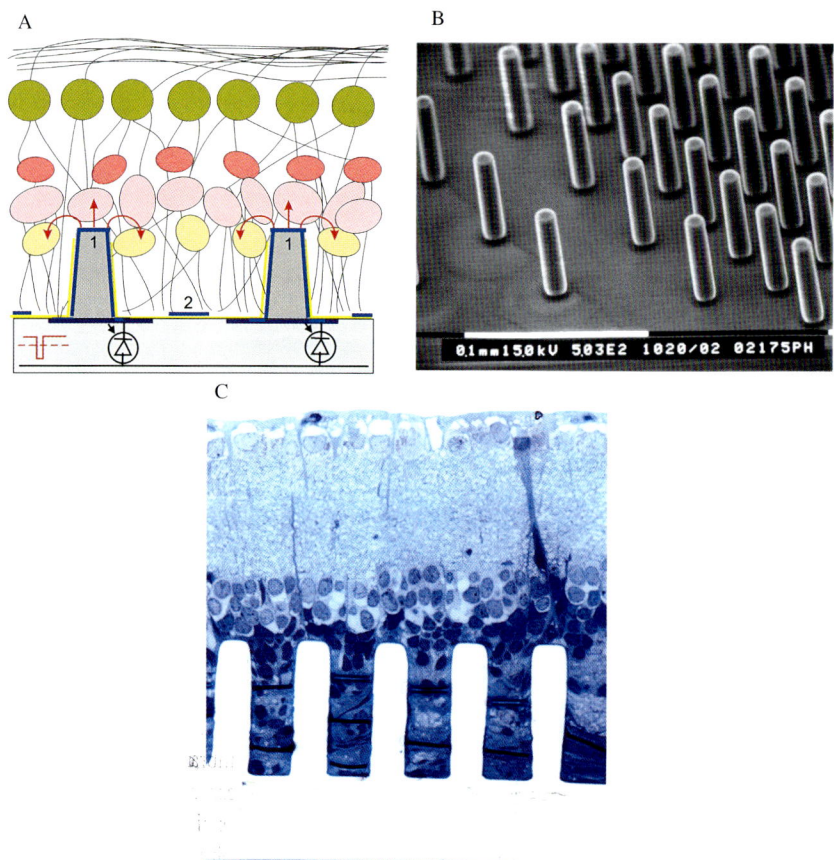

FIGURE 14.8. (a) Concept of protruding electrodes on the sub-retinal array penetrating deep into the retina after migration of the retinal cells into the empty spaces between the pillars. Penetration depth is set by the length of the pillars, which are insulated at the sides and exposed at the top. (b) Lithographically fabricated array of pillars of $10\,\mu m$ in diameter and $70\,\mu m$ in height having pillar spacing varying between 20 and $60\,\mu m$ (center-to-center) on various parts of the implant. (c) Lithographically fabricated 10-μm-wide pillars penetrating into the inner nuclear layer of the RCS rat retina 15 days after the implantation. Horizontal dark lines between the pillars are histological artifacts due to folding.

the stimulating signal in each pixel by projecting light from the goggles-mounted LCD display onto a retinal implant having an array of powered photodiodes, as diagrammatically shown in Figure 14.9. An image from the small video camera located on the patient's goggles is processed using a portable microcomputer. The processed image is displayed on the LCD micro-display similar to those used for "virtual reality" imaging systems (medical, military, etc.). The LCD display will emit pulsed near-infrared (IR) light (800–900 nm). The IR image

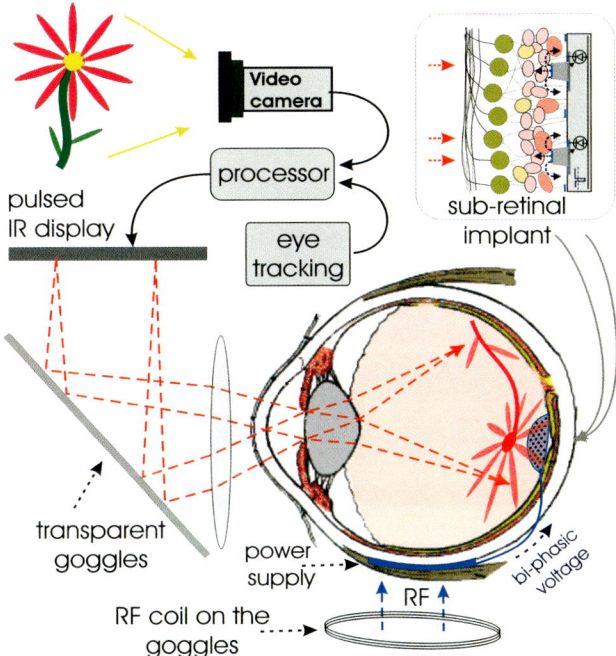

Figure 14.9. Diagram of the prosthetic system including video camera, image processing unit, and IR image projection from the goggles display onto the retina. Part of the image which is projected onto retinal chip activates its photo-sensitive stimulating pixels using inductively–coupled radiofrequency–driven pulsed power supply.

from the display is reflected from the transparent goggles and projected onto the retina using natural optical properties of the eye, as shown in Figure 14.9. The projected IR image is thus superimposed onto a normal image of the world observed through the transparent goggles. The retinal chip has an array of photosensitive elements converting the pulsed IR light into stimulating current in each pixel using an intraocular pulsed bi-phasic power supply, as described below. Advantages of this system as compared to current approaches to visual prosthesis include the following:

– The video display projects an image corresponding to the visual field of about 30°, which is much larger than the retinal chip (about 10°). With this system, the patient can use his natural eye movements in order to observe the larger field of view, rather than scanning it by moving his head-mounted video camera.
– Optical transmission of information from the LCD screen to all the pixels in the chip is conducted simultaneously, i.e. pixels are activated in a parallel fashion, and there is no need for serial decoding in the implant multiplexer, as it is done for the single emitter–receiver links (either optical [38] or radio frequency [5, 6]).

- The IR video display on the goggles can emit as much power as the eye can thermally tolerate thus providing a robust signal to each pixel at any level of ambient illumination.
- The implant-stimulated vision is provided simultaneously with the residual natural vision in the areas outside the retinal implant. The infrared projected image is not detected by photoreceptors. Conversely, the implant's response to natural visible light in the eye is negligible compared to the bright and pulsed infrared image.
- Intensity, duration, and repetition rate of the stimulating signal produced by the retinal chip can be controlled by the intensity, duration, and repetition rate of the light-emitting pixels in the LCD screen. These parameters can be adjusted without need for any changes in the retinal chip itself. This feature provides flexibility in optimization of the stimulation parameters and image-processing algorithm, which might have to be adjusted for each patient.
- This projection system can be used for both epiretinal and subretinal implants.

As described above, the stimulation current for an electrode of 10 μm in diameter is on the order of 2 μA. The photodiode converts photons into electric current with efficiency of up to 0.6 A/W, thus 3.4 μW of light power will be required for activation of one pixel. If light pulses are applied for 0.5 ms at 25 Hz, the average power will be 42 nW/pixel. With 18,000 pixels on the chip, the total light power irradiating an implant will be 0.75 mW.

The LCD screens used in video goggles emit light into a wide angle, and only a small fraction of it (typically $<1\%$) reaches the retina, while most of it is absorbed by the sclera and iris. In addition, only a small part of the retina (about 5%) is covered by an implant. To provide 1 mW of light on a 3 mm retinal implant, the LCD goggles should in total emit about 2 W of light power! This is certainly not practical.

This problem can be resolved by addressing both aspects of the loss of light: (1) providing a collimated illumination, and (2) activating only a small part of the screen – that which is projected onto the implant, position of which will be constantly monitored with a tracking system. The pulsed illumination for the implant will be provided by the near-IR LED array or laser diode illuminating the LCD with collimated beam. A condenser lens directs the main axis of the diodes into the center of the eyeball. Assuming no magnification between the screen and the retina, the diameter of the light spot on the pupil will be $D = d_{chip} + \alpha L$, where d_{chip} is the implant size, $\alpha = 8°$ is the divergence of the LED beam, and $L = 17$ mm is the distance between the implant and the pupil. With these assumptions the spot of light on iris can be as small as 5.3 mm in diameter. To provide for a eye scanning range of 30°, a spot size on the iris should be actually a little larger – about 9 mm in diameter, thus with the pupil of 3 mm in diameter, 10% of light will be transmitted into the eye, thus only 10 mW of power would be required from the LCD screen.

Tracking System

Since the eye is frequently moving, the proper activation of the display requires information about position of the retinal chip at each moment. With the angular size of the retinal implant being 10°, 1° of precision in tracking will suffice for the purpose of limiting the amount of IR light on the retina by emitting only from a part of the display which is projected on the implant. Such precision can be provided with one of the standard pupil tracking systems, which are capable of detecting the direction of gaze with accuracy of up to 0.25°.

For a more advanced image processing the pixel-specific adjustments might be required, which requires a significantly more precise link between the display and the implant. To be able to address a specific pixel on the moving implant, precision of tracking should not exceed the pixel size (20 μm, or 0.06°). During fixation, the human eye drifts at an average angular velocity of 0.5 deg/s, up to 2 deg/s [39]. Such precision and tracking rate can be achieved with retinal tracking [40]. However, during saccades, large ballistic movements, the eye can move with a velocity of hundreds of degrees per second, and tracking most probably will be lost at these moments. On the other hand, normal visual perceptions greatly decrease during saccades, especially sensitivity to motion, so real-time image processing will not be required at these rates.

The retinal tracking system can monitor positions of a few reference points on the retinal implant. The reference points reflect (or emit) light back through the pupil and are imaged onto the tracking array that is positioned in the plane conjugated to the LCD display. For tracking of the location and the torsional orientation of the retinal chip only 2 reference points on its surface should be sufficient. More reference points may provide higher reliability and precision in localization of the chip. Reference points on the implant can be made as back-reflectors, for example, in the shape of 3-sided pyramidal indentations with highly reflective walls. They also may be made as small LEDs emitting a wavelength or temporal pattern different from that emitted by the image display. This will allow for discrimination between the emission by the reference points and the scattering from elsewhere in the eye.

Optoelectronic Implant Design

As described above, proximity of retinal cells in the inner nuclear layer to the stimulating electrodes can be achieved by promoting cellular migration into the sub-retinal implant. One possible design of a sub-retinal photosensitive stimulating array that takes advantage of this effect is shown in Figures. 14.7 and 14.8. A wafer of about 15–25 μm in thickness is divided into separate photosensitive pixels similarly to a CCD array. Each pixel is a biased photodiode which converts local light intensity into bi-phasic charge-balanced pulses using a commonbi-phasic power line. Current in the negative phase of the waveform is proportional to the light intensity. Current in the positive phase passes through the diodes freely and provides the charge balance compensation for the electrodes. In experiments with such circuits, we verified preservation of the charge balance on Pt electrodes with precision better than 0.01% independently of the light intensity in individual pixels. The stimulating bi-phasic pulse conducted through the photo-diodes is applied to either the inner electrode in the cavity (1) or to the pillar electrode exposed at its tip (1), while the return electrode (2) is common to all pixels. The current from the inner electrode in each chamber is concentrated inside the aperture and thus the cells located near this bottleneck will be affected by electric field the most. The stimulation zone can be extended up and around the aperture by positioning the return electrode (2) away from the edge of the aperture.

As shown in Figure 14.3, the optoelectronic prosthesis having 18,000 pixels with electrodes of $10\,\mu m$ in diameter can consume about $0.2\,mW$ of power if the target cells are located within $20\,\mu m$ from the array. This power can be provided by a pulse bi-phasic voltage generator driven by the inductive RF transmission of energy from the coil located on the goggles into a coil located under the conjuctiva or inside the eye [41]. Power supply will provide positive voltage between the pulses and the negative voltage during the light pulse, as diagrammatically shown in Figure 14.9.

Location-Dependent Image Processing

The central part of the macula (fovea) does not contain bipolar and ganglion cells. Photoreceptors in this area radiate their synaptic connections outside the fovea (Henle's fibers), to a distance of about $0.5\,mm$ in diameter. Thus if the retinal chip covers the macula, the image should be processed so that stimuli are delivered to the bipolar or ganglion cells outside the foveola. This means that an image projected onto the retinal stimulating array should have a black spot in the foveola (since there are no cells to stimulate in that area), and the rest of the macular area should be stretched in a radial pattern matching the retinal organization in the macula. The fact that retinal architecture is non-uniform around the center of the macula necessitates image processing that depends on position of the foveola relative to the projected image, i.e. it depends on direction of gaze.

There are several additional reasons necessitating the position-sensitive image processing for the visual chip controlled by the optical projection system:

– Various pixels in the array may have different impedances (due to the tissue growth or electrode contamination) or different distances from the cells and thus may require different pulse intensity.
– The retina has a complex system of image processing involving intertwined patterns of "ON" and "OFF" cells with large receptive fields having center-surround organization. For correct transmission of the image there might be a need for the stimulation pattern matching this cellular organization. In addition, there might be a need for temporal variation of the stimulation of the neighboring pixels [42]. (We must state, however, that many visual neurons have rectified responses, reacting similarly to either sign, and thus a number of aspects of visual processing might not be substantially affected.

For example, many ganglion cells are of the "ON-OFF" type, responding to either increases or decreases in light intensity. Additionally, in the retinal pathway that distinguishes moving objects from background motion, signals are rectified, and exchanging black for white does not change the firing patterns of object motion–sensitive retinal ganglion cells [43]. Adjustment of the image-processing algorithm for humans would be performed based on communication with the patient [44].

Summary. (a) High resolution in retinal stimulation cannot be achieved unless very close proximity (on the order of cellular size) between the electrodes and the target cells will be established along the whole interface of the implant with the retina, (b) For normal visual perception, the image should not be dissociated from the eye movements, and (c) The image processing between the camera and the implant should depend on the implant location, i.e. direction of gaze. The system described in this article includes: (1) an optically controlled implant enabling delivery of visual information related to the natural eye movements, (2) position-sensitive image processing, and (3) techniques for bringing retinal neurons into required proximity with stimulus elements.

Acknowledgments. Authors would like to thank Professor Michael Mirkin from Queens College at CUNY for consultation on issues of electrochemistry and Professor Stephen Baccus from the Department of Neurobiology at Stanford for discussions on issues of image processing. Authors also would like to thank Yev Freyvert (UC Berkeley) for his help with fabrication and assembly of the implants.

Funding was provided in part by the Medical FEL program (Air Force Office of Scientific Research) and by VISX Corp.

References

1. Marc, R. E. and B. W. Jones (2003). "Retinal remodeling in inherited photoreceptor degenerations." *Mol Neurobiol* **28**(2): 139–147.
2. Marc, R. E., B. W. Jones, et al. (2003). "Neural remodeling in retinal degeneration." *Prog Retin Eye Res* **22**(5): 607–655.
3. Humayun, M. S., E. de Juan, et al. (1999). "Pattern electrical stimulation of the human retina." *Vision Research* **39**(15): 2569–2576.
4. Humayun, M. S. (2003). *Clinical Trial Results with a 16-Electrode Epiretinal Implant in End-Stage RP Patients.* The First DOE International Symposium on Artificial Sight, Fort Lauderdale, FL, Department of Energy.
5. Rizzo, J. F., 3rd, J. Wyatt, et al. (2003). "Methods and perceptual thresholds for short-term electrical stimulation of human retina with microelectrode arrays." *Invest Ophthalmol Vis Sci* **44**(12): 5355–5361.
6. Rizzo, J. F., 3rd, J. Wyatt, et al. (2003). "Perceptual efficacy of electrical stimulation of human retina with a microelectrode array during short-term surgical trials." *Invest Ophthalmol Vis Sci* **44**(12): 5362–5369.
7. Humayun, M. S., J. D. Weiland, et al. (2003). "Visual perception in a blind subject with a chronic microelectronic retinal prosthesis." *Vision Research* **43**(24): 2573–2581.

8. Smith, G. and D. A. Atchison (1997). The eye. *The Eye and Visual Optical Instruments*. Cambridge, Cambridge University Press: 291–316.

9. Margalit, E., M. Maia, et al. (2002). "Retinal prosthesis for the blind." *Survey of Ophthalmology* **47**(4): 335–356.

10. Margalit, E., J. D. Weiland, et al. (2003). "Visual and electrical evoked response recorded from subdural electrodes implanted above the visual cortex in normal dogs under two methods of anesthesia." *Journal of Neuroscience Methods* **123**(2): 129–137.

11. Sachs, H. G., K. Kobuch, et al. (2000). "Subretinal implantation of electrodes for acute in vivo stimulation of the retina to evoke cortical responses in minipig." *Investigative Ophthalmology & Visual Science* **41**(4): S102–S102.

12. Stett, A., W. Barth, et al. (2000). "Electrical multisite stimulation of the isolated chicken retina." *Vision Research* **40**(13): 1785–1795.

13. Zrenner, E., F. Gekeler, et al. (2001). "Subretinal microphotodiode arrays to replace degenerated photoreceptors?" *Ophthalmologe* **98**(4): 357–363.

14. Palanker, D., A. Vankov, et al. (2005). "Design of a high resolution optoelectronic retinal prosthesis." *Journal of Neural Engineering* **2**: S105–S120.

15. Palanker, D., P. Huie, et al. (2004). "Migration of retinal cells through a perforated membrane: implications for a high-resolution prosthesis." *Invest Ophthalmol Vis Sci* **45**(9): 3266–3270.

16. Palanker, D., P. Huie, et al. (2004). *Attracting retinal cells to electrodes for high-resolution stimulation*. SPIE, Ophthalmic Technologies, San Jose, CA, SPIE, vol. 5314.

17. Coppola, D. and D. Purves (1996). "The extraordinarily rapid disappearance of entopic images." *Proc Natl Acad Sci USA* **93**(15): 8001–8004.

18. Coburn, B. (1989). "Neural modeling in electrical-stimulation." *Critical Reviews in Biomedical Engineering* **17**(2): 133–178.

19. Yang, X. L. and S. M. Wu (1997). "Response sensitivity and voltage gain of the rod and cone bipolar cell synapses in dark-adapted tiger salamander retina." *Journal of Neurophysiology* **78**(5): 2662–2673.

20. Berntson, A. and W. R. Taylor (2000). "Response characteristics and receptive field widths of on-bipolar cells in the mouse retina." *Journal of Physiology-London* **524**(3): 879–889.

21. Malmivuo, J. and R. Plonsey (1995). Hodgkin-Huxley membrane model. *Bioelectromagnetism*. New York, Oxford University Press: 74–93.

22. Hibino, M., H. Itoh, et al. (1993). "Time courses of cell electroporation as revealed by submicrosecond imaging of transmembrane potential." *Biophysical Journal* **64**(6): 1789–1800.

23. Greenberg, R. J., T. J. Velte, et al. (1999). "A computational model of electrical stimulation of the retinal ganglion cell." *Ieee Transactions on Biomedical Engineering* **46**(5): 505–514.

24. Jensen, R. J., O. R. Ziv, et al. (2005). "Thresholds for activation of rabbit retinal ganglion cells with relatively large, extracellular microelectrodes." *Investigative Ophthalmology & Visual Science* **46**(4): 1486–1496.

25. Aurian-Blajeni, B., X. Beebe, et al. (1989). "Impedance of hydrated iridium oxide electrodes." *Electrochimica Acta* **34**(6): 795–802.

26. Weiland, J. D., D. J. Anderson, et al. (2002). "In vitro electrical properties for iridium oxide versus titanium nitride stimulating electrodes." *Ieee Transactions on Biomedical Engineering* **49**(12): 1574–1579.

27. Hesse, L., T. Schanze, et al. (2000). "Implantation of retina stimulation electrodes and recording of electrical stimulation responses in the visual cortex of the cat." *Graefes Archive for Clinical and Experimental Ophthalmology* **238**(10): 840–845.
28. Bard, A. J. and L. R. Faulkner (2001). *Electrochemical Methods: Fundamentals and Applications*. New York, John Wiley & Sons.
29. Oldham, K. B. (2004). "The RC time constant at a disk electrode." *Electrochemistry Communications* **6**(2): 210–214.
30. Rohsenow, W. M., J. P. Hartnett, et al. (1985). Chapter 4. *Handbook of Heat Transfer Fundamentals*, McGraw-Hill: 164.
31. Bron, M., J. Radnik, et al. (2002). "EXAFS, XPS and electrochemical studies on oxygen reduction catalysts obtained by heat treatment of iron phenanthroline complexes supported on high surface area carbon black." *Journal of Electroanalytical Chemistry* **535**(1–2): 113–119.
32. Hammerle, H., K. Kobuch, et al. (2002). "Biostability of micro-photodiode arrays for subretinal implantation." *Biomaterials* **23**(3): 797–804.
33. Mozota, J. and B. E. Conway (1983). "Surface and bulk processes at oxidized iridium electrodes .1. Monolayer stage and transition to reversible multilayer oxide film behavior." *Electrochimica Acta* **28**(1): 1–8.
34. Klein, J. D., S. L. Clauson, et al. (1989). "Morphology and charge capacity of sputtered iridium oxide-films." *Journal of Vacuum Science & Technology a-Vacuum Surfaces and Films* **7**(5): 3043–3047.
35. Meyer, R. D., S. E. Cogan, et al. (2001). "Electrodeposited iridium oxide for neural stimulation and recording electrodes." *Ieee Transactions on Neural Systems and Rehabilitation Engineering* **9**(1): 2–11.
36. Robblee, L. S., J. Mchardy, et al. (1980). "Electrical-stimulation with Pt electrodes .5. The effect of protein on Pt dissolution." *Biomaterials* **1**(3): 135–139.
37. Rose, T. L. and L. S. Robblee (1990). "Electrical-stimulation with Pt electrodes .8. Electrochemically safe charge injection limits with 0.2 Ms pulses." *Ieee Transactions on Biomedical Engineering* **37**(11): 1118–1120.
38. Eckmiller, R., R. Hünermann, et al. (1999). "Exploration of a dialog-based tunable retina encoder for retina implants." *Neurocomputing* **26–27**: 1005–1011.
39. Skavenski, A. A., R. M. Hansen, et al. (1979). "Quality of retinal image stabilization during small natural and artificial body rotations in man." *Vision Research* **19**(6): 675–683.
40. Hammer, D. X., R. D. Ferguson, et al. (2003). "Compact scanning laser ophthalmoscope with high-speed retinal tracker." *Applied Optics* **42**(22): 4621–32.
41. Kendir, G. A., W. T. Liu, et al. (2005). "An optimal design methodology for inductive power link with class-E amplifier." *Ieee Transactions on Circuits and Systems I-Regular Papers* **52**(5): 857–866.
42. Baruth, O., D. Neumann, et al. (2003). "Pattern encoding and data encryption in learning retina implants." *Investigative Ophthalmology & Visual Science* **44**(suppl.2): U701–U701.
43. Olveczky, B. P., S. A. Baccus, et al. (2003). "Segregation of object and background motion in the retina." *Nature* **423**(6938): 401–408.
44. Asher, A., Segal, W. A., Baccus, S. A., Yaroslavsky, L. P., Palanker, D. V. (2007). "Image processing for a high-resolution optoelectronic retinal prosthesis". IEEE Transactions on Biomedical Engineering, in print.

15
Computational Modeling of Electromagnetic and Thermal Effects for a Dual-Unit Retinal Prosthesis: Inductive Telemetry, Temperature Increase, and Current Densities in the Retina

Stefan Schmidt[1], Carlos J. Cela[1], Vinit Singh[1], James Weiland[2],
Mark S. Humayun[2] and Gianluca Lazzi[1]

[1]*Department of Electrical and Computer Engineering,*
North Carolina State University
[2]*Doheny Retina Institute, Keck School of Medicine,*
University of Southern California

Abstract: Recent advances in electromagnetic and thermal modeling of a dual-unit retinal prosthesis are presented. The focus is on the latest computational methods to quantify electrical and thermal deposition in the human tissue with the ultimate goal of addressing safety concerns and optimizing the overall performance of the system. A Partial Inductance Method (PIM) is used for the computation of the electrical coupling parameters of the radiating and receiving telemetry coils. Results for the inductive coil coupling are presented and different coil geometries are compared. Further, a finite difference method for the solution of a bio-heat equation is used to compute the temperature increase caused by the implanted electronics and the electromagnetic absorption due to the external power and data telemetry link. Temperature increases due to the implanted microchip, coils, and stimulating electrode array are shown. Finally, our computational approach based on a multi-resolution impedance method is used for the computation current spread in the human tissue. Results are presented showing variations of current spread in the retina and eye due to different electrode array geometries and placement configurations.

Introduction

Retinitis Pigmentosa (RP) and Age-related Macular Degeneration (AMD) are retinal diseases that cause a slowly progressing loss of vision due to a degeneration of the light sensitive cells (rods and cones) in the retina. In the end state, RP and AMD can lead to complete blindness due to the loss of photoreceptors, while the connected ganglion and bipolar cells in the retina as well as the optical nerve remain largely intact. The aim of a number of retinal prostheses efforts is to partially restore vision to patients with severe cases of RP and AMD by stimulating the ganglion cells with electrical pulses emitted through an array of electrodes attached to the retina. Various clinical trials have already demonstrated the stimulation of visual sensation using single electrodes and electrode arrays. In light of these successful results, research is focused on increasing the number of electrodes in the array, which is likely to increase the capability to perform useful tasks for patients receiving the implant.

While a number of efforts toward the realization of a retinal prosthesis are ongoing [1–3], we will focus on a dual-unit prosthesis approach consisting of an external unit, with a camera and wireless transmitter, and a unit implanted in the eye consisting of a wireless receiver, implanted microelectronics, and a stimulating electrode array [4, 5]. The wireless transmission of data and power to a chronically implanted prosthesis is necessary to avoid percutaneous wire connections. Biocompatibility and safe operation of the prosthesis components has to be maintained with minimal deposition of electromagnetic power and heat, setting stringent design conditions.

This chapter is organized in three sections, respectively addressing aspects of the inductively coupled telemetry link, the thermal heat simulations of the retinal prosthesis components, and simulations of the retinal electrode array.

Inductively Coupled Links for a Dual-Unit Retinal Prosthesis

In the dual-unit retinal prosthesis under consideration here, like in numerous other biomedical applications, inductive coupling is the preferred method for transcutaneous power transfer. Inductive links can carry data to and from implanted biomedical devices without the need of wires piercing the skin, therefore reducing the risk of infection. The effectiveness of the inductive link and compliance with safety standards are of critical importance for these applications. In general, skin mobility and variations in the thickness of subcutaneous fatty tissue can cause misalignment of the coils, leading to a change of transmission characteristics. In the retinal prosthesis, eye movement can obviously cause substantial changes in the relative positions of external and internal coils. Thus, numerical studies are necessary to assess the performance of the coupling between external and internal coils in advance to clinical trials. Different coil geometries are considered here as examples for the retinal prosthesis under consideration. However, the methods discussed and presented are general and applicable to many different coil configurations and applications.

There have been several approaches to the analysis and design of inductively coupled transcutaneous links, with the goal of minimizing misalignment effects and maximizing the coupling efficiency [6–10]. In many cases, an analytical static approximation, based on the Partial Inductance concept [11], can be used successfully to calculate the mutual and self-inductance of coupled coils. The PIM can only provide an estimate for the coupling between external coils and coils implanted inside the eye of a human head model since it ignores the presence of the human tissue. However, at low frequencies it is very suitable for maximizing the coupling efficiency of inductive links and observing the effects of implant motion and misalignment. This method provides a very simple and efficient free-space analysis of inductively coupled wire traces and, due to its simplicity and capabilities, will be briefly illustrated here.

Partial Inductance Method

The inductive interaction between conductors carrying currents is caused by electrodynamic effects, which take place concurrently: currents flowing through conductors create magnetic fields (Ampere's Law); time-varying magnetic fields create induced electric fields (Faraday's Law). The inductive coupling of complex geometric structures and open loops can be calculated using the Partial Inductance concept [11–13]. Figure 15.1 illustrates two inductively coupled current loops.

The inductance between two wire loops i and j is be defined as $L_{ij} \equiv \frac{\phi_{ij}}{I_j}$, where the mutually coupled flux is $\phi_{ij} = \frac{1}{a_i} \oint_i \int_{a_i} \mathbf{A}_{ij} \cdot d\mathbf{l}_i da_i$. $\mathbf{A}_{ij} = \frac{\mu}{4\pi} \frac{I_j}{a_j} \oint_j \int_{a_j} \frac{d\mathbf{l}_j da_j}{r_{ij}}$ is the magnetic vector potential in loop i due to the current in loop j. Thus, the mutual inductance between two wires can be written as

$$L_{ij} = \frac{\mu}{4\pi a_i a_j} \int_{a_i} \int_{a_j} \oint_{l_i} \oint_{l_j} \frac{d\mathbf{l}_j \bullet d\mathbf{l}_i}{r_{ij}} da_j da_i \tag{1}$$

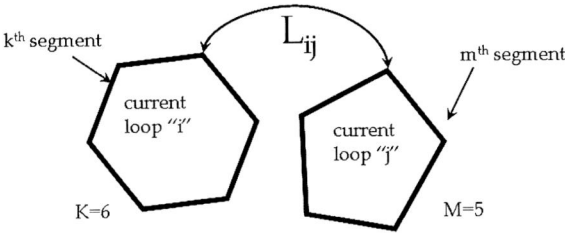

FIGURE 15.1. Decomposition of current loops into segments of partial inductances.

In the PIM, the integral along the complete paths of wire loops, dl_i and dl_j, respectively, is partitioned into linear wire segments as shown in Figure 15.1. Thus, the inductance of Eq. (1) can be written as a sum of partial inductances

$$L_{ij} = \sum_{k=1}^{K} \sum_{m=1}^{M} \frac{\mu}{4\pi} \frac{1}{a_i a_j} \int_{a_k} \int_{a_m} \int_{b_k}^{c_k} \int_{b_m}^{c_m} \frac{dl_k \bullet dl_m}{r_{ij}} da_k da_m = \sum_{k=1}^{K} \sum_{m=1}^{M} L_{km}^{P} \qquad (2)$$

The partial inductance integral L_{km}^{P} can be calculated in a closed form for the self-inductance of a cylindrical wire segment (where k = m) [14], as

$$L_{kk}^{P} = 2 \left(l \log \frac{l + \sqrt{l^2 + r^2}}{r} - \sqrt{l^2 + r^2} + \frac{1}{4} + r \right) \qquad (3)$$

where l is the wire segment length and r is the wire radius. The mutual inductance between two cylindrical wire segments (k, m) is approximated as the mutual coupling of two filamentary currents flowing along the longitudinal axis of cylindrical wires segments. The equation for the mutual coupling of two straight filaments placed in any desired position can be found, for example, in [13]. The equations for the self- and mutual inductances of wire segments can be evaluated very efficiently, and in this fashion the total loop inductance can be computed very quickly even for a large number of wire segments.

Coil Models and Coupling Computation

For an optimal coupling efficiency, the external telemetry and power transfer coil for the retinal prosthesis considered here must be placed very close to the eye, as illustrated in Figure 15.2. Further, the axis of external and internal coils should be aligned. As an example, Figure 15.3 shows the geometry of external and internal coils that can be used for a retinal prosthesis. The smaller internal coil has two layers, each having 9 turns of 40/44 copper litz wire. The larger external coils also has two layers, each having 10 turns of 165/46 copper litz wire.

Using the PIM, the mutual coupling of the coils shown in Figure 15.3 was computed as a function of the distance between the two. In Figure 15.4 the quasi-static results, obtained with the PIM, are compared with coupling measurements at 1 MHz, obtained using a vector network analyzer. The figure demonstrates that there is good agreement between the results obtained from the PIM and the measurements at low frequencies.

In another set of PIM simulations, shown in Figure 15.5, the mutual coupling for two different external coils are compared. Here, the external coils both have a total number of 20 turns; the coil used in Figure 15.5(a) has only one layer of 20 turns, while the coil used in Figure 15.5(b) has two layers of 10 turns each. The internal coil considered here only has 10 turns on one layer. The figures show the mutual coupling between external and internal coils for a range of distances between them and a range of angles of rotation of the eyeball that causes a misalignment of the axes of the coils. The figures illustrate that the external coil only has one layer, but the same number of turns has nearly 50% better coupling

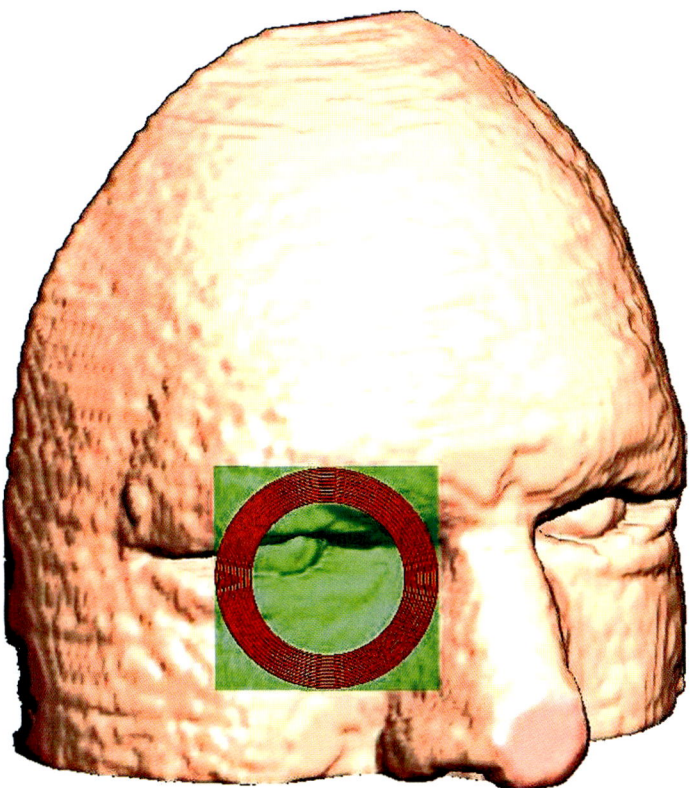

FIGURE 15.2. Illustration of the human head model and placement of the external telemetry coil.

over the two-layer coil at the same distance and coaxial alignment ($\theta = 0$). The two-layer coil, however, is less sensitive to misalignment due to eye rotation ($\theta \neq 0$).

Thermal Modeling

One of the key requirements to prevent potential tissue damage caused by an electronic implant – such as the retinal prosthesis – is to limit the thermal elevation induced in the human body due to the operation of the implant itself. While the extra-ocular components of the retinal prosthesis are not expected to have any direct thermal effects on the tissues of the eye, the intra-ocular components will cause temperature increase due to power dissipation. Further, the wireless telemetry link between the external (primary) coil and the internal (secondary) coil will cause electromagnetic power deposition in the tissues, which in turn could lead to temperature increase. While sources of heat cannot

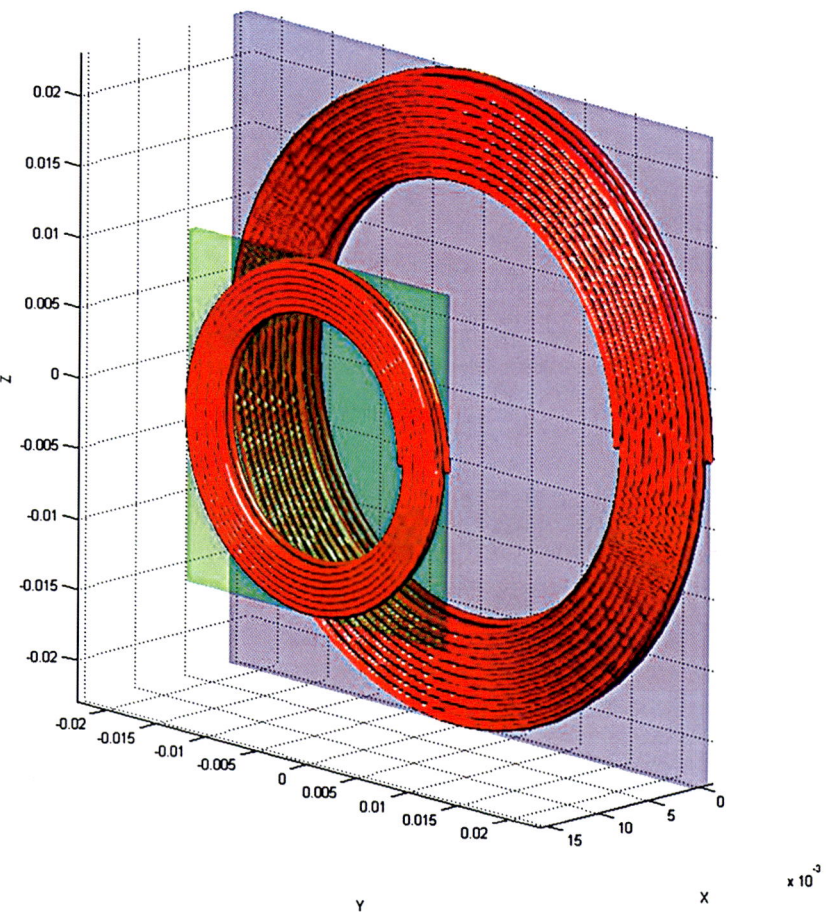

FIGURE 15.3. An example of internal and external coil geometries for a dual-unit retinal prosthesis.

be removed completely, measures can be taken to minimize the heat due to a biomimetic device such as the retinal prosthesis.

Bio-Heat Equation Formulation

In general, the temperature variation in a generic body can be described by a partial differential equation (PDE) of conduction, given by $C\rho\frac{\partial T}{\partial t} = \nabla \cdot (K\nabla T) \pm S$, where C is the specific heat $\left[\frac{J}{kg\,°C}\right]$, ρ is the mass density of the material $\left[\frac{kg}{m^3}\right]$, K is the thermal conductivity $\left[\frac{J}{m\,s\,°C}\right]$, T is the temperature [$°C$], and S is a heat source (positive) or sink (negative) $\left[\frac{W}{m^3}\right]$. To adapt this equation to the tissues in the human body and the regulating action of blood, Pennes incorporated terms to describe the warming effect of the basal metabolism and the regulatory

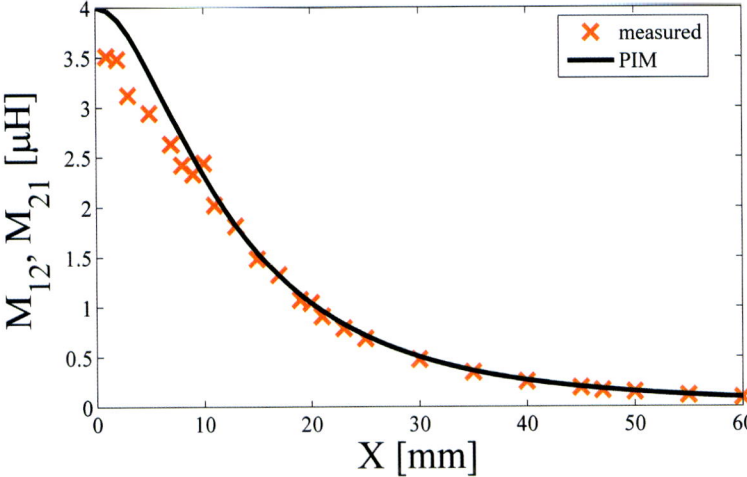

FIGURE 15.4. Simulated and measured results for the mutual coupling between the two coils in Figure 15.3.

influence of blood on tissue temperatures [15]. Pennes' basic Bio-Heat Equation (BHE) can therefore be written as

$$C_\rho \frac{\partial T}{\partial t} = \nabla \cdot (K \nabla T) + A - B(T - T_b) \tag{4}$$

where A is the rate of heat production due to metabolic processes per unit volume, B is the blood perfusion constant, and T_b is the temperature of blood (assumed constant at $T_b = 37.0\,^{\circ}\mathrm{C}$). As can be seen from Eq. (4), the temperature variation with respect to time depends on a number of tissue properties, which would lead to wide varying temperatures within the human body. Also, while the metabolic rate is a source of heat, the blood perfusion constant B is a measure of how well a particular tissue is permeated with blood, which is reflected in its temperature. The term $B(T - T_b)$ acts as a source of heat if $T > T_b$, and as a sink if $T < T_b$.

Electromagnetic Deposition

If a biological object is exposed to RF energy (e.g. the wireless telemetry system), E- and H-fields penetrate the tissues leading to power deposition. This power deposition leads to temperature increase per unit time given by $\frac{\partial T}{\partial t} = \frac{SAR}{C}$, where SAR is the Specific Absorption Rate [16]. The SAR is defined as the time rate of incremental energy $\left(SAR = \frac{d}{dt}\frac{dW}{dm}\right)$, dissipated in a material incremental mass dm, contained in a volume element dV of given mass density ρ, i.e. $dm = \rho\, dV$. The SAR is related to magnitude $|E|$ of the electric field by $SAR = \frac{\sigma}{\rho}|E|^2$, where σ is the electrical conductivity of the tissue. Thus, once we know the E-field distribution in the tissues, we can obtain the SAR, from which we can obtain

A

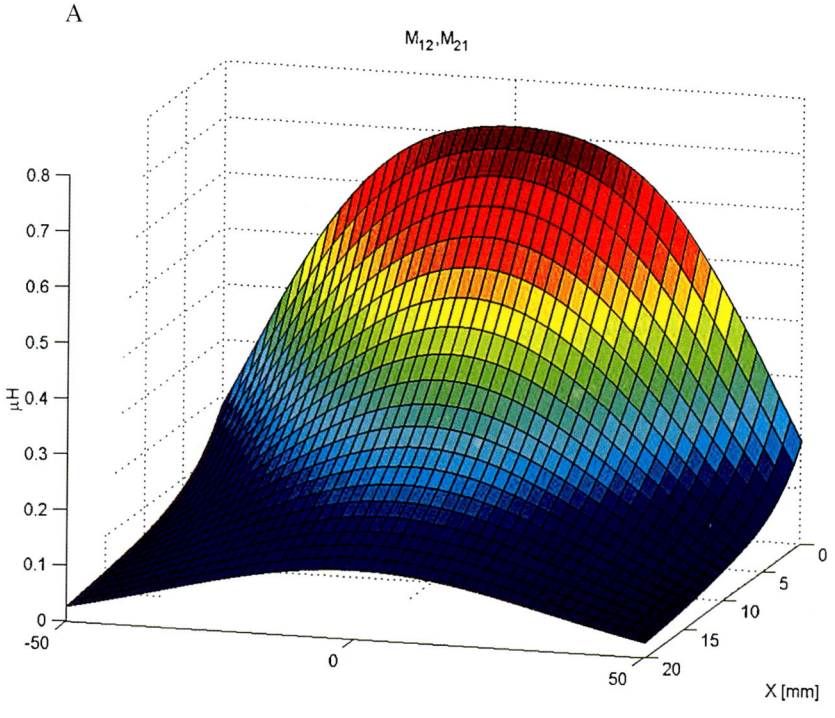

B

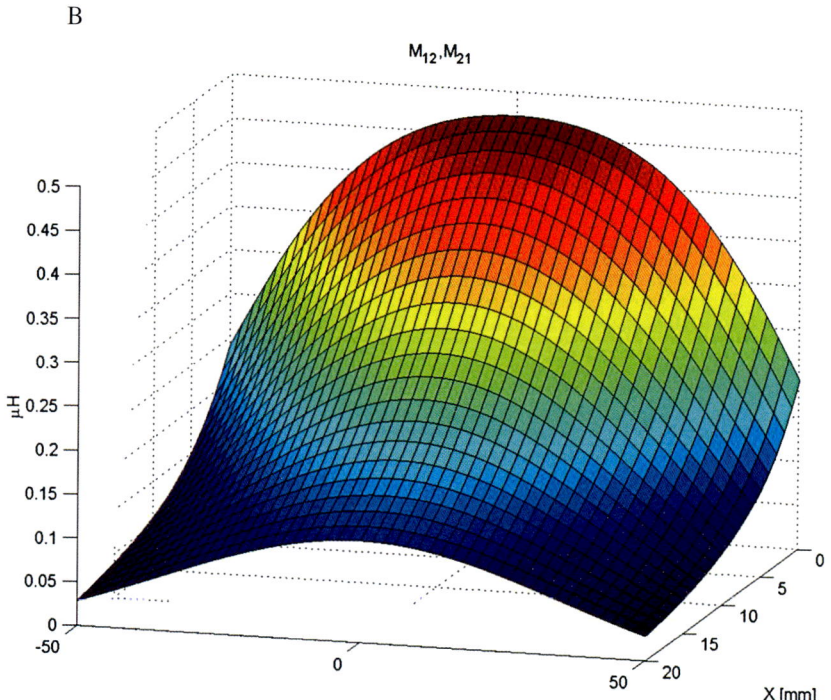

FIGURE 15.5. Mutual coupling inductances for (a) one-layer external coil, and (b) two-layer external coil.

the heat generated per unit volume per unit time, which can be incorporated in the BHE as a source.

Joule Heating

When considering electronic implants within the tissue, power dissipation within the electronic circuits has to be considered. Assuming the heat dissipation per unit volume of an implant is P_i, it can be incorporated in the BHE as a source. Thus, the complete BHE governing heat conduction within a body volume under RF exposure and with internal electronic components can be expressed as

$$C_\rho \frac{\partial T}{\partial t} = \nabla \cdot (K \nabla T) + A - B(T - T_b) + \rho \text{SAR} + P_i \qquad (5)$$

Computational Method

The BHE given in Eq. (5) is a PDE and can be solved using a variety of computational methods. An efficient computational solution of Eq. (5) can be obtained by means of 3D finite difference methods, such as that in [17], which is based on the conservation of thermal energy $Q^{i,j,k}$ within a cell at the 3D coordinates (i, j, k). Heat generated by the sources is added algebraically to the heat flowing into a cell, which is equal to the energy increment in the cell. The total heat generation (in Joules) within the cell is therefore $Q_G = Q_A + Q_B + Q_{\text{implant}} + Q_{EM}$, where

1. Metabolic heat: $Q_A^{i,j,k} = A^{i,j,k} \, \delta^3 \Delta t$
2. Blood perfusion: $Q_B^{i,j,k} = B^{i,j,k} \left(T^{i,j,k} - T_B \right) \delta^3 \Delta t$
3. Joule heating: $Q_{\text{implant}} = P_{\text{implant}} \Delta t$
4. EM power deposition: $Q_{EM}^{i,j,k} = \rho^{i,j,k} \, \text{SAR}^{i,j,k} \, \delta^3 \, \Delta t$

For simplicity, the cells have been assumed to be cubes of size δ. The flow of heat into the cell via conduction is calculated using the analogy of heat flow to current flow where the equivalent of Ohm's Law $(V = IR)$ is Fourier's Law $(\Delta T = Q' R_{th})$. Here, Q' is the rate of heat transfer and R_{th} is the thermal resistance. For the derivation of this analogy, the reader is referred to [18].

Let $K^{i\pm1,j,k}$, $K^{i,j\pm1,k}$, and $K^{i,j,k\pm1}$ be the thermal conductivity of the tissue cells adjacent to cells $K^{i,j,k}$. The series thermal resistance between point $K^{i,j,k}$ and $K^{i+1,j,k}$ can be written as $R_{th} = \frac{1}{2\delta K^{i,j,k}} + \frac{1}{2\delta K^{i+1,j,k}}$. The conduction heat flow into a cell during the time interval Δt can be written as $Q_C = Q' \, \Delta t$. Thus, conduction from just one cell is given by

$$Q_C^{(i+1,j+k)\to(i,j,k)} = (T^{i+1,j,k} - T^{i,j,k}) \left(\frac{1}{K^{i,j,k}} + \frac{1}{K^{i+1,j,k}} \right) \delta \Delta t \qquad (6)$$

Similarly, conduction heat contributions from the other five adjacent cells are added to obtain the total heat energy increment, $Q_{\text{total}} = Q_C + Q_G$. The temperature increment from time $t_n = n \cdot \Delta t$ to $t_{n+1} = (n+1) \cdot \Delta t$ is then given by

$$Q_{\text{total}}^{i,j,k} = \rho^{i,j,k} C^{i,j,k} \frac{\partial T}{\partial t} \delta^3$$

$$\approx \rho^{i,j,k} C^{i,j,k} \frac{T_{n+1}^{i,j,k} - T_n^{i,j,k}}{\Delta t} \delta^3 \qquad (7)$$

Thus, with the knowledge of boundary values, initial conditions, and properties of the tissues, we can use an explicit finite difference method to solve the BHE (5).

The boundary condition between human body tissue and air is obtained by using the concept of continuity of heat flow normal to the skin surface. This is expressed as the convective boundary condition, $K \frac{\partial T}{\partial \hat{n}} = H_a(T - T_a)$, where $\hat{n}$ signifies the unit normal to the skin surface, H_a is the convective heat coefficient of the ambient (air), and T_a is the ambient temperature. Boundaries truncating the model within the body are modeled using Dirichlet boundary conditions (boundary condition of the first kind), i.e. the temperature at the boundaries is assumed to be constant.

The electromagnetic power deposition due to the wireless link for power and data transfer can be calculated using the Finite-Difference Time-Domain (FDTD) method, as done in [19], and included if needed in Eq. (5). However, if the design of the wireless link is such that the electromagnetic absorption is within limits established by international safety standards, such as IEEE and ANSI, the influence of the SAR on the temperature increase is minor compared to the other sources of heat, and is neglected in the BHE computation.

To obtain the initial temperature values within the head, the BHE is solved for the 3D head model without any external factors (SAR, P_{implant}). In the retinal prosthesis case, for practical purposes, only the region around the eye is needed since the temperature increase is largely confined around the region of the implant. Therefore, after the initial (basal) temperature is obtained with a large model of the human body, only a subsection of this is needed for the computation of the temperature increase due to the implant. Thus, only smaller volumes around the eye region were used, for example, in [19] to compute the thermal increase due to the retinal prosthesis implant.

Tissues Properties

For the purpose of obtaining computational results of the temperature increase due to the retinal prosthesis system, a 1mm resolution 3D head model based on the Visible Human Project of the National Library of Medicine [20, 21] was used. A number of other models derived from MRI scans of volunteers are in general widely available and can also be used for the purpose. The thermal parameters (C,K,A,B) for the tissues can be obtained as described in [16, 22], while the dielectric properties at the frequency of the telemetry system can be obtained from [23–26]. Both the metabolic rate (A) and the blood perfusion constant (B) describe thermoregulatory mechanisms of the human tissues. The head models used have a skin–air and a cornea–air interface, each characterized by a different convection coefficient H_a ($H_a = 10.5\frac{W}{m^2}$ for skin–air, $H_a = 20\frac{W}{m^2}$ for cornea–air). Previous simulations [19] showed that using a constant $H_a = 10.5\frac{W}{m^2}$ for all tissue-to-air interfaces has a negligible effect on the accuracy.

A very important aspect of modeling the eye is the relationship between the retina and the choroid tissue. The choroid is a highly vascularized tissue, i.e. it contains a significant amount of blood vessels. Therefore, the choroid is modeled with the dielectric properties of blood. However, during simulation, the temperature of the choroid is allowed to vary, as opposed to being held constant at T_B.

Extensive studies have indicated that the high blood circulation density between the choroid and the retina is responsible for the retinal temperature regulation. To this extent, the retinal tissues are assumed to be perfused with the "choroidal blood", i.e. instead of T_B the regulating temperature is $T_{choroid}$. To get a deeper understanding of the choroidal effects on the retina, the reader is referred to [27–29].

Component Modeling and Thermal Elevation Results

Stimulator Chip

Gosalia et al. [19] previously obtained thermal elevation results using the explicit finite difference method to implement the BHE. Different positions (in the center of the vitreous cavity and at the anterior of the eye between the ciliary muscles) and sizes ($4 \times 4 \times 0.5$ mm and $6 \times 6 \times 1$ mm) were considered. The chip was covered by an insulating encapsulation with a uniform thickness of 0.5 mm, and was allowed to dissipate 12.4 mW over its entire volume (excluding the insulation). The thermal conductivities of the chip and the insulation were assumed to be constant and uniform over their volume and equal to 60 J/(m.s.°C) and 30 J/(m.s.°C), respectively. As those simulations showed, placing the chip in the anterior region, or increasing its size, reduced the computed temperature increase in the vitreous humor and the retina. The complexity of the relation between temperature increase and physical characteristics of the implanted electronics is not limited to their position or size; for example, it can also involve the power distribution characteristics and material properties of biocompatible materials used for insulating the chip. To complement the study in [19], here we show as an example the effect on the temperature increase in the eye of the thermal conductivity K of the insulation surrounding the microchip. Four cases have been considered, representing different constant K or graded K values for the chip insulation.

Figure 15.6a gives a 3D view of the stimulator chip implanted in the eyeball. Figure 15.6b shows a horizontal cross section of the same model (at 0.25 mm resolution), with the chip located in the middle of the vitreous humor. Figure 15.6b is a tissue-rich model, with different colors indicating different

A

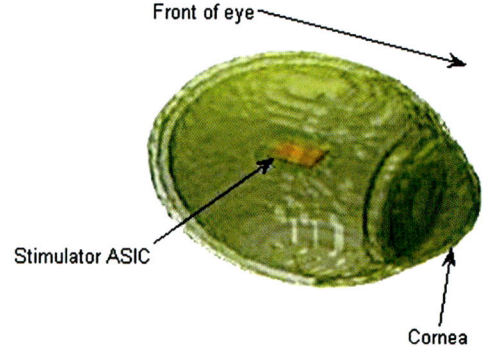

Front of eye

Stimulator ASIC

Cornea

B

Cornea
(front of eye)

VITREOUS HUMOR

Chip

Outer
Insulation

Inner Insulation

FIGURE 15.6. (a) Location of stimulator ASIC (chip) in eyeball. (b) Cross section of a 0.25 mm resolution model. A graded-K coating arrangement is shown in the figure.

tissue types. The dotted line indicates the axis along which the thermal profiles for different insulator configurations are shown in Figure 15.7.

Figure 15.7 gives the temperature profile parallel to the axis of the eye, through the insulator. All the simulations were performed for 22 minutes of real-world time, during which the temperatures obtain their steady-state values.

As expected, the maximum temperature rise ($\Delta T = 1.657\,°C$) within the chip/insulator package takes place for configuration with uniform $K_{INSULATOR} = 0.2\,W/m°C$, while the minimum rise ($\Delta T = 0.986\,°C$) takes place for uniform $K_{INSULATOR} = 30\,W/m°C$. The graded coating configuration partially restricted

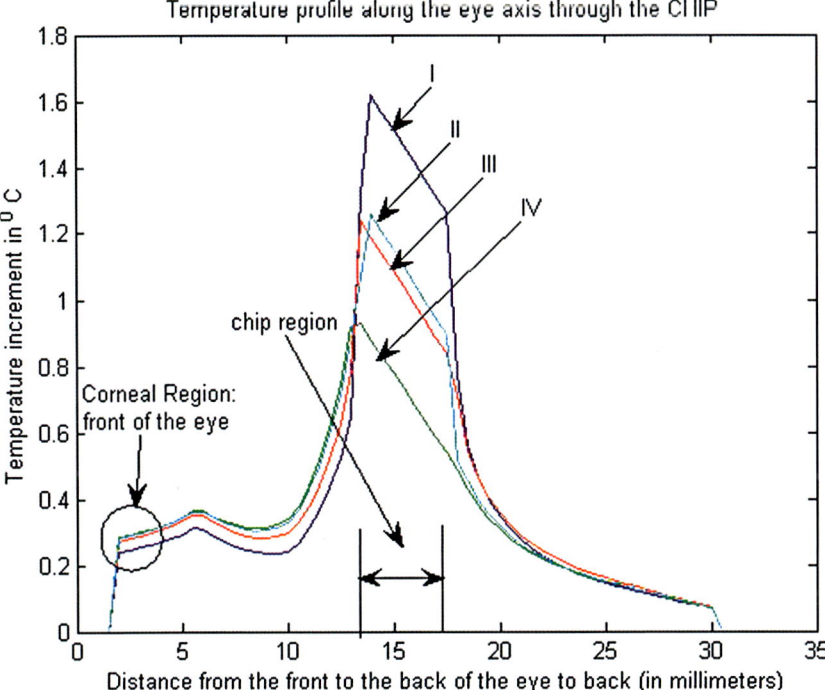

FIGURE 15.7. Profiles of the temperature increase along the axis of the eye going through the chip. (I) Only one insulator coating of $K = 0.2\,\text{W/m.}^\circ\text{C}$; (II) Graded insulator coating, $K_{INNER} = 0.2\,\text{W/m.}^\circ\text{C}$; $K_{OUTER} = 30\,\text{W/m.}^\circ\text{C}$; (III) Graded insulator coating, $K_{INNER} = 30\,\text{W/m.}^\circ\text{C}$; $K_{OUTER} = 0.2\,\text{W/m.}^\circ\text{C}$; (IV) Single Uniform coating, $K = 30\,\text{W/m.}^\circ\text{C}$.

the flow of heat from the chip, leading to intermediate temperature increments in the chip/insulator region and the surrounding parts. For high-K insulator, heat flows freely outward from the chip. Consequently, the temperature rise in the chip is lower, and the tissues at far away points have higher thermal increment. For low-K insulator, the restricted heat induces comparatively higher temperature rise within the chip/insulator region, and the relatively faraway tissues are almost unaffected.

Implanted Coil

The heat dissipated by the receiving coil for the telemetry system in the retinal prosthesis can contribute to the thermal increase in the surrounding tissues. As an example, we have considered the effect of two possible locations of the receiving coil: (1) outside the eye, surrounding the eyeball, and (2) implanted in the eye, in place of the lens. In both cases, the axis of the coil is identical to the axis of the eyeball. Figure 15.8 illustrates the position of the coils relative to the eyeball.

The power dissipated in the coil was calculated using the total resistance of the coil and the current that is assumed to flow in the coil itself. Table 15.1 compares

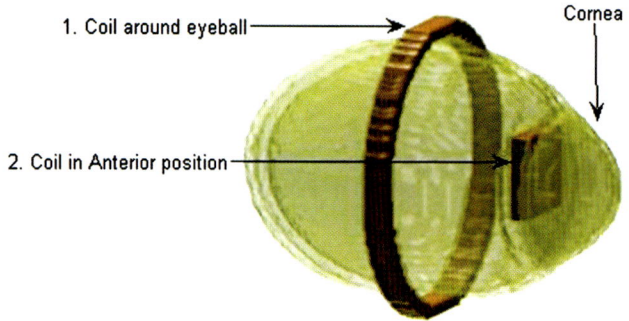

1. Coil around eyeball

Cornea

2. Coil in Anterior position

FIGURE 15.8. The location of the two receiving-coil positions considered for thermal simulations. Note that both have been assumed to be tightly wound four-turn coils.

TABLE 15.1. Maximum temperature increase in various tissues of the human body due to the power dissipation of the receiving coil for the wireless telemetry. "Anterior" indicates the results with the receiving coil implanted in place of the lens, while "Surrounding" implies that the receiving coil is of $\sim 25\,mm$ diameter and surrounds the eyeball.

Tissue	Maximum temperature rise (°C)	
	Anterior	Surrounding
Retina	0.025	0.358
Skin	0.089	1.433
Fat	0.152	1.222
Muscle	0.409	1.045
Cornea	0.240	0.105
Vitreous Humor	0.415	0.314

the maximum temperature increments caused by the two considered coil configurations on some of the tissues, for a given power dissipation ($546\,\mu W$). Further details can be found in [30].

Electrode Array

The effect of the power dissipation of the electrode array can also be quantified. Figure 15.9a and 15.9b are horizontal and vertical cross sections, respectively, showing the position on the retina, at the back of the eyeball, of an 8×8 stimulating electrode array, in a 0.125 mm resolution head model. This model was extracted from the larger, lower resolution (0.25 mm) eye model shown in Figure 15.6b.

The current injected in each electrode is biphasic in nature, with a pulse that for illustrative purposes can be assumed of equal width for the cathodic and anodic phase. Figure 15.10 shows the power dissipation pattern in each electrode for two different current waveforms.

A

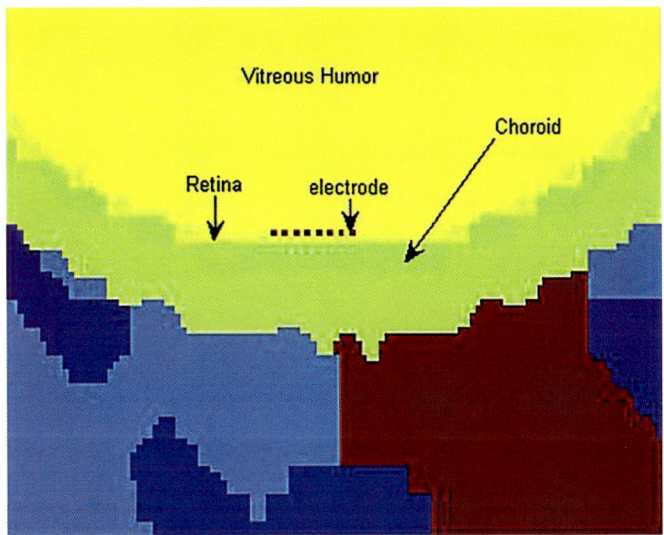

B

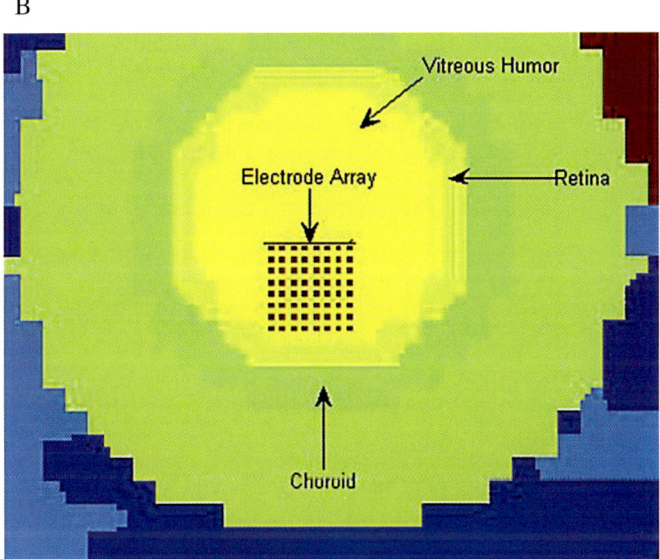

FIGURE 15.9. Cross section of the thermal model showing configuration of the electrode array on the retina. (a) Perpendicular to electrode array. (b) Parallel to electrode array.

For the given current injection values and periodicity, the temperature increment due to the power dissipated in the electrode array is lower than 0.01 C. The low thermal elevation is expected since the power dissipation per unit volume (~ 158 Watts/m^3) is extremely small and zero most of the time. The

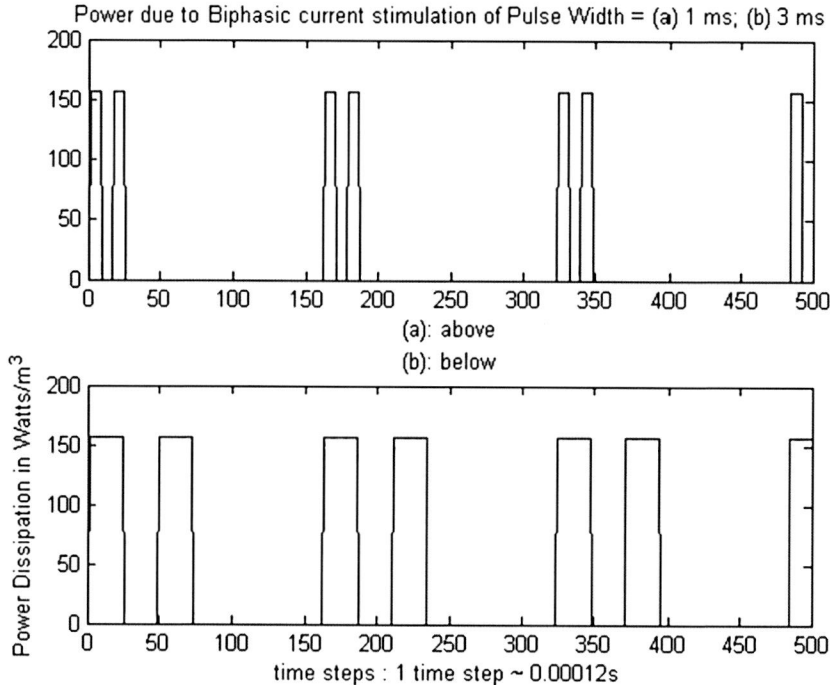

FIGURE 15.10. Power dissipation by a single platinum electrode for an injected current of ~ 617 μA at pulse width = (a) 1ms, (b) 3 ms; and pulse repetition rates of 50 Hertz.

temperature increase in the tissue due to the flow of the current can be calculated in a similar fashion, and it is currently under consideration.

Computation of Electric Current Densities in the Retina

When designing the stimulating electrode array, it is useful to predict the electric current densities and potentials induced in the various retinal layers by the stimulating electrodes. Determination of currents and potentials is important for achieving optimal stimulation while avoiding risks of tissue damage due to excessive current [31].

Current density simulations can answer questions regarding:

- optimal electrode geometry and current return placement
- safe maximum amount of current to inject for a given configuration
- efficiency of current injection for particular setting
- aspects of safety of implanted device for particular configurations, related to current circulation through living tissue.

Due to the very low frequencies used for the stimulation of the retina, the simulation of the resulting current densities can be approximated through static or quasi-static methods. The "Impedance Method" [32–34] or its dual, the "Admittance Method" [35, 36] are two computationally efficient methods that can be used to calculate the resulting current densities in 2D or 3D. In the Impedance Method, the anatomical model of the retina is discretized and approximated as an equivalent impedance mesh, built of lumped circuit elements. This approach reduces the problem to a solution of a linear system of equations. The final result of the simulation are the approximate current density vectors at the centers of each of the voxels forming the anatomical model, as well as the electric potential at each vertex of each voxel in reference to the model's ground. Before applying the Impedance Method, a suitable discrete retinal model needs to be produced. For a successful simulation, the model must be anatomically correct and have a resolution that is sufficient to describe the electrical characteristics of the retinal layers. Further, the low frequency impedance of each tissue must be known if the tissue itself is to be modeled as a "bulk" material with given dielectric properties.

The following sections briefly describe some aspects of high-resolution discrete model generation, the basics of the Impedance Method, the challenges involved in such high-resolution modeling, and illustrative results.

Layered Retina Model

In order to understand the interaction of the current injected by the electrode array and the ganglion cell layer of the retina, the retinal models must be anatomically correct and sufficiently detailed. One such model is the Visible Human model, available at a resolution of 1 mm [21]., While acceptable for tissues surrounding the eye, a 1mm resolution is grossly insufficient to describe the anatomical characteristics of the retina. Thus, models of the retina with accurate descriptions of all its layers [37, 38] have been artificially generated through mathematical descriptions or obtained from anatomical atlases, and these refined models were used to replace retinal tissue in the original 1mm resolution model.

Figure 15.11 shows a 2D model of the retina and electrode array created with a software application that was written to generate discretized models of the human eye – or other anatomical structures – at arbitrary resolution starting from geometrical descriptions. In our simulations we use current sources to provide a stimulus to the retina. In addition to organic structures, the model needs to include the electrodes used for excitation, the current return, and a marker for the ground potential, as shown in Figure 15.12.

Impedance Method

The idea behind the impedance method [32] is to convert a physical system to a discretized model, and combine that information with the electrical impedance of each material to obtain an equivalent electric circuit, built of lumped circuit

A

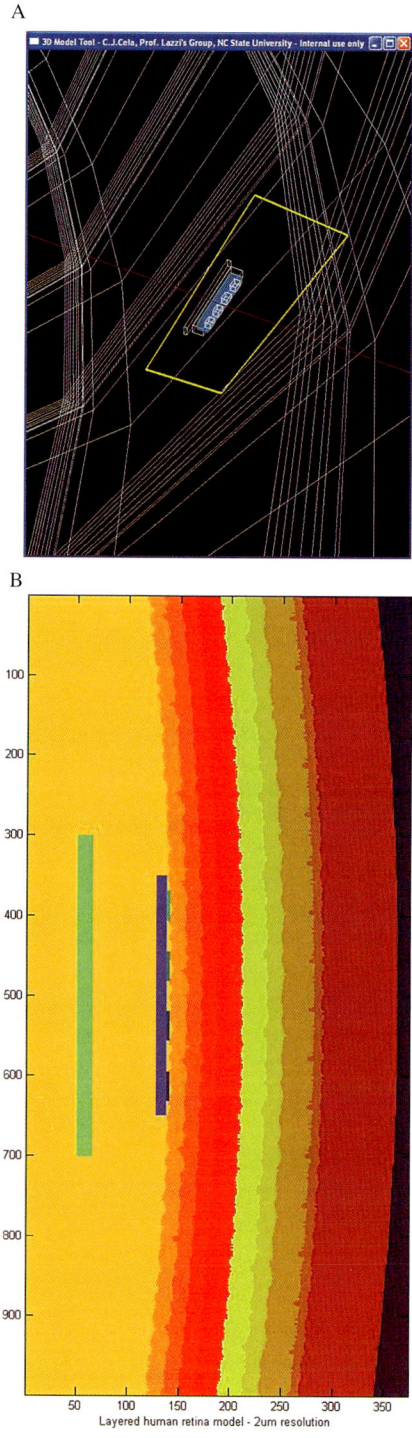

B

FIGURE 15.11. (a) Geometric model of human retina and linear electrode array. (b) 2D slice across the Z-plane of resulting model discretized at 2μm resolution.

A

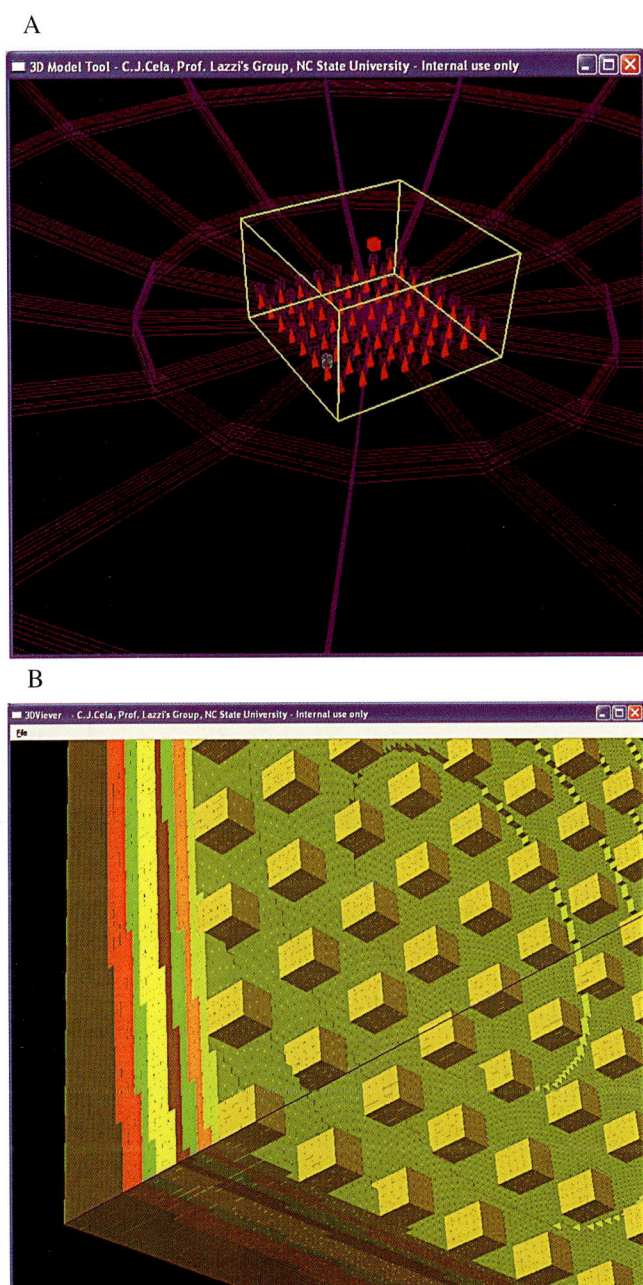

B

FIGURE 15.12. (a) 3D geometric retina model with array of electrodes. (b) View of matching uniformly discretized section, 10μm resolution; insulator between electrodes is not shown.

components. This equivalent electric circuit can be represented as a linear system, and then be solved using Ohm's Law.

Starting from the discretized model, the solution of the system will result in knowing the electric potential at each vertex of each voxel relative to the model's ground as well as the current density vector at the center of each voxel. The method can be applied to 2D or 3D discretized models, formed of either uniform or different sized cells or voxels. The method for uniform resolution cells can be found in [32] and it will be briefly summarized here.

The first step is to convert each voxel of the model to its equivalent impedance network formed of lumped circuital elements. Since the model is linear, each of the three orthogonal directions (X, Y, and Z) is considered separately, and the final impedance network for the voxel is obtained combining all the resulting components.

Considering a single voxel of the discretized model (Figure 15.13), it is possible for example to approximate the impedance seen by a current flowing in the X direction by lumping the impedance of each of the sub-volumes into a circuit element, knowing that the equivalent resistance will be

$$R = \frac{L}{W\,H}\rho \qquad (8)$$

and the equivalent capacitance will be

$$C = \frac{W\,H}{L}\varepsilon_0\varepsilon_r \qquad (9)$$

where ε_0 is the permittivity of free space. ρ and ε_r are the resistivity and the relative electric permittivity, respectively, of the voxel's material along the X axis. W, H, and L are the respective width, height, and length of the sub-voxel volume.

For each voxel, this process is performed in all three orthogonal axes, as shown in Figure 15.14. Then, the resulting lumped elements are combined to form the equivalent impedance network that approximates the voxel electrical

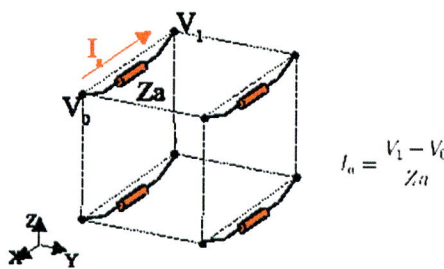

FIGURE 15.13. Voxel sub-volumes used to calculate lumped circuit elements in X direction. The resistor in parallel with the capacitor will represent the low-frequency impedance of one sub-volume.

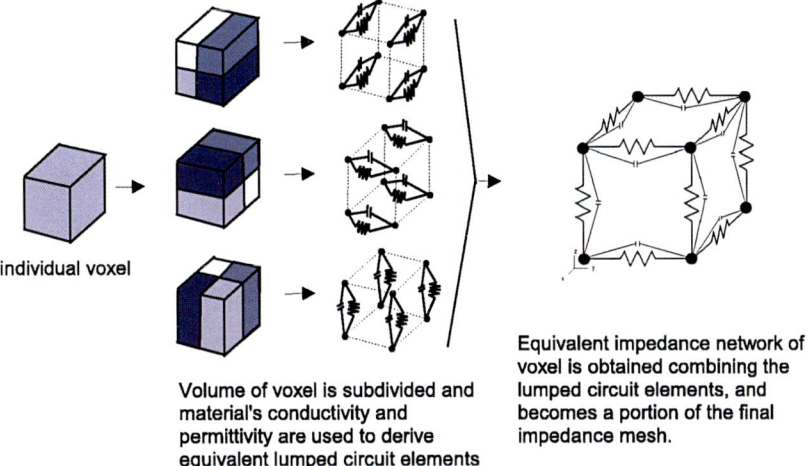

individual voxel

Volume of voxel is subdivided and material's conductivity and permittivity are used to derive equivalent lumped circuit elements

Equivalent impedance network of voxel is obtained combining the lumped circuit elements, and becomes a portion of the final impedance mesh.

FIGURE 15.14. Process for generating equivalent impedance network of a voxel.

behavior. Contributions from all voxels in the model are then combined to form the equivalent impedance network that represents the entire model.

Note that the resultant electrical circuit will have nodes coinciding with the vertices of the model's voxels. One node is assigned the ground potential. External stimuli can then be modeled as current sources, connected between two nodes, one belonging to an electrode and the other to the current return. The equivalent impedance matrix for the circuit is then built and the system solved for voltages and currents. The X, Y, and Z components of the voxels' total current density can then be calculated by considering the area normal to the sub-volume that each lumped impedance represents. Finally, the total current density vector for each voxel is calculated by adding the contributions of all the sub-volumes in all three axes, as shown in Figure 15.15.

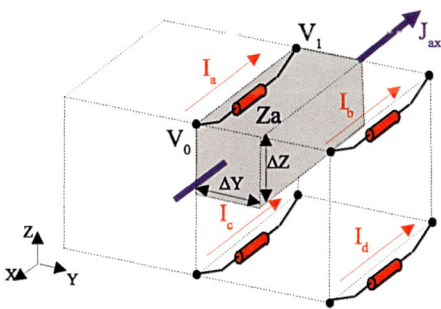

FIGURE 15.15. Current density calculation with the impedance method.

It should be noted that the method could incorporate arbitrary lumped impedances at any point of the impedance mesh to account for contact capacitances or any other effects that can be described by means of localized impedances. Furthermore, the method can be characterized by a "multi-resolution" grid, where the size of the cells is chosen to fit the model to be simulated with fine resolution in regions of interest and relatively coarse resolution elsewhere [34].

Computational Challenges

As mentioned earlier, the size of the computational space can easily become too large to be handled with uniform cells. An approach to processing large models is to use a multi-resolution impedance method [33], where larger voxels are used in homogeneous regions of the model, and small voxels are used in regions with material boundaries, where detail is needed. In retinal models, the use of multi-resolution models can reduce the number of voxels from 30% to over 80% depending on the level of detail in the model. Building an optimized multi-resolution model is a complex process. Figure 15.16 shows a simplified way of creating a multi-resolution model from a uniformly discretized model. As electric properties change at material boundaries, it is important to keep small voxels at the boundaries to minimize numerical errors.

To process larger models, there are additional techniques that can be used. In some cases simulations can be performed in 2D instead of 3D. While those results are often qualitative in nature, preliminary results from simulations performed by our group indicate that some configurations allow the scaling of data obtained using 2D simulations into 3D values with an acceptable error margin. In particular, current density values taken far from the current return and close to the symmetry plane of the electrode array, in configurations with electrode arrays larger than 4×4, can be scaled well from 2D to 3D.

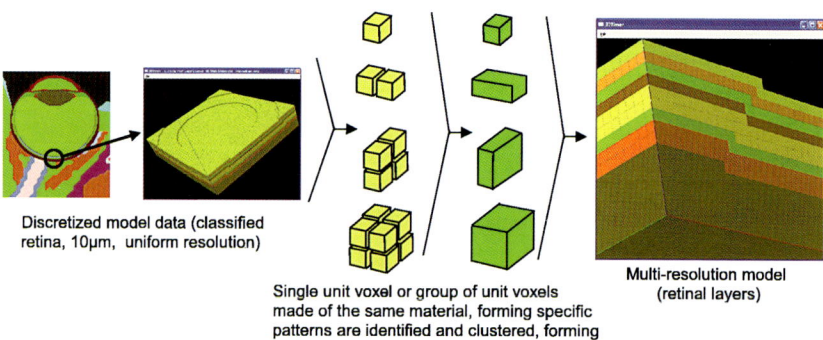

Discretized model data (classified retina, 10µm, uniform resolution)

Single unit voxel or group of unit voxels made of the same material, forming specific patterns are identified and clustered, forming multi-resolution voxels whenever possible.

Multi-resolution model (retinal layers)

FIGURE 15.16. Procedure to obtain a multi-resolution model.

Some circuit techniques can be used to reduce the size of the model's equivalent electric circuit. For instance, if the impedance value between two nodes is very small in comparison with surrounding impedance values, it can be approximated with a value of zero. While this generates a small error in the calculation, it also lumps two circuital nodes together, reducing the rank of the resulting impedance matrix by one.

In addition to optimizing the model representation, numerical methods used can be tuned to use less computing resources. In general, since in this type of simulation the resulting transfer function of the linear system can be represented as a large sparse symmetric matrix, and taking advantage of the fact that sparse matrices can be stored in compact data structures, computing storage space can be saved by solving the linear system using iterative methods that keep the matrix sparse.

Results

There are several different types of information that can be obtained using a current spread simulation through the impedance method or similar methods. One matter of interest is to relate the current density recorded in the ganglion cell layer of the retina with the particular electrode array geometry and intensity of injected currents. Further, the effect that the location of the electrode arrays' current return has on the ganglion cell layer excitation pattern may be studied with this type of simulation. Models with resolution fine enough to describe the geometrical characteristics of actual retinal cells can also be developed. Results below show how current spread simulations can provide information about the response to excitation by stimulating electrode arrays.

The current spread simulation provides the quantitative data needed to understand what would be the excitation pattern – and thus possibly the visual pattern – induced by a particular electrode arrangement and activation pattern.

The simulation results in Figure 15.17 show a transversal cut of a retinal model excited by an electrode array, and the resulting current densities inside the ganglion cell layer of the retina. The 4×4 electrode array is composed of electrodes measuring 75μm per side and is backed by a dielectric material. Individual electrodes are 75μm apart. Each electrode is injecting a current of 200μA. The resolution of the model is 2μm.

The location of the current return changes the current path and affects the current density in the ganglion cell layer. As in the previous simulation, Figure 15.18 shows a transversal cut of a retinal model excited by an electrode array, and the resulting current densities inside the ganglion cell layer of the retina. The setup of the model is the same as the previous simulation, with the exception of the current return placement, which is not centered over the electrode array in this case.

A

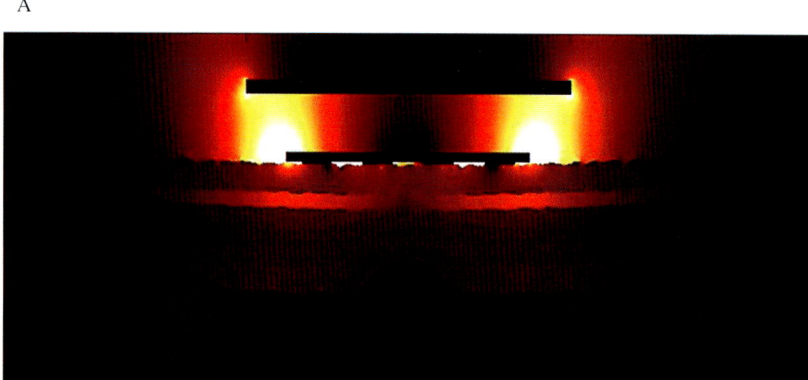

3D section log colorplot - Case 1, 200 uA per electrode

B

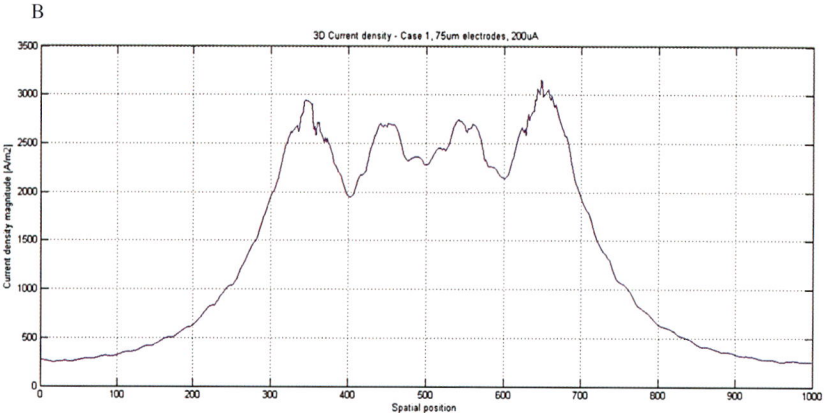

FIGURE 15.17. Impedance method simulation results: (a) logarithmic colorplot represents current densities in slice of 3D retinal model (lighter = more current). (b) Line plot showing current densities for a cross section through the ganglion cell layer. The effect of the current injected by each individual electrode can be seen as a peak in the current density value, aligned with the spatial position of each electrode.

Knowing the current densities in the retinal tissue could help understanding how effectively the implant will operate. It also helps answering a number of questions regarding the design of implantable electrode arrays, including determining a proper location for the implant to be placed, measuring efficiency of different shapes and sizes of electrodes to be used, calculating the optimal inter-electrode spacing, finding a convenient location for the current return, and verifying that the implant will provide safe levels of current to the surrounding tissue.

A

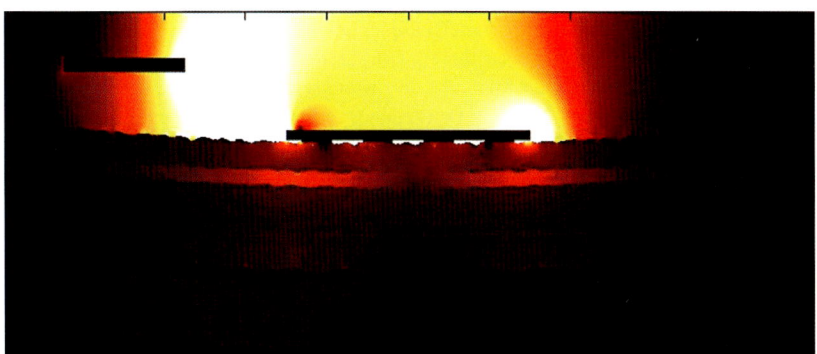

3D section log colorplot - Case 3, 200 uA per electrode

B

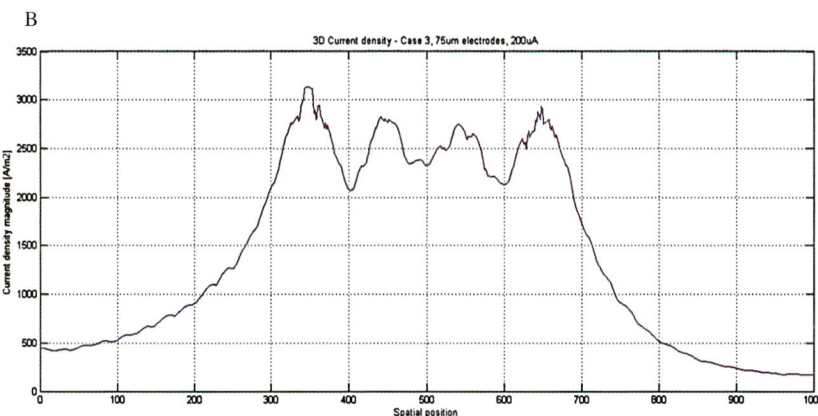

FIGURE 15.18. Impedance method simulation results: (a) logarithmic colorplot represents current densities in slice of 3D retinal model (lighter = more current). Note the higher current densities closer to the current return (top). (b) Line plot showing current densities for a cross section through the ganglion cell layer. While the four peaks can be differentiated, current density is higher in the leftmost peak, which is closer to the current return.

References

1. J. Wyatt and J. Rizzo, "Ocular implants for the blind," *IEEE Spectrum*, vol. 33, no. 5, pp. 47–53, 1996.
2. A. Y. Chow, M. T. Pardue, G. A. Peyman, and N. S. Peachey, "Development and application of subretinal semiconductor microphotodiode array," in *Vitreoretinal Surgical Techniques*, F. A. Peyman, S. A. Meffert, M. D. Conway, and F. Chou, Eds. London, UK: Martin Dunitz, 2001, pp. 575–578.
3. E. Zrenner, "Will retinal implants restore vision?," *Science*, vol. 295, no. 5557, pp. 1022–1025, February 8, 2002.

4. M. S. Humayun, J. D. Weiland, B. Justus, C. Merrit, J. Whalen, D. Piyathaisere, S. J. Chen, E. Margalit, G. Fujii, R. J. Greenberg, E. J. de Juan, D. Scribner, and W. Liu, "Towards a completely implantable, light-sensitive intraocular retinal prosthesis," presented at Proceedings of the 23rd Annual International Conference of the IEEE Engineering in Medicine and Biology Society, 2001.

5. M. S. Humayun, E. de Juan Jr., J. D. Weiland, G. Dagnelie, S. Katona, R. Greenberg, and S. Suzuki, "Pattern electrical stimulation of the human retina," *Vision Research*, vol. 39, no. 15, pp. 2569–2576, 1999/7 1999.

6. N. d.-N. Donaldson and T. A. Perkins, "Analysis of resonant coupled coils in the design of radio frequency transcutaneous links," *Medical & Biological Engineering & Computing*, vol. 21, no. 5, pp. 612–627, Sept. 1983.

7. F. C. Flack, E. D. James, and D. M. Schlapp, "Mutual inductance of air-cored coils: Effect on design of radio-frequency coupled implants," *Medical & Biological Engineering*, vol. 9, no. 2, pp. 79–85, March 1971.

8. D. C. Galbraith, M. Soma, and R. L. White, "A wide-band efficient inductive transdermal power and data link with coupling insensitive gain," *IEEE Transactions on Biomedical Engineering*, vol. 34, no. 4, pp. 265–275, April 1987.

9. C. R. Neagu, H. V. Jansen, A. Smith, J. G. E. Gardeniers, and M. C. Elwenspoek, "Characterization of a planar microcoil for implantable microsystems," *Sensors and Actuators A: Physical*, vol. 62, no. 1–3, pp. 599–611, July 1997.

10. M. Soma, D. C. Galbraith, and R. L. White, "Radio-frequency coils in implantable devices: Misalignment analysis and design procedure," *IEEE Transactions on Biomedical Engineering*, vol. 34, no. 4, pp. 276–282, April 1987.

11. A. E. Ruehli, "Inductance calculations in a complex integrated-circuit environment," *IBM Journal of Research and Development*, vol. 16, no. 5, pp. 470–481, Sept. 1972.

12. E. B. Rosa, "The self and mutual inductances of linear conductors," *Bulletin of the Bureau of Standards*, vol. 4, no. 2, pp. 301–344, 1908.

13. F. W. Grover, *Inductance Calculations: Working Formulas and Tables*. New York: D. Van Nostrand, 1946.

14. E. B. Rosa and F. W. Grover, "Formulas and tables for the calculation of mutual and self-inductance [revised]," *Bulletin of the Bureau of Standards*, vol. 8, no. 1, pp. 1–237, 1 January 1912.

15. H. H. Pennes, "Analysis of tissue and arterial blood temperatures in the resting human forearm," *Journal of Applied Physiology*, vol. 1, no. 2, pp. 93–122, 1 August 1948.

16. G. Lazzi, S. C. DeMarco, W. Liu, J. D. Weiland, and M. S. Humayun, "Computed SAR and thermal elevation in a 0.25-mm 2D model of the human eye and head in response to an implanted retinal stimulator – part II: results," *IEEE Transactions on Antennas and Propagation*, vol. 51, no. 9, pp. 2286–2295, 2003.

17. P. Bernardi, M. Cavagnaro, S. Pisa, and E. Piuzzi, "Specific absorption rate and temperature elevation in a subject exposed in the far-field of radio-frequency sources operating in the 10–900-MHz range," *Biomedical Engineering, IEEE Transactions on*, vol. 50, no. 3, pp. 295–304, 2003.

18. D. Poulikakos, *Conduction Heat Transfer*. Englewood Cliffs, N.J.: Prentice-Hall, 1994.

19. K. Gosalia, J. Weiland, M. Humayun, and G. Lazzi, "Thermal elevation in the human eye and head due to the operation of a retinal prosthesis," *Biomedical Engineering, IEEE Transactions On*, vol. 51, no. 8, pp. 1469–1477, 2004.

20. M. J. Ackerman, "The Visible Human Project," *Proceedings of the IEEE*, vol. 86, no. 3, pp. 504–511, 1998.

21. "Dosimetry Models," *ftp://starview.brooks.af.mil/EMF/dosimetry_models/*.
22. S. C. DeMarco, G. Lazzi, W. Liu, J. D. Weiland, and M. S. Humayun, "Computed SAR and thermal elevation in a 0.25-mm 2D model of the human eye and head in response to an implanted retinal stimulator – part I: models and methods," *Antennas and Propagation, IEEE Transactions On*, vol. 51, no. 9, pp. 2274–2285, 2003.
23. C. Gabriel, R. J. Sheppard, and E. H. Grant, "Dielectric properties of ocular tissues at 37 degrees C," *Physics in Medicine and Biology*, vol. 28, no. 1, pp. 43–49, January 1983.
24. C. Gabriel, S. Gabriel, and E. Corthout, "The dielectric properties of biological tissues: I. Literature survey," *Physics in Medicine and Biology*, no. 11, pp. 2231–2249, 1996.
25. S. Gabriel, R. W. Lau, and C. Gabriel, "The dielectric properties of biological tissues: II. Measurements in the frequency range 10 Hz to 20 GHz," *Physics in Medicine and Biology*, no. 11, pp. 2251–2269, 1996.
26. S. Gabriel, R. W. Lau, and C. Gabriel, "The dielectric properties of biological tissues: III. Parametric models for the dielectric spectrum of tissues," *Physics in Medicine and Biology*, no. 11, pp. 2271–2293, 1996.
27. J. T. Ernest, "Choroidal Circulation," in *Retina*, S. J. Ryan, Ed., 2nd ed. St. Louis: Mosby, 1994, pp. 76–80.
28. P. W. V. Gurney, 'Is Our "Inverted" Retina Really "Bad Design"?,' in *Creation Ex Nihilo*, vol. 13, 1999, pp. 37–44.
29. L. M. Parver, C. Auker, and D. O. Carpenter, "Choroidal Blood-Flow as a Heat Dissipating Mechanism in the Macula," *American Journal of Ophthalmology*, vol. 89, no. 5, pp. 641–646, 1980.
30. G. Lazzi, "Thermal Effects of Bioimplants," *to appear in Engineering in Medicine and Biology Magazine*, 2005.
31. T. I. C. o. N.-I. R. P. (ICNIRP), "Guidelines for Limiting Exposure to Time-Varying Electric, Magnetic, and Electromagnetic Fields (up to 300 GHz)," *Health Physics*, vol. 74, no. 4, pp. 494–522, April 1998.
32. O. P. Gandhi, J. F. DeFord, and H. Kanai, "Impedance Method for Calculation of Power Deposition Patterns in Magnetically Induced Hyperthermia," *IEEE Transactions on Biomedical Engineering*, vol. BME-31, no. 10, pp. 644–651, October 1984.
33. M. Eberdt, "A multi-resolution meshing scheme for the impedance method," North Carolina State University. 2001, pp. viii, 73 leaves.
34. M. Eberdt, P. K. Brown, and G. Lazzi, "Two-dimensional SPICE-linked multi-resolution impedance method for low-frequency electromagnetic interactions," *IEEE Transactions on Biomedical Engineering*, vol. 50, no. 7, pp. 881–889, July 2003.
35. D. W. Armitage, H. H. LeVeen, and R. Pethig, "Radiofrequency-induced hyperthermia: computer simulation of specific absorption rate distributions using realistic anatomical models," *Physics in Medicine and Biology*, vol. 28, no. 1, pp. 31–42, January 1983.
36. P. K. Brown, "A three-dimensional multi-resolution admittance method for low-frequency bioelectromagnetic interaction," in *Electrical and Computer Engineering Thesis*. Raleigh: North Carolina State University 2005.
37. C. J. Karwoski, D. A. Frambach, and L. M. Proenza, "Laminar profile of resistivity in frog retina," *Journal of Neurophysiology*, vol. 54, no. 6, pp. 1607–1619, December 1985.
38. R. W. Rodieck, "The Primate Retina," in *Comparative Primate Biology*, vol. 4, G. Mitchell and J. Erwin, Eds. New York: A. R. Liss, 1986, pp. 203–274.

16
Microstimulation with Chronically Implanted Intracortical Electrodes

Douglas McCreery

Neural Engineering Program
Huntington Medical Research Institutes

Abstract: Stimulating microelectrodes that penetrate into the brain afford a means of accessing the basic functional units of the central nervous system. Microstimulation in the region of the cerebral cortex that subserve vision may be an alternative, or an adjunct, to a retinal prosthesis, and may be particularly attractive as a means of restoring a semblance of high-resolution central vision. There also is the intriguing possibility that such a prosthesis could convey higher order visual percepts, many of which are mediated by neural circuits in the secondary or "extra-striate" visual areas that surround the primary visual cortex. The technologies of intracortical stimulating microelectrodes and investigations of the effects of microstimulation on neural tissue have advanced to the point where a cortical-level prosthesis is at least feasible. The imperative of protecting neural tissue from stimulation-induced damage imposes constraints on the selection of stimulus parameters, as does the requirement that the stimulation not greatly affect the electrical excitability of the neurons that are to be activated. The latter is especially likely to occur when many adjacent microelectrodes are pulsed, as will be necessary in a visual prosthesis. However, data from animal studies indicates that these restrictions on stimulus parameter are compatible with those that can evoke visual percepts in humans and in experimental animals. These findings give cause to be optimistic about the prospects for realizing a visual prosthesis utilizing intracortical microstimulation.

Introduction

The feasibility of a visual prosthesis to restore vision to a blind person using intracortical microstimulation (ICMS) in the visual cortex was first studied in a 42-year-old woman who had been totally blind for 22 years secondary to glaucoma [1]. Thirty-eight microelectrodes were implanted in the right visual cortex, near the occipital pole, for a period of 4 months. Visual percepts reported as small spots of light, called phosphenes, were produced by 34 of the 38 implanted microelectrodes. Threshold currents for phosphene generation with trains of biphasic pulses were as low as 1.9 A, and most of the microelec-trodes had thresholds below 25 A. The phosphenes' brightness could be modified with stimulus amplitude, frequency, and pulse duration. The phosphenes did not flicker during the stimulation and ended quickly when the stimulation was terminated. The apparent size of phosphenes ranged from a "pin-point" to a "nickel" (20-mm-diameter coin) held at arm's length. Distinct phosphenes could be elicited by microelectrode spaced as closely as 500 m, suggesting that a prosthesis based on ICMS might restore vision with good spatial detail.

The Anatomy and Physiology of the Visual System, as they Relate to a Cortical Visual Prosthesis

The axons of the optic nerve and tract project onto the lateral geniculate nucleus of the thalamus, which in turn projects in an orderly fashion onto the striate cortex at the posterior (occipital) pole of the brain. This "visuotopic" projection from the retina onto the striate cortex creates a map of the corresponding half of the visual field in the cortex. The macula lutea, the portion of the retina that mediates high-resolution vision and thus the perception of spatial detail, is represented in the posterior part of the striate cortex, while more peripheral regions of the retina (the visual field) are represented more anteriorly (Figure 16.1). The macula occupies only a small portion of the retina but is represented by a disproportionately large region of the cerebral cortex, which is commensurate with its role in the perception of fine spatial detail; this "cortical magnification factor" reflects the high density of photoreceptors in and around the macula. The small size of the macula (a few mm in diameter) may make it difficult to develop a retinal prosthesis that can deliver electrical stimulation into this region with sufficient spatial detail so as to convey to a blind person a facsimile of the high-resolution vision that this region subserves in a sighted person. However, due to the cortical magnification factor, the macula is represented by many square centimeters of cerebral cortex. Further, since this region is located at the extreme posterior pole of the brain, surgical access is relatively easy, so a cortical prosthesis that incorporates a large number of intracortical microelectrodes may offer the best prospects for restoring useful central vision to blind person. However, the topology of the cortical projection of more peripheral regions of the visual field is much less favorable for a cortical-level prosthesis, since they project onto cortical regions deeper within the central sulcus between the cerebral hemispheres and into the depths of the calcarine sulci of both hemispheres.

There also is the intriguing possibility that such a prosthesis could convey higher order visual percepts, many of which are mediated by neural circuits in the "secondary" or "extra-striate" visual areas that surround the primary visual cortex. For example, the perception of the speed and direction of a moving object in the visual field appears to be mediated in the middle temporal cortex, usually designated as visual area MT or V5 [2]. The high spatial selectivity afforded by ICMS is well suited to access this neuronal circuitry. Thus microstimulation

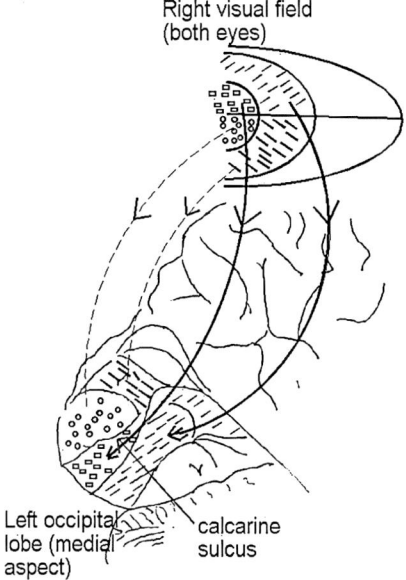

Right visual field
(both eyes)

Left occipital
lobe (medial
aspect)

calcarine
sulcus

FIGURE 16.1. Diagrammatic representation of the visuotopic projection of the retina onto the human striate cortex, which occupies a large part of the occipital lobe. Central vision is represented in a large region at the posterior pole of the brain, whereas peripheral vision is represented more anteriorly (Adapted from [28]).

with 10 A current pulses in area MT of a Rhesus monkey appeared to impart a sense of directed motion to an object in the visual field [3, 4]. However, when the stimulus current was increased to 80 A (and thus was able to activate neurons throughout a greater volume of tissue surrounding the microelectrode), The monkey's performance indicated that the percept of motion in a particular direction was much more ambiguous. These observations suggested that the percept of directed motion was mediated by direct activation of a highly localized population of neurons in area MT.

Microelectrodes for Chronic Intracortical Microstimulation

If a neural prosthesis based on ICMS is to restore a useful facsimile of central vision, or is to convey higher order visual perception into the extra-striate cortex, it must include a large number of microelectrodes and also must accommodate the somewhat irregular geometry of the cerebral cortex. The local curvature and irregularities of the brain can be accommodated by implanting a large number of arrays, each with a small footprint. Figure 16.2a shows such an intracortical microstimulating array that was fabricated in our laboratory [5, 6]. The microelectrodes extend from an epoxy superstructure that is 3 mm in diameter, and can be of various lengths. In addition to the 16 working microelectrodes, the array contains 3 longer stabilizing pins, which help to prevent torsion and traction

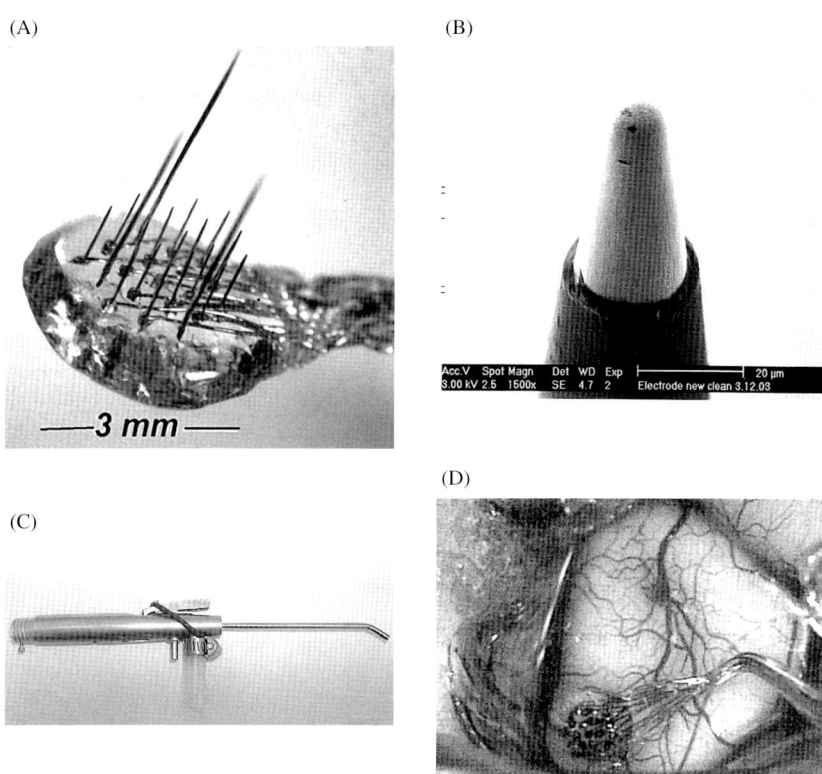

FIGURE 16.2. (a) An array of 16 iridium intracortical microelectrodes extending from an epoxy superstructure. (b) A scanning electron micrograph of one of the iridium microelectrodes. The Parylene-C insulation has been ablated from the tip region, revealing the iridium metal which constitutes the microelectrode's active surface. (c) An inserter tool for implanting the microstimulating arrays. Prior to being inserted into the brain, the microelectrodes are protected within the tool's barrel. (d) A microelectrode array immediately after implantation into the cerebral cortex of a cat.

in the cable from dislodging the newly implanted array from the brain. This feature has proved to be especially useful when implanting multiple arrays. In the version shown, the microelectrodes are spaced approximately 380 m apart. The individual microelectrodes (Figure 16.2b) are formed from pure iridium wire by electrolytic etching. The shafts then are insulated with approximately 3 m of Parylene-C, a biocompatible insulating material, which then is ablated from the electrode's tip region using a finely focused ultraviolet laser. This allows good control of the amount of exposed metal, which constitutes the electrode's working surface. For our studies of microstimulation in the cat cochlear nucleus and cerebral cortex, and in the visual cortex of the rhesus monkey, we have used electrodes with surface areas of $0.001–0.002\,\mathrm{mm}^2$, or $1000–2000\,\mathrm{m}^2$ [5, 7–10]. The tips of these microelectrodes are quite blunt with radii of curvature of $4–5 \times 10^{-6}\,\mathrm{mm}$ (4–5 m). This blunted shape promotes a more even distribution of the stimulus current over the electrode's working surface [11] and it also appears to minimize the risk of injury to the microvessels of the brain [8, 12].

The electrode shafts are composed of iridium, a metal that is especially useful for microelectrodes that must function throughout the life of the patient. An iridium electrode's working surface can be oxidized ("activated") to form a hydrous oxide film. Then, during the microstimulation, this oxide can shuttle between valence states, and thereforeefficiently transfer charge from the electrode into the brain [13, 14]. Another important property of activated iridium is its extremely low rate of dissolution during stimulation, even when the electrode is operated at moderately high current densities; an important property of a prosthesis that must function for the life of the patient.

Figure 16.2c shows an inserter tool that we have developed for implanting microstimulating arrays into the human cochlear nucleus, but which also is suitable for implanting the arrays into the cerebral cortex [5, 7, 10]. Prior to being inserted into the brain, the microelectrodes are protected within the tool's barrel so that the tip of the barrel can be placed in contact with the brain at the intended point of implantation. The electrodes then are injected into the brain at a moderately high velocity (approximately 1.5 m/sec). The bend at the end of the barrel facilitates access to brain surfaces that are partially obstructed by juxtaposing structures. Figure 16.2d shows an array immediately after implantation into the cerebral cortex of a cat. We also have used this tool to implant multiple microelectrode arrays into the visual cortex of 3 rhesus monkeys [7, 10]. The tool can be constructed so that the end of barrel is angled at up to about 70° (In Figure 16.2b, the angle is approximately 40°). This feature will be valuable when implanting the microelectrode arrays into the portion of the human visual cortex that lies within the central sulcus near the brain's occipital pole, between the two cerebral hemispheres.

Other technologies have been developed that embody the principle of the "floating" array of multiple penetrating microelectrodes, most notably the Utah Intracortical Array of 100 microelectrodes [15, 16], shown in Figure 16.3. The Utah array has been used primarily for long-term recording for cortical neurons [17], but with appropriate modifications, it could be the technology of choice for a cortical visual prosthesis, particularly if, or where, a high spatial density of electrodes is desirable [18].

Tissues Responses to Chronically Implanted Microelectrodes

With any neural prosthesis, the risk of injury to the tissue at the site of implantation must be carefully evaluated. In the case of the intracortical electrode array, there is an opportunity for tissue injury during implantation as the microelectrodes penetrates down into the highly vascular brain tissue, and also during the subsequent period of residence in the brain. There also is the opportunity for injury from the electrical stimulation itself. The micrograph in Figure 16.4a

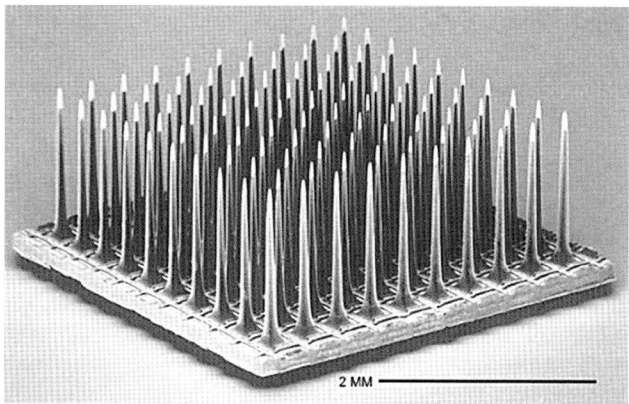

FIGURE 16.3. The Utah Intracortical array. It contains 100 microelectrodes, and is an example of a technology for efficiently fabricating high-density microelectrode arrays.

shows the footprint in a cat's cerebral cortex left by the 18 iridium shafts of an array of the type depicted in Figure 16.2a. The array had been implanted for 30 days. The histologic section was cut perpendicular to the axis of the electrode shafts at a depth of approximately 0.5 mm below the array's superstructure and below the surface of the brain, and shows the tracks of the 16 working electrodes (T) and those of the 3 longer stabilizing shafts (S).

The state of the neural tissue shown in Figure 16.4a is typical of our findings at the implant sites of these arrays. In spite of the numerous small blood vessels permeating the tissue (some of which appear similar to the electrode tracks), we have rarely seen evidence of vascular injury within the array's footprint. We presume that as the electrodes are being inserted into the brain, their blunt tips push the vessels aside rather than severing them. Figure 16.4b shows a histologic section through the tip site of one of the microelectrodes from the same array that had been subjected to 8 hours of electrical stimulation. Figure 16.4c shows the tip site of an unpulsed microelectrode. The particulars of the stimulation regimen are described below. A conspicuous feature of the stimulated site is the aggregate of inflammatory cells (seen as irregular elongated profiles) around the tip site [19].

In Figures 16.4b, c, there are normal-appearing neurons surrounding all of the pulsed and the unpulsed tip sites. However, prolonged electrical stimulation does convey a risk of injury to nearby neurons. There are several mechanisms by which electrical stimulation might inflict tissue injury. A detailed discussion of this topic is beyond the scope of this chapter, but the interested reader is referred to the review by the author [20]. The propensity for neural damage is affected strongly by the interaction of two variables, the stimulus charge per phase and the stimulus charge density. In most cases, the stimulus waveform is configured so that the positive and negative phases inject equal amounts of charge. Charge per phase is the charge injected by the electrode during each presentation of either the positive or the negative (anodic or cathodic) phases of the stimulus current.

(A)   (B)

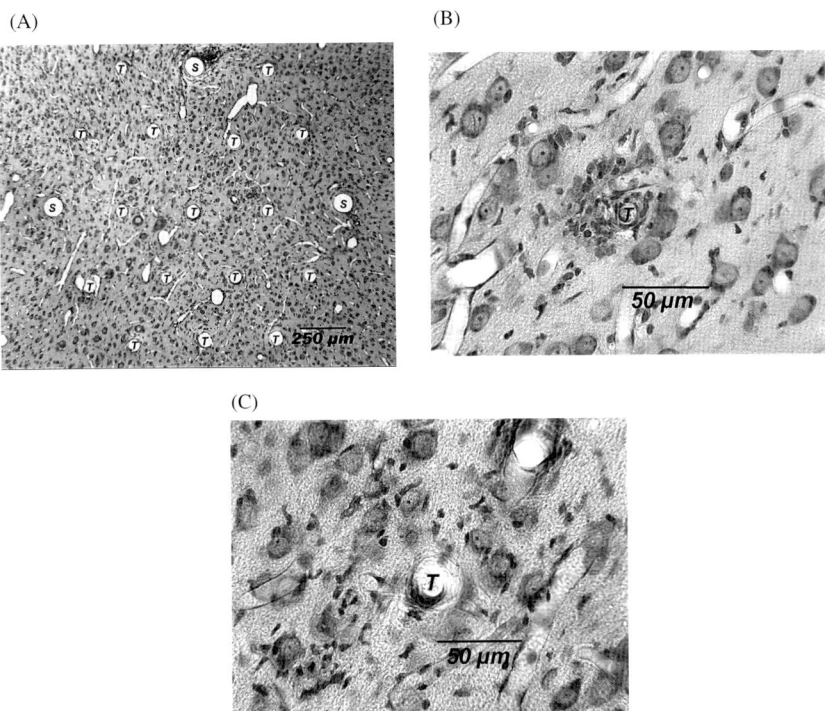

(C)

FIGURE 16.4. (a). The footprint in a cat's cerebral cortex left by the 18 iridium shafts of an array of the type depicted in Figure 15.2A, implanted for 30 days. (b) A histologic section through the tip site of one of the microelectrodes that have been subjected to 8 hours of electrical stimulation at .004 μC/ph, and at a charge density of 200 μC/cm^2. (c) A histologic section through the tips site of an unpulsed microelectrode.

A stimulus waveform that is often used for neural stimulation is composed of a pair of cathodic and anodic current pulses of constant amplitude and so charge per phase is simply $I \times d$, where "I" is the stimulus current pulse amplitude and "d" is the duration of the pulse comprising the anodic or cathodic phase of the stimulus. The unit of electric charge is the Coulomb (C) which corresponds to 1 ampere of current for 1 second. For microstimulation, the most common units are microcoulombs (10^{-6} C), and nanocoulombs (10^{-9} C). Charge density per phase is charge per phase divided by the electrode's surface area. It usually is expressed as microcoulombs per square cm of electrode Figure 16.5 shows how charge density and charge shows how charge density and charge per phase interact to determine the risk of neuronal injury. These data are drawn from several studies [21–24] but all employed electrodes implanted on the cat's cerebral cortex, or microelectrodes implanted within the cortex. Filled and open symbols represent combinations of charge density and charge per phase that did or did not cause injury to nearby neurons. It is apparent that charge density and charge per phase interact synergistically to determine the propensity for neural injury.

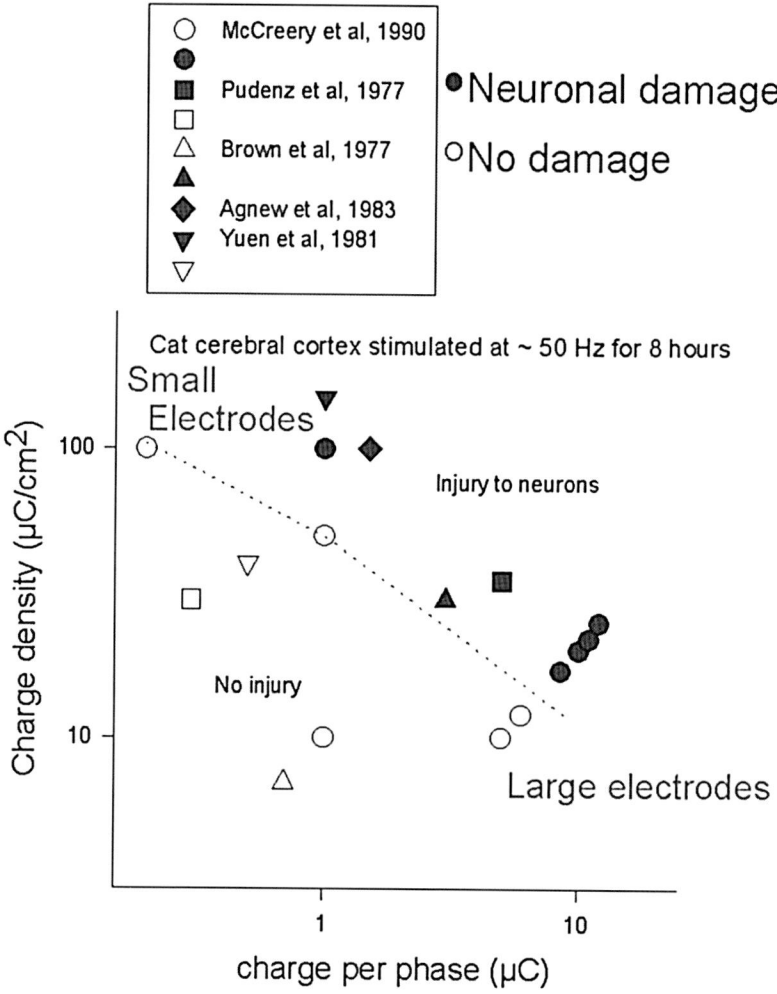

FIGURE 16.5. The relation between charge density, charge per phase and the induction of injury to neurons. The diagonal line indicates the approximate demarcation between the domains of damaging and the non-damaging stimulation.

Penetrating microelectrodes, with their small surface areas, are represented in the extreme upper left portion of the graph, and in this region charge density can be relatively high without injuring the nearby neurons. However, injury will occur if the combination of charge density and charge per phase is excessively high. Figure 16.6 shows a histologic section through the tip site of a microelectrode that underwent 8 hours of pulsing at 0.06 C/phase and at a charge density of approximately 3000 C/cm². There are no recognizable neurons within about 70 m of the tip site. Fortunately, these stimulus parameters are far in excess of what would be required for a visual prosthesis employing ICMS. The microelectrodes depicted in Figure 16.4 were evaluated with a regimen that is more appropriate for

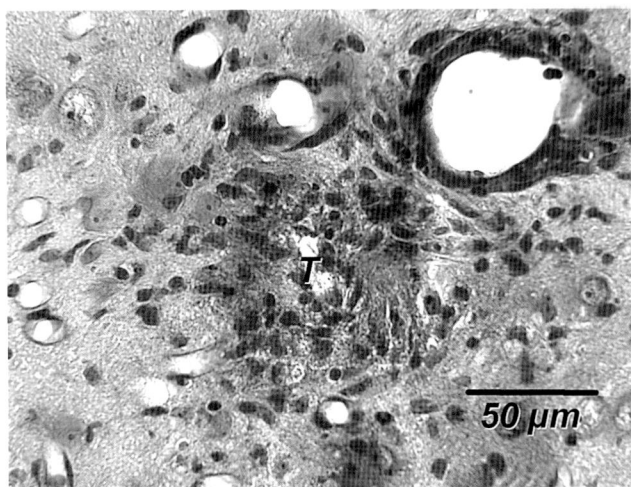

FIGURE 16.6. A histologic section through the tip site of a microelectrode that have been subjected to 8 hours of electrical stimulation at .06 μC/ph, and at a charge density of 3000 μC/cm². There is marked infiltration of inflammatory cells into the tissue surrounding of the tip site (T), and no recognizable neurons near the tip.

ICMS. Eleven of the 16 microelectrodes underwent 8 hours pulsing at a charge density of .004 C/ph, and at a charge density of 200 C/cm². This continuous stimulus was delivered at a rate of 50 pulses per second (50 Hz). Two days after the end of stimulation, the cat was deeply anesthetized and the brain tissue was examined for evidence of tissue injury. All histologically normal neurons within 100 m of each of the electrode tip sites were counted in three adjacent histologic sections through each of the tip sites, and the distance of each neuron from the center of the tip site was recorded. In this animal, the average density of normal-appearing neurons around the pulsed and unpulsed electrodes (Figures 16.4b, c) differed by less than 5% between 30 and 100 m from the center of the tip site.

The selection of stimulus parameters must be constrained by the imperative of not causing neural injury. However, additional constraints are imposed by the requirement that the stimulation does not markedly perturb the physiological properties of the neurons to be stimulated, and in particular, that the stimulation does not induce large changes in their electrical excitability. If a cortical visual prosthesis is to restore useful vision to a blind person, numerous microelectrodes will have to be pulsed either simultaneously or in an appropriate temporal sequence that is yet to be determined. Prolonged neural stimulation may induce a persisting reduction of the electrical excitability of neurons that are close to the electrodes. In the cerebral cortex, this phenomena of stimulation-induced depression of electrical excitability, or "SIDNE", [6, 9, 12] appears to be exacerbated when many closely spaced microelectrodes are pulsed [9]. SIDNE can occur in the absence of histologically detectable neural injury [12], but it remains an issue in the design of stimulus protocols since recovery of neuronal excitability after cessation of the stimulation typically requires many days, and when SIDNE is present, the functionality of the prosthesis obviously will be degraded and/or unstable.

We have observed persisting but ultimately reversible SIDNE in the feline cochlear nucleus and in the cerebral cortex after prolonged microstimulation [6, 9, 12], so it may be a general manifestation of the response of neurons to highly localized protracted stimulation. We investigated SIDNE in the sensorimotor cortex of the cat [9], a site which, while not identical to the primate visual cortex, is a convenient model because many of its neurons project into the corticospinal tract (the "pyramidal tract"), a bundle of axons that traverses from the cortex into the spinal cord and is sufficiently compact to be accessible with a single recording microelectrode. Arrays of 16 iridium microelectrodes similar to Figure 16.2a were implanted chronically into the sensorimotor cortex of adult cats for at least 40 days, and a recording electrode was implanted into the pyramidal tract (Figure 16.7a). Neuronal responses characteristic of single pyramidal tract axons ("unit-like responses" or ULRs) were recorded during ICMS (Figure 16.7b). Each trace was generated by averaging the response to 2048 consecutive intracortical pulses. The negative peak of the ULRs is indicated by (*). The graph's abscissa is the latency after the start of the $150\,\mu s$/phase biphasic stimulus pulse. The number near the right edge of each averaged trace signifies the amplitude of the $150\,\mu s$ intracortical stimulus pulse. The threshold of this ULR is 8 A. When the intracortical microelectrodes were not pulsed except to determine the electrical threshold of the ULRs, the electrical threshold of the ULRs was very stable for at least 7 hours (Figure 16.7c). Note that the threshold of most of the ULRs was below 12 A (1.8 nC, with the 150 s stimulus pulses used in the study). The abscissa is the ULR's initial threshold and the ordinate is the threshold after 7 hours, during which the microelectrodes were not pulsed. The broken lines represents a change of one stimulus level, and is the limit of accuracy for the determination of the unit's threshold. The numbers adjacent to some circles indicates multiple ULRs at those coordinates. As discussed above, eight hours of continuous pulsing of the intracortical microelectrodes at 4 nC/ph and 50 pulses per second did not induce histologically detectable neural damage, and when these stimulus parameters were applied for 7 hours to only 1 of the 16 microelectrodes in the intracortical array, there was elevation of the electrical threshold of only 1 of 18 ULRs evoked from these pulsed microelectrodes (Figure 16.7d). The data were acquired from 5 cats in which only the single intracortical microelectrode from which the ULRs were evoked was pulsed continuously for 7 hours, at $26.5\,\mu A$ (4 nC/ph) and at 50 Hz. The threshold of all but one of the ULRs was unchanged. However, when all 16 microelectrodes were pulsed for 7 hours at 4 nC/ph, the threshold of most of the ULRs became markedly elevated (Figures 16.8a, b), and the SIDNE was perhaps even more severe when the 16 microelectrodes were pulsed sequentially (at 50 Hz per microelectrode) (compare Figures 16.8a and 16.8b).

We postulate that reversible SIDNE is caused by the prolonged high-rate neuronal activity that is induced by the stimulation, and when the stimulus amplitude is sufficiently high so as to produce significant overlap in the regions

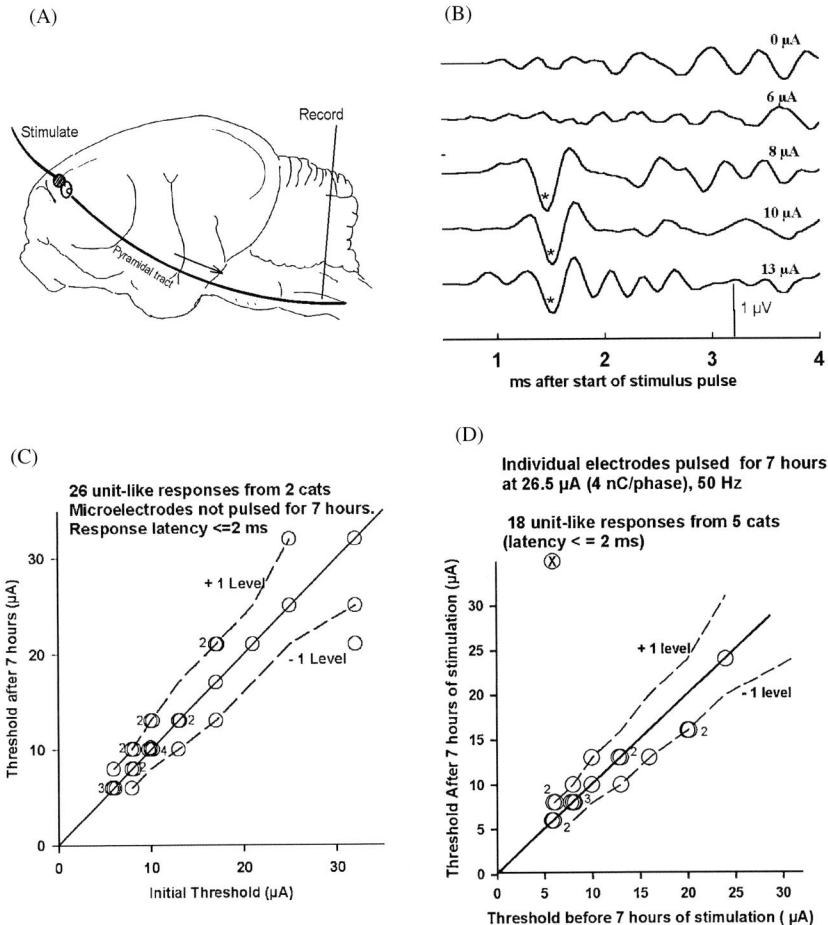

FIGURE 16.7. (a) The scheme used to study stimulation-induced change in neuronal excitability in the cat's cerebral cortex. The array of stimulating microelectrodes is implanted into the sensorimotor cortex and a single recording electrode is implanted though the cerebellum and into the pyramidal tract. (b) A "unit-like response" (ULR) evoked by microstimulation in a cat's postcruciate gyrus and recorded from the pyramidal tract. (c) The thresholds of 26 ULRs (circles) from 2 cats. (d) The effects of 7 hours of continuous stimulation on the threshold of 18 ULRs (From [9]McCreery et al, 2002, with permission of the publisher).

of effective stimulation from adjacent microelectrodes, sequential stimulation will evoke multiple responses from neurons in this overlapping region, forcing the neurons to fire at a very high rate. When the stimulus amplitude was lower (e.g. 1.8 nC/ph, Figures 16.8c, d), then sequential pulsing of the 16 microelectrodes (Figure 16.8c) induced much less SIDNE than simultaneous pulsing of the 16 electrodes (Figure 16.8d).

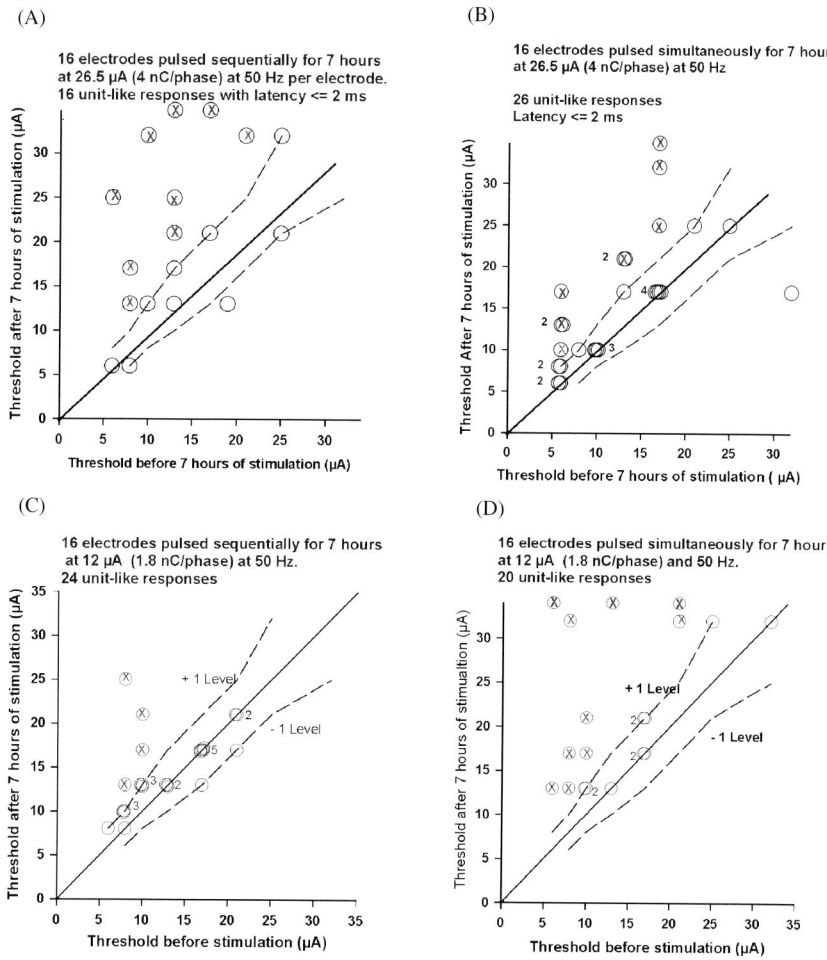

FIGURE 16.8. (a) The thresholds of 26 ULRs from one cat, in which all 16 intracortical microelectrodes were pulsed sequentially for 7 hours, at 4 nC/phase and 50 Hz per electrode. The ULRs whose threshold had increased during the 7 hours of stimulation are indicated by "X". (b) The thresholds of 18 ULRs from the same cat, in which the 16 microelectrodes were pulsed simultaneously for 7 hours at 4 nC/phase and 50 Hz. The stimulation sessions represented in A and B were conduced 20 days apart, when neuronal excitability had recovered. (c) The thresholds of 19 ULRs from another cat, in which the 16 microelectrodes were pulsed sequentially for 7 hours at 1.8 nC/phase and 50 Hz per electrode. (d) The thresholds of 22 ULRs from the same cat, in which the 16 microelectrodes were pulsed simultaneously for 7 hours at 1.8 nC/phase and 50 Hz (From [9]McCreery et al, 2002, with permission from the publisher).

In Figure 16.8c, the only neurons that exhibited SIDNE were those whose threshold at the start of the 7 hours of stimulation was less than the amplitude of the 7-hour test stimulus, suggesting that the SIDNE is related to the activation of those particular neutons. However, the fact that 7 hours of simultaneous pulsing at 1.8 nC caused a greater effect on the thresholds of these low-threshold ULRs suggests that the effects of the prolong stimulation on neuronal excitability are determined not only by the forced activation of the individual neurons, but also by the total number of neurons activated. Simultaneous pulsing of adjacent microelectrodes will induce neuronal activity over a greater distance than will sequential pulsing, due to summation of the stimulus from each electrode. While the quantitative details certainly will differ for microelectrode arrays with different interelectrode spacing, or when pulsed with different parameters, or when implanted into different regions of the cerebral cortex (visual vs. somatosensory cortex), and perhaps even when implanted into different species (humans vs. domestic cat), it is likely that SIDNE in a cortical visual prosthesis could be minimized by using a stimulation protocol that minimizes interaction between the microelectrodes (and thus also maximizing the spatial resolution). Sequential pulsing of the microelectrodes at an amplitude for which there is minimum overlap of the effective stimulus from each electrode would meet these requirements.

While SIDNE obviously will be an issue in the selection of the stimulus parameters to be used with a visual prosthesis employing ICMS, it is gratifying that we were able to deliver sequential ICMS for many hours through many adjacent microelectrode at an amplitude of 1.8 nC/ph and 12 A, without inflicting histologically detectable neural injury and without inducing detectable SIDNE in most of the low-threshold neurons near the microelectrodes. This stimulus lies within the range that was shown to induce phosphenes in a human volunteer with microelectrodes implanted in the visual cortex (1.9–25 A; [1]) which produced directed saccadic eye movements when delivered into the deep layers of the primary visual cortex of a rhesus monkey (10 A or less; [2, 25]) and which apparently is appropriate to convey a percept of motion to an object in the visual field when delivered via a microelectrode in cortical area MT (10 A; [3]).

In our laboratory, microelectrode arrays similar to the example shown in Figure 16.2a have been implanted into the cat cerebral cortex for up to 2 years [5], and have remained functional over that interval. Most of our experience with microstimulating arrays implanted for long intervals is derived from our program to develop an auditory prosthesis for person with severe hearing loss, and who are unable to benefit from a cochlear implant. We have implanted more than 50 arrays into the cats' cochlear nucleus, for up to 7 years [8, 12, 26, 27]. The cochlear nucleus receives its input from the inner ear via the auditory nerve, and is the first stage of processing of auditory information in the central nervous system. These animal studies have yielded valuable insights into the histologic and physiologic response of the neural elements surrounding penetrating microelectrodes that have been implanted for long intervals. Figure 16.9a shows a histologic section through the site of the tip of one of four microelectrodes that had been implanted in a cat's cochlear nucleus for 2588 days [8]. The histologic section was cut through the microelectrode's track at an oblique angle, so the microelectrode's tip site appears elongated. The tips site is encapsulated with a sheath of connective tissue approximately 20 m in thickness. Beyond this capsule, the fibrous network of nerve-cell processes, the "neuropil" appears normal. The neurons themselves (some indicated as "N") also appear normal. In the section, some of these neurons are within 70 m of the tip site. At the time that these stimulating microelectrodes were implanted in the cochlear nucleus, a recording electrode was implanted near the cat's inferior colliculus, a structure in the brainstem to which the neurons of the cochlear nucleus project. This electrode is too large to record the electrical

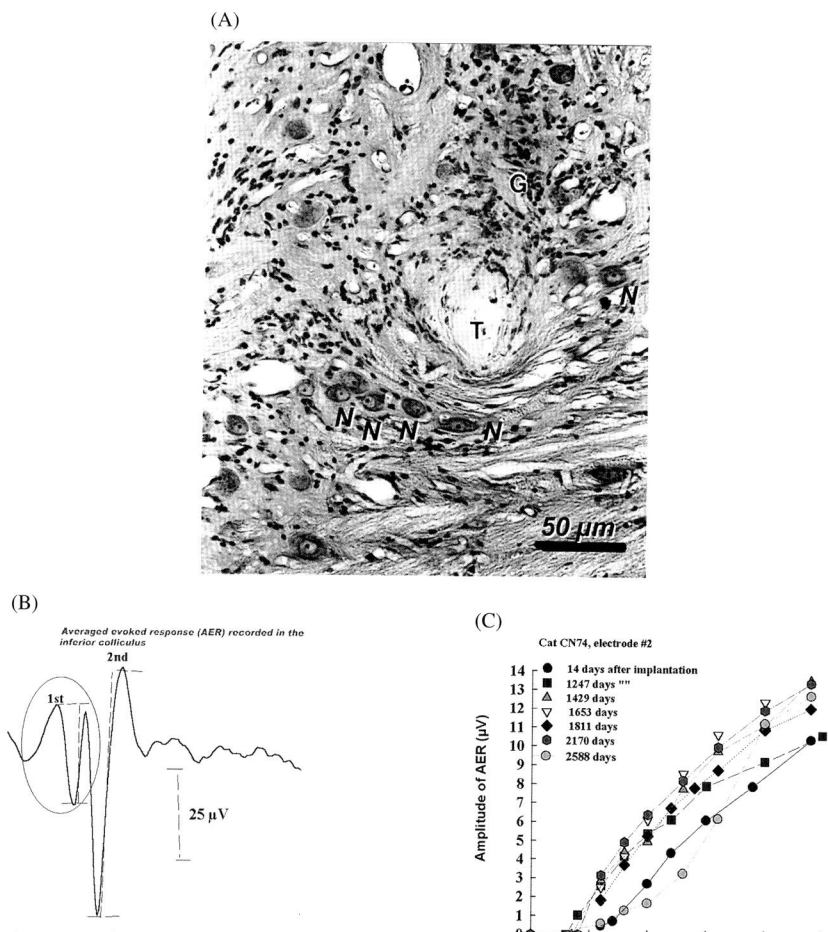

FIGURE 16.9. (a) A histologic section through the site of the tip of one of 4 microelectrodes that had been implanted in a cat's cochlear nucleus for 2588 days. (b) An averaged evoked response recorded in the cat's inferior colliculus while stimulating in the cochlear nucleus with the microelectrode whose tips site is shown in A. The first and second components of the response are indicated. (c) Plots of the amplitude of the early component of the AERs evoked from the microelectrode whose tip site is shown in (a).

activity of individual neurons but it does record a manifestation of the partly synchronized electrical activity of the population of nerve cells that is activated by the microstimulation in the cochlear nucleus. Figure 16.9b shows an example of this averaged evoked response (AER), which was generated by averaging the responses to 2048 successive presentation of the stimulus in the cochlear nucleus. Figure 16.9c shows plots of the amplitude of the early component of the AERs that were evoked from the microelectrode whose tip site is shown in Figure 16.9a. The graph's abscissa is the amplitude of the 150 s stimulus current pulses. Throughout the 7 years, the threshold of the neuronal response was quite constant (approximately 5 to 8 A) and the growth of the response with increasing stimulus amplitude also was very constant. This stability of the neuronal response is consistent with the good condition of the neural elements surrounding the tip site (Figure 16.9a).

Conclusions

Certainly, many hurdles remain to in order develop a visual prosthesis based on ICMS, and to determine its place amongst the alternative approaches to a visual prosthesis. At this time, there is not even solid evidence that the topographic projection of the retina onto the primary visual cortex can be exploited so as to convey to the patient the percept of a complex shape. However, these issues now are being investigated in animal studies (3, 26). In other animal studies, we have shown that with an appropriate choice of stimulus parameters, it should be possible to safely activate the small populations of neurons in the visual cortex that mediate either simple visual percepts such as phosphenes or some higher-order percepts, including the perception of motion by an object in the visual field. The blunt-tipped activated iridium microelectrodes used in our studies have been shown to evoke neuronal responses that are stable over many years in the brain, and the small arrays of these microelectrodes are compatible with an approach of "tiling" the surgically accessible portions of the visual cortex with numerous closely spaced arrays. Certainly, the intracortical microelectrodes and the associated implantable electronic components will continue to evolve, as will technique for safely and efficiently implanting large numbers of microelectrode into the cortex.

Acknowledgments. We thank Clarence Graham, Jesus Chavez, David Minik, Nijole Kulevitute, and Alfred Tirado for able technical assistance. Mr. Leo Bullara was responsible for much of the design of the electrode arrays developed at HMRI. We thank Edna Smith and the animal care staff for excellent care of the animals and Cheryl Long provided secretarial assistance.

The animal studies conducted at HMRI were approved by the Animal Care & Use Committee of HMRI, and were performed under the guidelines set forth in the Guide to Care and Use of Laboratory Animals (1996 edition). This work was supported in part by research grant RO1-NS40690-01A1 and by contracts NO1-NS-8-2388, NO1-NS-5-2324, and NO1-NS-5-2324 from the National Institutes of Health.

References

1. Schmidt EM, Bak MJ, Hambrecht FT, Kufta CV, O'Rourke DK, and Vallabhanath P (1996) Feasibility of a visual prosthesis for the blind based on intracortical microstimulation of the visual cortex. *Brain* 119: 507–522.
2. Salzman CD, Murasugi CM, Britten KH, and Newsome WT (1992) Microstimulation in visual area MT: effects on direction discrimination performance. *J Neurosci* 12: 2331–2355.
3. Murasugi CM, Salzman CD, and Newsome WT (1993) Microstimulation in visual area MT: effects of varying pulse amplitude and frequency. *J Neurosci* 13: 1719–1729.
4. Shadlen MN and Newsome WT (2001) Neural basis of a perceptual decision in the parietal cortex (area LIP) of the rhesus monkey. *J Neurophysiol* 86: 1916–1936.
5. Liu X, McCreery DB, Carter RR, Bullara LA, Yuen TG, and Agnew WF (1999) Stability of the interface between neural tissue and chronically implanted intracortical microelectrodes. *IEEE Trans Rehabil Eng* 7: 315–326.
6. McCreery DB, Bullara LA, and Agnew WF (1986) Neuronal activity evoked by chronically implanted intracortical microelectrodes. *Exp Neurol* 92: 147–161
7. Bradley DC, Troyk PR, Berg JA, Bak M, Cogan S, Erickson R, Kufta C, Mascaro M, McCreery D, Schmidt EM, Towle VL, and Xu H. (2005)Visuotopic mapping through a multichannel stimulating implant in primate V1. *J Neurophysiol* 93: 1659–1670. Epub 2004 Sep 1651
8. McCreery DB, Yuen TGH, and Bullara LA Physiologic and Histologic effects of prolonged microstimulation in the feline ventral cochlear nucleus. Conference on Implantable Auditory Prostheses, Asilomar CA (Abstr.), 2001.
9. McCreery DB, Agnew WF, and Bullara LA (2002) The effects of prolonged intracortical microstimulation on the excitability of pyramidal tract neurons in the cat. *Ann Biomed Eng* 30: 107–119.
10. Troyk P, Bak M, Berg J, Bradley D, Cogan S, Erickson R, Kufta C, McCreery D, Schmidt E, and Towle VA (2003) Model for intracortical visual prosthesis research. Artif Organs 27: 1005–1015.
11. McIntyre CC and Grill WM (2001) Finite element analysis of the current-density and electric field generated by metal microelectrodes. *Ann Biomed Eng* 29: 227–235.
12. McCreery DB, Yuen TG, and Bullara LA (2000) Chronic microstimulation in the feline ventral cochlear nucleus: physiologic and histologic effects. *Hear Res* 149: 223–238.
13. Beebe X and Rose TL (1988) Charge injection limits of activated iridium oxide electrodes with 0.2 msec pulses in bicarbonate buffered saline. *IEEE Trans Biomed Eng* BME-35: 494–495.
14. Robblee LS, Lefko J, and Brummer SB (1983) Activated Ir: An electrode suitable for reversible charge injection in saline solution. *Journal of the Electrochemical Society* 130: 731–733.
15. Maynard EM, Nordhausen CT, and Normann RA (1997) The Utah intracortical Electrode Array: a recording structure for potential brain-computer interfaces. *Electroencephalogr Clin Neurophysiol* 102: 228–239

16. Normann RA, Maynard EM, Rousche PJ, and Warren DJ (1999) A neural interface for a cortical vision prosthesis. *Vision Res* 39: 2577–2587.
17. Nordhausen CT, Maynard EM, and Normann RA (1996)Single unit recording capabilities of a 100 microelectrode array. *Brain Res* 726: 129–140.
18. Maynard EM (2001) Visual prostheses. *Annu Rev Biomed Eng* 3: 145–168
19. Yuen TG, Agnew WF, McCreery D, and Bullara L (1998) Accumulation of lymphocytes elicited by microstimulation of the cat's cerebral cortex. *Society for Neuroscience (abstr.)*.
20. McCreery D Tissue reaction to electrodes: The problem of safe and effective stimulation of neural tissue. In: *Neural Prosthesis: Theory and Practice*, edited by Horch KW and Dhillon GS: World Scientific Publishing; River Edge, NJ, 2004, pp. 592–607.
21. Agnew WF, Yuen TG, and McCreery DB (1983) Morphologic changes after prolonged electrical stimulation of the cat's cortex at defined charge densities. *Exp Neurol* 79: 397–411.
22. McCreery DB, Agnew WF, Yuen TG, and Bullara L (1990) Charge density and charge per phase as cofactors in neural injury induced by electrical stimulation. *IEEE Trans Biomed Eng* 37: 996–1001.
23. Pudenz RH, Agnew WF, Yuen TG, Bullara LA, Jacques S, and Shelden CH (1977) Adverse effects of electrical energy applied to the nervous system. *Appl Neurophysiol* 40: 72–87.
24. Yuen TG, Agnew WF, Bullara LA, Jacques S, and McCreery DB (1981) Histological evaluation of neural damage from electrical stimulation: considerations for the selection of parameters for clinical application. *Neurosurgery* 9: 292–299
25. Salzman CD, Britten KH, and Newsome WT (1990) Cortical microstimulation influences perceptual judgements of motion direction. *Nature* 346: 174–177.
26. McCreery DB, Yuen TG, Agnew WF, and Bullara LA (1992) Stimulation with chronically implanted microelectrodes in the cochlear nucleus of the cat: histologic and physiologic effects. *Hear Res* 62: 42–56.
27. McCreery DB, Yuen TG, Agnew WF, and Bullara LA (1997) A characterization of the effects on neuronal excitability due to prolonged microstimulation with chronically implanted microelectrodes. IEEE Trans Biomed Eng 44: 931–939.
28. Chapter 11, "Central Visual pathways". In: *Neurosciences (1997)*, edited by Purves D, Augustine, Fitzpatrick D, Katz LC, LaMantia A and McNamara JO. Sunderland, MA: Sinauer Associates, Inc.

17
A Tissue Change After Suprachoroidal-Transretinal Stimulation with High Electrical Current in Rabbits

Kazuaki Nakauchi[1], Takashi Fujikado[1], Akito Hirakata[2] and Yasuo Tano[3]

[1] Department of Applied Visual Science, Osaka University Medical School
[2] Department of Ophthalmology, Kyorin University School of Medicine
[3] Department of Ophthalmology, Osaka University Medical School

Abstract: Purpose: To investigate the safety range of current by suprachoroidal-transretinal stimulation (STS) using a high-current continuous stimulation.

Method: Sclerotomy was performed at the area just beneath the visual streak of rabbits and the platinum (Pt) electrode (diameter: $100\,\mu m$ or $200\,\mu m$) embedded in silicone plate was attached on the fenestrated sclera. Return electrode was placed in the vitreous cavity. Retina was stimulated by biphasic pulses (anodic first, duration: 0.5 msec, frequency: 20 Hz) with a current ranged from 1 to 3 mA continuously for an hour. The rabbit eyes were enucleated immediately after microscopic fundus observation, fixated with glutar-aldehide, embedded in paraffin and stained with hematoxylin-eosin.

Result: For $100\,\mu m$ electrode, no histological change was observed with a current of 1 mA, but retinal change was observed with a current of 1.5 mA.

For $200\,\mu m$ electrode, no histological change was observed with a current of 1.5 mA but with a current of 2.0 mA, retinal change was observed. The residual scleral thickness was 50–$100\,\mu m$.

Conclusion: The results of acute experiment suggested that a relatively large amount of current was able to be injected with STS method without tissue damages. In the next step, an experiment with chronic stimulation is needed to verify the safety of STS method.

Introduction

Artificial retina is one of the promising treatment to restore vision for blind people and several approaches have been developed. Epi-retinal or sub-retinal type electrodes have been promoted nowadays but they are directly attached to fragile retinal tissues. So we have developed indirect stimulating method of artificial retina that an electrode is inserted into the scleral pocket. And we already reported that our original stimulating method 'suprachoroidal-transretinal stimulation (STS)' was able to elicit electrical evoked potential (EEP) in acute experiment in rabbits [1]. Then chronic implanting method was developed and EEP responses were recorded without attenuation for 3 weeks [2]. As is widely known, although biphasic charge balanced pulse is used, the neural tissue will be damaged if excessive amount of electrical current is injected [3, 4]. In our STS method, electrode does not contact directly with retina, so it is principally safer to inject current compared with other method. In this study, we investigated the safety range of current by STS using a high-current continuous stimulation. And we studied the tissue change induced by STS with a current beyond the safety limit.

Material and Methods

Animals: Ten Dutch-belted rabbits were used. Twenty eyes were used for stimulation experiment (contents are referred to Table 17.1) and six eyes of them were for histological study. The rabbits were anesthetized with ketamine (50 mg/kg) and xylazine (20 mg/kg) cocktail injection. In need of additional anesthesia, half of first bolus was injected. The procedures used on the animals conformed to the Institutional Guidelines of Osaka University and the ARVO Resolution on the Use of Animals in Ophthalmic Research.

Electrodes: Stimulating electrode is shown in Figures 17.1a, b. A single Pt wire electrode (diameter: $100 \mu m$, $200 \mu m$, polyurethane coating) which was embedded in silicone plate ($2 \times 5 \times 0.3 mm$, manufactured by Unique Medical, Osaka, Japan) was used. And the electrode wire was bent perpendicularly at the tip, with $50 \mu m$ protruding from the plate surface. The surface area (including top and side) of electrode was $2.36 \times 10^{-4} cm^2$ for $100 \mu m$ electrode, and $6.28 \times 10^{-4} cm^2$ for $200 \mu m$ electrode. For vitreous return electrode, Pt wire (diameter: $100 \mu m$, polyurethane coating) was used. The tip 3 mm was bent perpendicularly, and the top 2 mm was uncoated.

TABLE 17.1. Tissue change by STS, These data derived from 20 eyes with macroscopic observation, 11 eyes were used for $100 \mu m$ electrode stimulation, and 9 eyes were used for $200 \mu m$ electrode stimulation. A fraction, for example, $+4/4$ means every time damaged of 4 times stimulation. Six eyes of them (with 1 mA, 1.5 mA, 2 mA of both electrode) were observed with light microscope. (ND; not done)

mA	1	1.5	2	2.5	3
$100 \mu m$ electrode Tissue change	$-2/2$	$+4/4$	$+3/3$	$+1/1$	$+1/1$
$200 \mu m$ electrode Tissue change	$-2/2$	$-2/2$	$+4/4$	ND	$+1/1$

(A) A Photo of stimulating electrodes

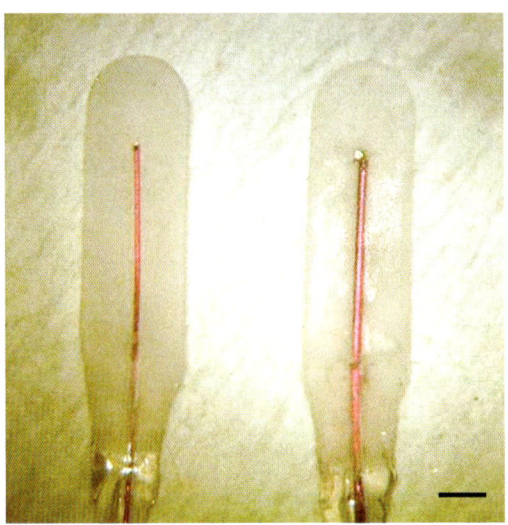

(B) A schema of stimulating electrodes

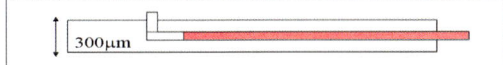

300μm

FIGURE 17.1. (a) A photo of stimulating electrodes. Left: 100 μm electrode. Right: 200 μm electrode. Silicone plate size is $2 \times 5 \times 0.3$ mm. Bar = 1 mm. Dark-colored wire part is coated with polyurethane. (b) A schema of stimulating electrode. An electrode wire is protruding 0.05 mm from silicone plate. The surface area of electrode is 2.36×10^{-4} cm^2 for 100 μm electrode, and 6.28×10^{-4} cm^2 for 200 μm electrode (including top and lateral area).

Surgery: Sclerotomy was performed at the area just beneath the visual streak of rabbits as thin as possible until choroid was observed. And the stimulating electrode (100 μm or 200 μm) was attached on the fenestrated sclera with 8-0 Vicryl and 5-0 Dacron suture. Return electrode was placed in the vitreous cavity at the ora serrata. One electrode was used on one eye for one-time stimulation, and for secondary stimulation, eye and electrode was changed.

Stimulation: The schema of STS is shown in Figure 17.2. The retina over the electrode was stimulated by biphasic pulse (anodic first, duration: 0.5 msec, frequency: 20 Hz) with a current ranged from 1 to 3 mA continuously for an hour. The stimulus shape was made with signal processor (SEN-7203, Nihon-Kohden, Shinjyuku, Japan) and generated by the isolator (WPI-A365, WPI, Sarasota, USA). The voltage between stimulating and return electrode during stimulation was measured with oscilloscope (TPS-2014, Tektronics, Beaverton, USA).

A schema of STS

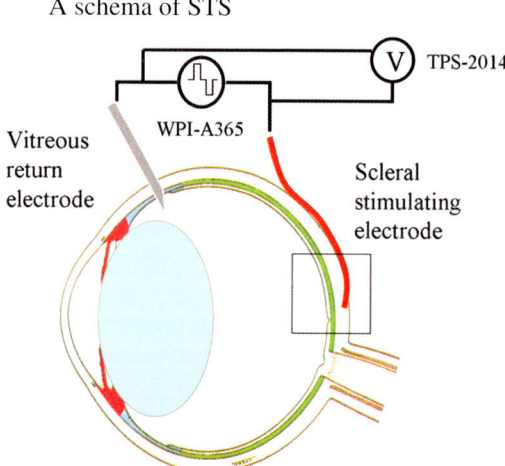

FIGURE 17.2. A schema of STS. A transretinal electric current is injected from the stimulating electrode to the vitreous electrode. A biphasic charge-balanced current is generated by isolator (WPI-A365), and the voltage between active and return electrode is monitored by oscilloscope (TPS-2014). A square surrounded place is dissected for histological study.

After stimulation, funduscopy was performed to study whether there was a tissue change or not.

Histology: When both eyes stimulation and funduscopy were over, the rabbit eyes were enucleated and marked the scleral dent where the protruded electrode was attached. The eyes were fixated with 2.5% glutaraldehyde (0.1M phosphate buffer, Ph: 7.4), embedded in paraffin and sectioned with 3 μm thickness around the marked scleral site with an interval of 20 μm and staining with hematoxylin-eosin.

Results

The results of tissue change observed with funduscopy are shown in Table 17.1. For 100 μm electrode, no tissue change was observed with a current of 1 mA (n = 2); however, retinal change was observed with a current of 1.5 mA (n = 4), 2 mA (n = 3), 2.5 mA (n = 1), and 3 mA (n = 1).

For 200 μm electrode, no tissue change was observed with a current of 1 mA (n = 2) and 1.5 mA (n = 2); however, retinal change was observed with a current of 2 mA (n = 4) and 3 mA (n = 1). With the eyes examined with a current of 1 mA, 1.5 mA, and 2 mA of theses 3 parameters for both electrodes, one preparation was made for observation with light microscope. Histological examination showed the residual scleral thickness was 50–100 μm in all preparations (n = 6).

Representative examples of no histological change with low currents are shown in Figures 17.3a, c.

Histological changes of the retina by high current stimulation are shown in Figures 17.3b, d.

These changes included the following: enlargement of choroidal vessels, irregular arrangement of outer nuclear layer cells, nuclear condensation of inner nuclear layer cells, cell swelling (hypertrophy), and expansion of the extracellular space in the outer plexiform layer, inner plexiform layer, and nerve fiber layer.

The most damaged area in the retina was positioned on the central part of the electrode, where all layers were destroyed. Peripheral to the lesion, the retina was less damaged, but localized damage was observed more in the inner

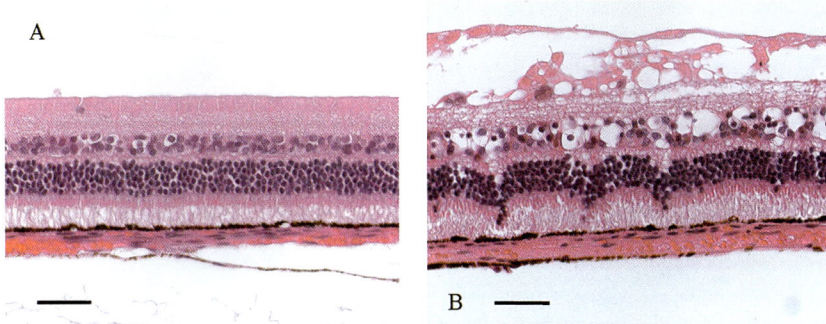

fig.3-A: 1mA injected tissue (φ100μm)   fig.3-B: 1.5mA injected tissue (φ100μm)

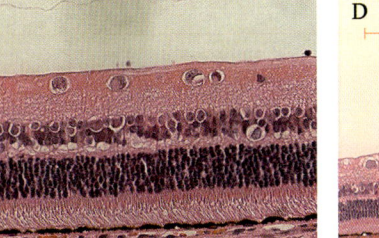 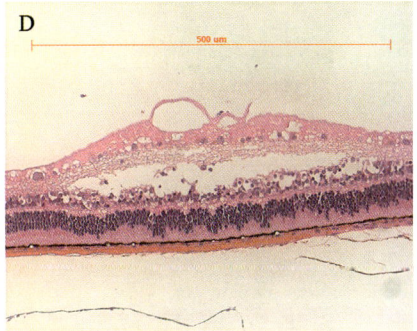

fig.3-C: 1.5mA injected tissue (φ200μm)   fig.3-D: 2mA injected tissue (φ200μm)

FIGURE 17.3. (a) The retina is not damaged by stimulation with a current of 1 mA (with 100 μm electrode). Bar = 50 μm. × 400. (b) The retina is damaged with a current of 1.5 mA (with 100 μm electrode). Bar = 50 μm × 400. All layers are damaged, but especially inner layer of the retina is affected severely. There is prominent cytoplasmic vacuolization of the neurons in the inner retina. (c) The retina is not damaged by stimulation with a current of 1.5 mA (with 200 μm electrode). Bar = 50 μm. × 400. An enlargement of choroidal vessels is observed. (d) The retina is damaged with a current of 2 mA (with 200 μm electrode) ×200. All layers are damaged, and inner nuclear layer is separated.

layers, especially the nerve fiber layer, than in the outer layers including RPE or photoreceptors.

Discussion

The Safety of Electrical Injection Limit of STS Method

We have reported previously that EEPs can be elicited from rabbits by the same 100 μm Pt-wire electrode with relatively small electricity, by a 0.5-ms-duration biphasic pulse stimulation with a current as small as 100 μA [2].

We also studied the safety of continuous 1–hour electrical stimulation (anodic first, duration; 0.5 msec, frequency; 20 Hz) by STS with a range of current between 0.5 and 1 mA. And we were not able to find any tissue change with that electrical current.

This study investigated the safe limit with continuous 1-hour stimulation. That was 1 mA with 100 μm electrode, and 1.5 mA with 200 μm electrode from result of funduscopycal change.

There was a wide safety range between current to elicit EEP (100 μA) and to make tissue damage (1.5 mA).

Previous studies have been reported about the cause and the threshold of electrode-induced neural damage. And one of main causes is thought to be electrode surface electrolysis or pH change, so the charge per phase or charge densities are used dominantly to describe a threshold [4]. This law may be correct in direct retinal stimulation, but in our non-direct retinal stimulation it is not clear whether the charge also regulates the retinal damage. In recently, some papers of epi- or sub-retinal electrode, a safe current of stimulation describes threshold as the other unit (e.g. W/cm2 or Voltage).

Next, the histological safe injection limit in our study was calculated by wattage or electrical charge density (using a simple formula: $P = I V$, $Q = I T$, where I - current, V - voltage, T - duration time, P - electrical power, Q - electrical charge).

For 100 μm electrode, a safe current of 1 mA (11 V was necessary to inject between active and return electrode) corresponded to 0.93 W/cm² or 2119 μC/cm², and harmful current of 1.5 mA (14 V) corresponded to 1.78 W/cm² or 3178 μC/cm².

For 200 μm electrode, safety 1.5 mA (12 V) corresponded to 0.57 W/cm² or 1194 μC/cm², and harmful 2 mA (17 V) corresponded to 1.08 W/cm² or 1592 μC/cm².

Reports by other groups concerning retinal (tissue) damage after electrical stimulation indicated that an epi-retinal type heater probe with 2.5 W/cm² caused immediate change of canine retina [5], and recent report about chronic stimulation by epi-retinal type electrode showed that threshold charge density of canine retina was 0.1 mC/cm² [6] and by sub-retinal type electrode with 2.5 V caused acute change of rabbit retina [7]. In the study of brain tissue, 800 μC/cm², 5.0 μC/phase (electrode; 65×10^{-6} cm²) was the safety limit with

7 hours of stimulation in cat [4]. The safety of electrical injection limit of STS method is higher than other retinal prosthesis, in voltage, wattage, and charge density. STS method has potential advantage to protect the fragile retina from the harmful effect of electricity, because electrodes are situated far from the retina.

What Causes this Retinal Degeneration?

Histologically, the tissue change in the damaged retina was atypical when compared with damage after photo-coagulation [8] or transscleral thermotherapy [9]. Although stimulating electrode was set at the sclera, the damage of RPE cells was minimal and inner retinal cells was more severe.

   Several causes are speculated for this retinal damage. Thermal burn, chemical burn, and blood flow disorder, or the mixture of these could be the causes. However, the mechanism of damaging retina with STS should be studied more precisely in future.

Conclusion

The results of acute experiments suggested that a relatively large amount of current could be injected with STS method without causing tissue damage. In the next step, an experiment with chronic stimulation is needed to verify the safety of STS method.

Acknowledgment. This work was supported by Health Science Research Grants from the Ministry of Health, Labor and Welfare, Japan.

References

1. Nakauchi K et al. (2005) Transretinal electrical stimulation by an intrascleral multichannel electrode array in rabbit eyes. Graefes Arch Clin Exp Ophthalmol. 243:169–174.
2. Nakauchi K et al. (2004) Effectiveness of transretinal electrical stimulation using chronically implanted intrascleral electrodes in rabbits. Invest Ophthalmol Vis Sci. 44: ARVO E-abstract 4185.
3. Donaldson NN and Donaldson PE. (1986) When are actively balanced biphasic ('Lilly') stimulating pulses necessary in a neurological prosthesis? Histological background; Pt resting potential; Q studies. Med Biol Eng Comput. 24:41–49.
4. McCreery DB et al. (1990) Charge density and charge per phase as cofactors in neural injury induced by electrical stimulation. IEEE Trans Biomed Eng. 37:996–1001.
5. Piyathaisere DV et al. (2003) Heat effects on the retina. Ophthalmic Surg Lasers Imaging. 34:114–120.
6. Guven D et al. (2005) Long-term stimulation by active epiretinal implants in normal and RCD1 dogs. J Neural Eng. 2:65–73.

7. Schwahn HN et al. (2001) Studies on the feasibility of a subretinal visual prosthesis: data from Yucatan micropig and rabbit. Graefes Arch Clin Exp Ophthalmol. 239:961–967.

8. Borges JM et al. (1987) A clinicopathologic study of dye laser photocoagulation on primate retina. Retina. 7:46–57.

9. Rem AI et al. (2003) Transscleral thermotherapy. Arch Ophthalmol. 121: 510–516.

18
Electrical Stimulation of Mammalian Retinal Ganglion Cells Using Dense Arrays of Small-Diameter Electrodes

Chris Sekirnjak[1], Pawel Hottowy[2], Alexander Sher[3], Wladyslaw Dabrowski[2], Alan M. Litke[3] and E. J. Chichilnisky[1]

[1] *The Salk Institute for Biological Studies*
[2] *Faculty of Physics and Applied Computer Science, AGH University of Science and Technology*
[3] *Santa Cruz Institute for Particle Physics, University of California*

Abstract: Current epiretinal implants contain a small number of electrodes with diameters of a few hundred microns. Smaller electrodes are desirable to increase the spatial resolution of artificial sight. To lay the foundation for the next generation of retinal prostheses, we assessed the stimulation efficacy of micro-fabricated arrays of 61 platinum disk electrodes with diameters 8–12 μm, spaced 60 μm apart. Isolated pieces of rat, guinea pig, and monkey retina were placed on the multi-electrode array ganglion cell side down and stimulated through individual electrodes with biphasic, charge-balanced current pulses. Spike responses from retinal ganglion cells were recorded either from the same or a neighboring electrode. Most pulses evoked only 1–2 spikes with short latencies (0.3–10 ms), and rarely was more than one recorded ganglion cell stimulated. Threshold charge densities for eliciting spikes in ganglion cells were typically below 0.15 mC/cm^2 for pulse durations between 50 and 200 μs, corresponding to charge thresholds of $\sim$ 100 pC. Stimulation remained effective after several hours and at frequencies up to 100 Hz. Application of cadmium chloride did not abolish evoked spikes, implying direct activation. Thus, electrical stimulation of mammalian retina with small-diameter electrodes is achievable, providing high temporal and spatial precision with low charge densities.

Introduction

Electrical stimulation of retinas in blind people has demonstrated the potential for direct excitation of retinal neurons as a means of re-establishing sight [1]. The retina in patients with advanced neurodegenerative diseases (such as retinitis pigmentosa or age-related macular degeneration) contains very few photoreceptors, but a substantial fraction of ganglion cells remain intact [2, 3]. Epiretinal implants specifically target surviving ganglion cells by positioning stimulating electrodes in close proximity to the inner surface of the retina.

In spite of recent successes, the present-day implants are but a small step toward restoring meaningful sight. Psychophysical studies indicate that foveal implants which create useful vision must contain a minimum of about 600 electrodes [4]. To achieve this number or greater, electrodes must be tightly packed, necessitating small stimulation sites. At present, the resolution is exceedingly crude and the density of electrodes per implant area is low: a typical epiretinal implant contains a few electrodes with diameters of several hundred microns, spaced hundreds of microns apart [1].

Useful artificial vision will require implants with hundreds or thousands of much smaller electrodes. Ideally, an advanced implant would devote one electrode to every ganglion cell and each electrode would be similar in size to the cell it is designed to stimulate (tens of microns). Little is known about the parameters which would permit reliable retinal stimulation with electrodes which approach cellular dimensions. When the electrode surface area is reduced, current and charge densities increase drastically, and high charge densities are known to cause tissue damage by electrochemical reactions [5–7].

A review of the pertinent literature reveals that the feasibility of stimulation with arrays of small electrodes in mammalian tissue has not been adequately tested. The majority of studies involving retinal stimulation have used needle-shaped probes such as platinum wires or concentric microelectrodes [8–10]. Others have attempted to utilize smaller stimulating microprobes with tip diameters of $25\,\mu m$ or less [11–14]. The geometry of such probes differs greatly from the planar disk electrode design developed for current epiretinal implants. Stimulation is always limited to a single stimulation site, prohibiting the study of stimulation using multiple electrodes, their interactions, and crosstalk effects. The use of multi-electrode arrays for retinal stimulation has been mainly limited to large electrodes with diameters between 100 and $1500\,\mu m$ [15–18]. While multi-electrode arrays with smaller electrodes have been utilized to selectively stimulate the axons of retinal ganglion cells [19, 20] and to stimulate the retina in the subretinal space [21, 22], no study has reported the use of electrodes with surface areas below $200\,\mu m^2$ to target mammalian ganglion cells.

The goal of this study was to establish current and charge thresholds for stimulation of rat, guinea pig, and primate retina using small electrodes with surface areas of $50–100\,\mu m^2$ (corresponding to diameters of $8–12\,\mu m$). In this manner, this study directly addresses the prevalent concerns about the usability of small electrodes for retinal prosthetics. Our two-dimensional multi-electrode arrays use planar disk microelectrodes very similar to those utilized in present epiretinal prosthetics, but smaller by 1–2 orders of magnitude.

Materials and Methods

Retinal Preparation

This study utilized retinal tissue from 17 adult rats, 4 guinea pigs, and one macaque monkey. The average body weight was $284 \pm 7\,\mathrm{g}$ for rats, $386 \pm 52\,\mathrm{g}$ for guinea pigs, and $4\,\mathrm{kg}$ for the macaque monkey.

Rodent eyes were enucleated after decapitation of animals deeply anesthetized with 10 mg/kg Xylazine and 50 mg/kg Ketamine HCl. Primate eyes were obtained from terminally anesthetized macaque monkeys (Macaca radiata) used by other experimenters. Immediately after enucleation, the anterior portion of the eye and vitreous were removed in room light and the eye cup placed in bicarbonate-buffered Ames' solution.

Pieces of retina 1–2 mm in diameter (Figure 18.1) were separated from the retinal pigment epithelium and placed flat on the electrode array, with the ganglion cell layer facing the array. The tissue was held in place by weighted nylon netting positioned gently over the array. The preparation was then mounted on a circuit board attached to an inverted microscope and continuously super-fused at room temperature with Ames' solution bubbled with 95% oxygen and 5% carbon dioxide. Enucleation, vitrectomy, and dissections were performed in normal laboratory lighting conditions, likely resulting in substantial retinal photobleaching.

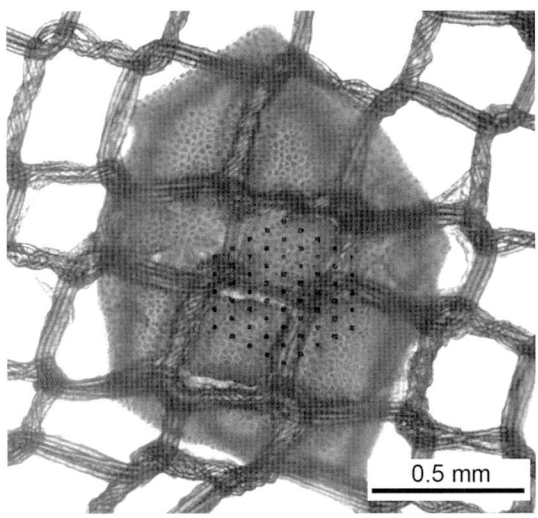

FIGURE 18.1. A retinal tissue piece on the stimulation/recording setup, photographed during an experiment. The hexagonal multi-electrode array is visible in the center. The nylon netting applies gentle pressure on the tissue.

In several experiments (10 cells), cadmium chloride (100–250 μM) was added directly to the perfusion solution to block synaptic transmission. An increase in spontaneous bursting activity was observed, confirming that the drug efficiently diffused into the ganglion cell layer.

Multi-Electrode Array

The array consisted of a planar hexagonal arrangement of 61 extracellular electrodes, approximately $0.5 \times 0.5\,mm^2$ in total size (Figure 18.2). These electrodes were used both to record action potentials extracellularly from ganglion cells [23, 24] and to apply current to the tissue for stimulation. The array was microfabricated on a glass substrate, with indium tin oxide leads and a silicone nitride insulation layer [25, 26]. Each electrode was formed by microwells (holes in the insulation layer) which were filled (electroplated) with platinum prior to an experiment. A platinum wire integrated into the array chamber served as indifferent ground, several millimeters away. All stimulations were performed using a monopolar configuration (electrode to distant ground).

Stimulation and Recording

Experiments were performed on a setup allowing for simultaneous recording of all 61 electrodes and stimulation on multiple electrodes. The array was connected to a circuit board containing two microchips which multiplexed the 61 signals

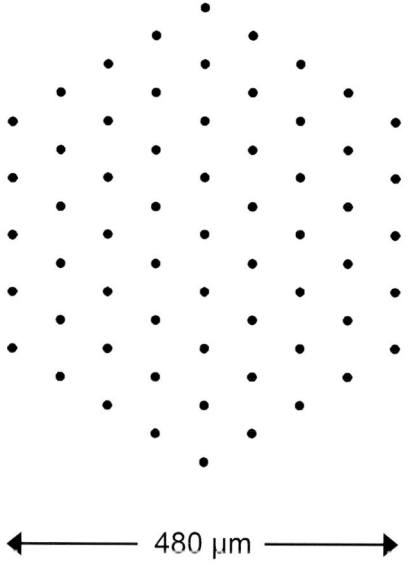

FIGURE 18.2. The hexagonal multi-electrode array. Electrode diameter varied between approximately 8 and 12 μm, with a fixed inter-electrode spacing of 60 μm.

and sent them to digitizing cards installed in a PC [25]. A dim level of illumination was maintained during the entire experiment (room lights or microscope illuminator). Recording and stimulation were controlled by interface software, and extracellular potentials were recorded from all 61 electrodes, digitized at 20 kHz [27], and stored for off-line analysis on a PC.

The stimulus pulse consisted of a negative square-wave current pulse of amplitude A and duration d, followed immediately by a positive pulse of amplitude $A/2$ and duration $2d$. Reported current values always refer to the negative phase amplitude A; pulse duration always refers to the duration d of the negative phase. All pulses were individually calibrated to produce biphasic stimuli with zero net charge. Stimulation frequency was varied between 0.25 and 100 Hz.

Many ganglion cells fire spontaneous spikes in isolated pieces of retina. Stimulation on a particular electrode was usually attempted only if spontaneous extracellular spikes could be recorded from that electrode. This approach guaranteed that the electrode was properly platinized and confirmed that ganglion cells in its vicinity were capable of firing action potentials. Typically, about half of the platinized electrodes on an array showed spontaneous activity from at least one cell.

Stimulation was attempted by using the lowest current settings and was then increased systematically if no response was seen. Threshold was defined as the current setting which produced a spike with nearly every stimulus pulse (approximately 95% of trials) during stimulation at 1–2 Hz.

Multi-electrode data were analyzed offline using Labview 6, Matlab 6, and Igor Pro 5. Means and group data were calculated in Microsoft Excel. Errors reported are standard errors of the mean (SEM), unless otherwise stated. Images were processed in Adobe Photoshop 7.

Digital Artifact Subtraction

To reveal spikes with latencies of less than 1 ms, we developed an artifact subtraction method, which takes advantage of the fact that near threshold some stimulation trials will result in evoked spikes and some will fail to evoke responses. Short-latency spikes obscured by the stimulation artifact (which typically lasted for several milliseconds) were made visible by increasing the stimulation current until a possible spike threshold was reached. Just below threshold, the recorded traces changed shape noticeably on about half of the stimulus trials (for example, a change in curvature or peak height), indicating that a possible spike was elicited. Traces containing such possible spikes were averaged and subtracted from the average over those traces without spikes. Subtracting the failures from the successful stimulations eliminated the artifact and cleanly revealed the recorded ganglion cell spike. In general, this method was successful only in recordings with exceptionally short stimulation artifacts.

Results

Evoked Spikes and Thresholds

Our objective was to elicit spikes in ganglion cells by passing current through electrodes on the multi-electrode array. Stimulation at individual electrodes with currents up to $3\,\mu A$ resulted in evoked spikes recorded at latencies between a few hundred μs and tens of ms. Of the 64 successfully stimulated ganglion cells, 45 were from rat, 11 from guinea pig, and 8 from monkey.

The majority of successful stimulation attempts yielded a single spike that often resembled the spontaneous spikes recorded at that particular electrode. In such experiments, the same electrode was used to stimulate and to record the response. Figure 18.3 shows three examples of spikes evoked in monkey, guinea pig, and rat retina, respectively. The average threshold current for 11 rat cells stimulated under identical conditions with 0.1 ms pulses was $0.66 \pm 0.03\,\mu A$, corresponding to a charge of $66\,pC$ and a charge density of $0.08\,mC/cm^2$. In all three species, typical charge densities at threshold were consistently below $0.35\,mC/cm^2$, an often-stated safe limit for stimulation with platinum electrodes [28]. Stimulation using low amplitudes such as these usually affected only cells in the immediate vicinity of the stimulation electrode: of 86 analyzed neighboring electrodes in 17 stimulation experiments, only two were an evoked spikes from a different cell observed. However, evoked spikes from different cells were frequently seen on neighboring electrodes when the current was increased several-fold, indicating current spread to locations about $60\,\mu m$ (the inter-electrode spacing) from the stimulation site.

Short-Latency Spikes

To reveal the earliest spike responses following stimulation onset, the stimulus artifact was digitally subtracted from recorded traces (see Section "Digital artifact subtraction"). Figure 18.4 shows an example of a short-latency response obscured by the artifact. Averaging and artifact subtraction revealed a spike at sub-millisecond latency which resembled the spontaneous spike recorded at the same electrode (inset). Of 13 subtracted short-latency spikes (latency 0.77 ± 0.15 ms; range 0.2–1.9 ms), 12 closely matched the spontaneous spikes observed on the same electrode.

Threshold-Duration Curves

Spike threshold was defined as the current amplitude which produced spikes at approximately 95% of trials and depended strongly on stimulus pulse duration. To quantify the relationship between threshold and duration, pulses with a range of durations between $50\,\mu s$ and 1 ms were used for stimulation. Figure 18.5 shows that in all three species, higher currents were required to evoke a spike when shorter pulses were used. No systematic species difference was observed.

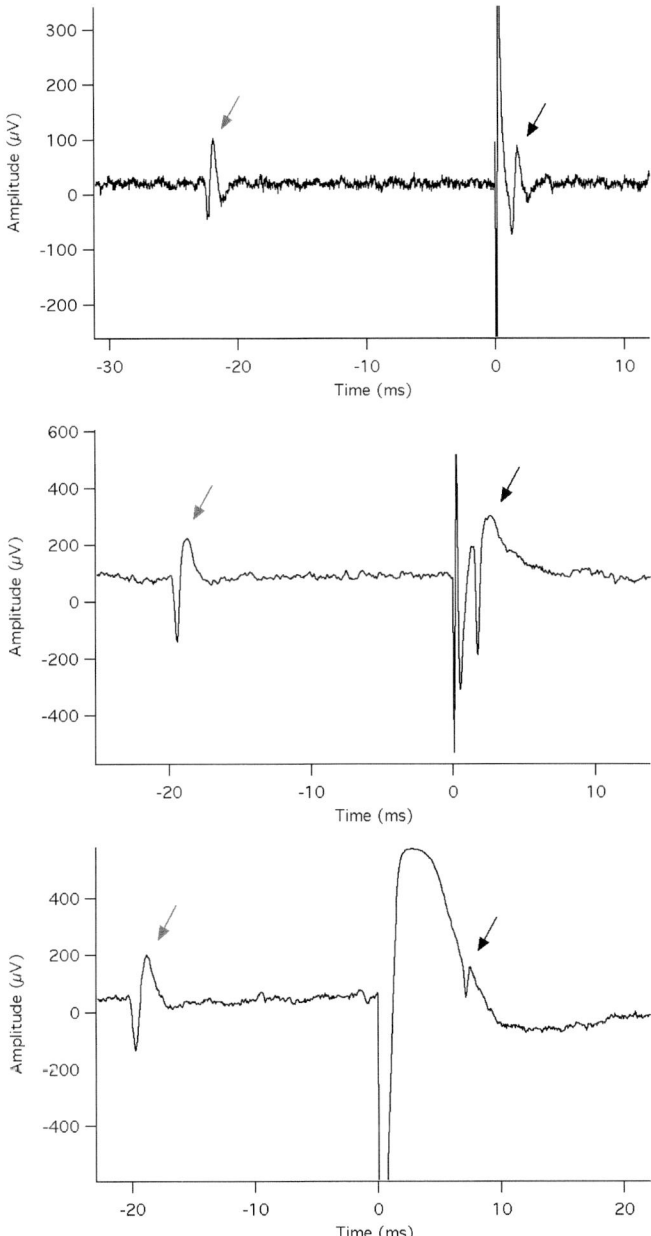

FIGURE 18.3. Examples of spikes evoked by electrical stimulation in three mammalian species. Top: monkey retina; middle: guinea pig retina; bottom: rat retina. Each example includes a spontaneous spike (gray arrow), a stimulus artifact (at time 0 ms), and a single evoked spike (black arrow). In the top two examples, the evoked spike is identical in shape to the spontaneous spike, while in the bottom trace it is distinct, indicating that a different cell was stimulated.

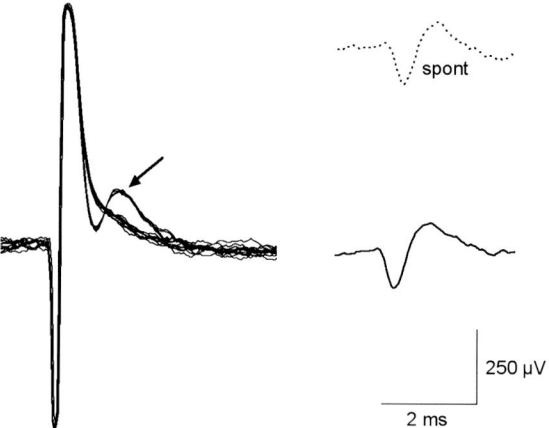

FIGURE 18.4. Artifact subtraction (rat retina). Left: overlay of 10 stimulation trials, some of which evoked a spike (arrow). Right: digitally subtracted spike with latency 0.55 ms. Top right: a spontaneous spike recorded from the same cell.

Exponential functions were fit to the data in Figure 18.5 by least-square regression and the resulting best-fit curves were used to estimate chronaxie (defined as the lowest stimulus duration required to elicit a response when using twice the minimum threshold amplitude). The average value for 12 cells was $196 \pm 66 \mu s$.

Simultaneous Stimulation

Functional artifical vision will likely require stimulation with many closely spaced electrodes. We tested the feasibility of spatial stimulation patterns by utilizing the multi-electrode array to stimulate at several electrodes at the same time. Figure 18.6 shows an example of simultaneous multi-site stimulation: spikes were evoked in two different ganglion cells by stimulating at two different electrodes with $0.6 \mu A$ pulses. Each electrode was first stimulated on its own to establish thresholds and latencies. The two evoked responses had similar but not identical threshold currents and different latencies (5 and 9 ms). Stimulation at one electrode did not evoke spikes at the neighboring site, indicating that the responses could be elicited independently of each other. Evoking independent spikes on multiple electrodes spaced $60 \mu m$ apart is thus feasible.

High Frequencies and Long Stimulation

Ganglion cells are capable of firing at frequencies of up to 100–200 Hz in response to visual stimuli [29]. High-frequency electrical stimulation is a requirement for retinal prosthetics, since sustained percepts of light must be

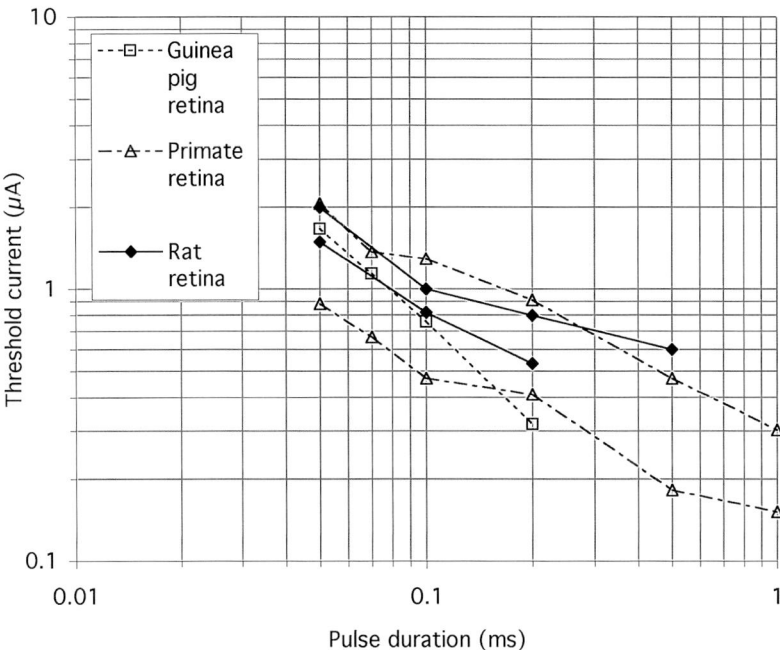

FIGURE 18.5. Threshold–duration relationships. Current necessary to evoke a spike is plotted as a function of pulse duration for 5 cells from three species.

generated. We used our setup to successfully stimulate 3 cells at frequencies up to 100 Hz. At such frequencies, spikes evoked at longer latencies (>2 ms) showed a dramatic decrease in response rate, while short-latency spikes could be evoked much more reliably.

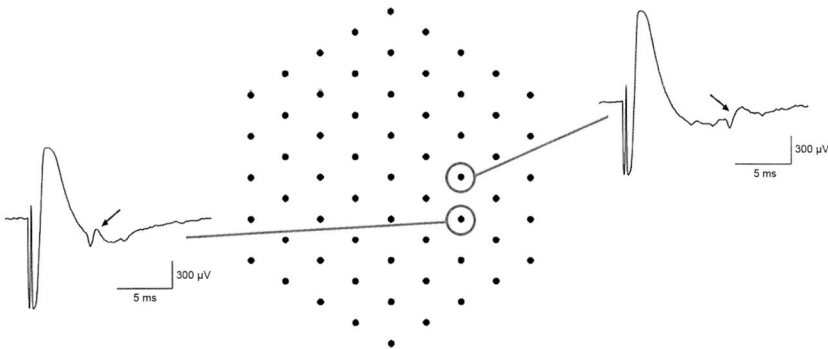

FIGURE 18.6. Simultaneous stimulation at multiple sites. Two nearby electrodes on the array were used to stimulate two ganglion cells independently of each other. The spike responses (arrows) are shown on the left and right, respectively.

Furthermore, human chronic retinal implants must be capable of delivering effective stimulation pulses over a period of many hours each day. To demonstrate that electrical stimulation remained functional after longer periods of time, two ganglion cells were stimulated continuously at 1–2 Hz for the longest duration that was experimentally feasible in our setup: spikes could still be evoked reliably after 4.5 hours, albeit with small threshold increases.

Miscellaneous Results

To ascertain that the applied electrical pulses directly stimulated ganglion cells, the calcium channel blocker cadmium chloride (100–250 μM) was applied to the bath solution to abolish synaptic transmission. In all 10 cells tested, evoked spikes were still observed after drug application, indicating that the observed spikes were not produced by mechanisms involving calcium-dependent synaptic transmission.

To establish upper limits for electrical stimulation using our multi-electrode arrays, several cells were stimulated with very high charge densities. Stimulation at currents and charge densities several-fold above threshold tended to disrupt the spontaneous and evoked spikes of ganglion cells, usually in a reversible manner: spike height decreased until spikes could no longer be detected. In many cells, spiking activity resumed after a period of several minutes. Local toxicity or heating effects were likely responsible for disrupting ganglion cell spiking, indicating that there is an upper ceiling for the optimal range of currents used in stimulation using small array electrodes.

Discussion

Electrical stimulation of rat, guinea pig, and primate retina with dense, small-diameter electrode arrays was achieved using low charge densities. Our multi-electrode arrays closely resemble those currently in use for human patient testing but contain much smaller electrodes at a much smaller electrode spacing. The purpose of this study was to elucidate the basic stimulation parameters so that the next generation of implants may incorporate a design using much smaller electrodes than are currently in use.

An important concern regarding implantable stimulation devices is their capability to deliver electrical current that is safe, yet efficient. As shown in this study, a single ganglion cell (at most a few cells) can be induced to fire a single spike (at most a few spikes) by using currents typically below 1 μA and with charge densities below 0.10 mC/cm^2. The literature is inconclusive about the safety of small electrodes for prosthetic implants: reported threshold values range from 0.16 mC/cm^2 [13] to 0.59 mC/cm^2 [14] for small-diameter microprobes, whereas 0.35 mC/cm^2 is the safe limit for stimulation with platinum electrodes [28]. We conclude that small-diameter electrode arrays (8–12 μm) do not require dangerously high charge densities to achieve reliable spiking in mammalian retina.

To determine the effectiveness of small-diameter electrodes, it would be of interest to record the neuronal activity in the vicinity of the stimulation electrode. However, stimulation and recording through the same small electrode brings about technical challenges. A large stimulus artifact obscures early spikes and others have reported that this artifact renders recording the response to stimulation near ganglion cell somas unfeasible [20]. While such technical limitations biased most of our observations to longer latency spikes (>1 ms), artifact subtraction performed on a few stimulated cells clearly revealed an early response at latencies of a few hundred μs. We speculate that most long latency stimulations also evoked an obscured short latency spike. Furthermore, since high-frequency stimulation tended to abolish long-latency spikes, we assume that the shortest-latency spikes will be the most significant during actual stimulation of human retina.

Small electrodes such as those used here typically only record spikes from cells located a short distance from the electrode. The shape of most evoked spikes resembled the spontaneous spike recorded at the recording electrode, indicating that the same nearby ganglion cell was stimulated. Taken together with the results of our neighbor-analysis, we conclude that currents around 1 μA usually do not activate cells outside a ~30 μm radius around the stimulating electrode.

The simultaneous stimulation experiment provided evidence for the feasibility of simultaneous stimulation with a fixed current pulse amplitude at multiple sites, such as would be necessary in a retinal prosthesis. It further demonstrates that two neighboring electrodes can stimulate two cells independently and without crosstalk. While only two electrodes were used here, an epiretinal implant would utilize hundreds or thousands of stimulation sites to achieve artificial sight. Our findings suggest that stimulation with high spatial resolution is feasible using small electrodes.

In summary, safe and effective stimulation of mammalian retina with small planar electrodes is clearly feasible with high temporal and spatial precision. Future generations of epiretinal prosthetics should aim toward a design incorporating electrodes whose size approaches cellular dimensions.

Acknowledgments. This research was supported by the Salk Institute Pioneer Postdoctoral Fellowship, Second Sight Medical Products Inc., McKnight Foundation NEI grant 13150 (EJC), NEI grant 12893 (SSMP), NSF Awards PHY-0245104 and PHY-0417175 (AML), Polish State Committee for Scientific Research, Project 3 T11E 011 27 (WD), and NATO Collaborative Linkage Grant (AML & WD).

References

1. Humayun M (2003) Clinical trial results with a 16-electrode epiretinal implant in end-stage RP patients. In: The First DOE International Symposium on Artificial Sight. Fort Lauderdale, FL: Department of Energy.
2. Santos A, Humayun MS, de Juan E, Jr., Greenburg RJ, Marsh MJ, Klock IB, Milam AH (1997) Preservation of the inner retina in retinitis pigmentosa. A morphometric analysis. Arch Ophthalmol 115:511–515.

3. Humayun MS, Prince M, de Juan E, Jr., Barron Y, Moskowitz M, Klock IB, Milam AH (1999) Morphometric analysis of the extramacular retina from postmortem eyes with retinitis pigmentosa. Invest Ophthalmol Vis Sci 40:143–148.

4. Cha K, Horch KW, Normann RA (1992) Mobility performance with a pixelized vision system. Vision Res 32:1367–1372.

5. Pollen DA (1977) Responses of single neurons to electrical stimulation of the surface of the visual cortex. Brain Behav Evol 14:67–86.

6. Brummer SB, Robblee LS, Hambrecht FT (1983) Criteria for selecting electrodes for electrical stimulation: theoretical and practical considerations. Ann N Y Acad Sci 405:159–171.

7. Tehovnik EJ (1996) Electrical stimulation of neural tissue to evoke behavioral responses. J Neurosci Methods 65:1–17.

8. Humayun M, Propst R, de Juan E, Jr., McCormick K, Hickingbotham D (1994) Bipolar surface electrical stimulation of the vertebrate retina. Arch Ophthalmol 112:110–116.

9. Weiland JD, Humayun MS, Dagnelie G, de Juan E, Jr., Greenberg RJ, Iliff NT (1999) Understanding the origin of visual percepts elicited by electrical stimulation of the human retina. Graefes Arch Clin Exp Ophthalmol 237:1007–1013.

10. Suzuki S, Humayun MS, Weiland JD, Chen SJ, Margalit E, Piyathaisere DV, de Juan E, Jr. (2004) Comparison of electrical stimulation thresholds in normal and retinal degenerated mouse retina. Jpn J Ophthalmol 48:345–349.

11. Dawson WW, Radtke ND (1977) The electrical stimulation of the retina by indwelling electrodes. Invest Ophthalmol Vis Sci 16:249–252.

12. Wyatt J, Rizzo JF, Grumet A, Edell D, Jensen RJ (1994) Development of a silicon retinal implant: epiretinal stimulation of retinal ganglion cells in the rabbit. Invest Ophthalmol Vis Sci 35:1380. ARVO abstract.

13. Rizzo JF, Grumet AE, Edell D, Wyatt J, Jensen R (1997) Single-unit recording following extracellular stimulation of retinal ganglion cell axons in rabbits. Invest Ophthalmol Vis Sci 38:S40.

14. Jensen RJ, Rizzo JF, 3rd, Ziv OR, Grumet A, Wyatt J (2003) Thresholds for activation of rabbit retinal ganglion cells with an ultrafine, extracellular microelectrode. Invest Ophthalmol Vis Sci 44:3533–3543.

15. Hesse L, Schanze T, Wilms M, Eger M (2000) Implantation of retina stimulation electrodes and recording of electrical stimulation responses in the visual cortex of the cat. Graefes Arch Clin Exp Ophthalmol 238:840–845.

16. Walter P, Heimann K (2000) Evoked cortical potentials after electrical stimulation of the inner retina in rabbits. Graefes Arch Clin Exp Ophthalmol 238:315–318.

17. Humayun MS, Weiland JD, Fujii GY, Greenberg R, Williamson R, Little J, Mech B, Cimmarusti V, Van Boemel G, Dagnelie G, de Juan E (2003) Visual perception in a blind subject with a chronic microelectronic retinal prosthesis. Vision Res 43: 2573–2581.

18. Rizzo JF, 3rd, Wyatt J, Loewenstein J, Kelly S, Shire D (2003) Methods and perceptual thresholds for short-term electrical stimulation of human retina with microelectrode arrays. Invest Ophthalmol Vis Sci 44:5355–5361.

19. Grumet A (1999) Electric stimulation parameters for an epi-retinal prosthesis. In: Department of Electrical Engineering and Computer Science, p 144: Massachusetts Institute Of Technology.

20. Grumet AE, Wyatt JL, Jr., Rizzo JF, 3rd (2000) Multi-electrode stimulation and recording in the isolated retina. J Neurosci Methods 101:31–42.

21. Zrenner E, Stett A, Weiss S, Aramant RB, Guenther E, Kohler K, Miliczek KD, Seiler MJ, Haemmerle H (1999) Can subretinal microphotodiodes successfully replace degenerated photoreceptors? Vision Res 39:2555–2567.
22. Stett A, Barth W, Weiss S, Haemmerle H, Zrenner E (2000) Electrical multisite stimulation of the isolated chicken retina. Vision Res 40:1785–1795.
23. Meister M, Pine J, Baylor DA (1994) Multi-neuronal signals from the retina: acquisition and analysis. J Neurosci Methods 51:95–106.
24. Chichilnisky EJ, Baylor DA (1999) Receptive-field microstructure of blue-yellow ganglion cells in primate retina. Nat Neurosci 2:889–893.
25. Litke AM (1998) The retinal readout system: an application of microstrip detector technology to neurobiology. Nucl Instrum Methods Phys Res A 418:203–209.
26. Litke AM, Chichilnisky EJ, Dabrowskic W, Grilloa AA, Grybosc P, S. K, Rahmand M, G. T (2003) Large-scale imaging of retinal output activity. Nucl Instrum Methods Phys Res A 501:298–307.
27. Litke AM (1999) The retinal readout system: a status report. Nucl Instrum Methods Phys Res A 435:242–249.
28. Brummer SB, Turner MJ (1977) Electrical stimulation with Pt electrodes: II-estimation of maximum surface redox (theoretical non-gassing) limits. IEEE Trans Biomed Eng 24:440–443.
29. Wandell BA (1995) Foundations of Vision, Sunderland, MA: Sinauer.

19
A Mechanism for Generating Precise Temporal Patterns of Activity Using Prosthetic Stimulation

Shelley I. Fried[1], Hain-Ann Hsueh[2] and Frank Werblin[3]

[1] *Vision Science*
[2] *Bioengineering*
[3] *Molecular & Cell Biology, UC Berkeley*

Introduction

Retinitis pigmentosa and age-related macular degeneration, two of the leading causes of blindness, result in more than 100,000 cases per year in the United States alone [1–4]. Each disease has many etiologies but they both lead to a degeneration of photoreceptors, resulting in a loss of visual function. Cells in the proximal retina remain morphologically intact [5–7], and electrical stimulation from electrodes placed close to the ganglion cell layer elicits light percepts in human patients [8, 9]. This indicates that the ganglion cells and their pathways to visual cortex remain viable, even in patients who have had no light perception for many years.

Several research groups are developing retinal prosthetic devices with the hope of providing meaningful visual information to blind patients [8–11]. To be effective, these prosthetic devices should generate patterns of activity that resemble patterns normally evoked by light. This means that the device should be capable of replicating the temporal properties of light-elicited spike trains from normal retina, while spatially confining the response to a focal region – ideally a single ganglion cell. In addition, there are approximately 12 different types of ganglion cells [12]; each population generates a unique spiking pattern that is carried to a distinct downstream visual site (in both cortical and non-cortical areas). This means that the prosthetic device must be capable of matching the appropriate spiking pattern to the correct cell type.

In this study, we focused on the first component of duplicating light responses – replicating the temporal patterns generated by light in normal retina. Maximum spike rates in normal retina can exceed 250 Hz [13], but not all cell types respond with an equal number and/or pattern of spikes, so the prosthetic device must be capable of generating a wide range of spike frequencies and patterns. Variations in spike frequency are also important within a single cell type, since modulations of spike frequency are used to code information about stimulus contrast. In addition, several studies now indicate that spike timing between multiple retinal neurons is often very precise [14, 15], suggesting that a prosthetic device may need to generate spikes with high temporal precision. Here, we have developed a paradigm to generate precise temporal patterns of spiking activity in retinal ganglion cells.

Methods

Animal Preparation

The care and use of animals followed all federal and institutional guidelines and all protocols were approved by the Animal Care and Use Committee, UC Berkeley. New Zealand white rabbits, $\sim 2.5\,$kg, were anesthetized with injections of xylazine/ketamine and subsequently euthanized with an intracardial injection of sodium pentobarbital. Immediately after death, the right eye was removed. All subsequent procedures were performed under dim red illumination. The front of the eye was removed, the vitreous eliminated and the eyecup dissected so that the visual streak and regions ventral were left intact – all other areas were discarded. Three rectangular pieces of eyecup each approximately $5 \times 7\,$mm were extracted and stored in oxygenated Ames medium (see below) prior to use. Storage times ranged from 15 minutes to 8 hours. Just prior to use, the retina was dissected from the eyecup and mounted, photoreceptor side down, to a 10 mm square piece of Millipore paper which was mounted with vacuum grease to the recording chamber ($\sim 1.0\,$ml volume). The Millipore paper had a 4 mm square hole in its centre which allowed light from below to be projected on to the photoreceptors. Retinas were superfused continuously at 7–10 ml/min with Ames medium (Sigma; pH 7.4, 36 °C), equilibrated with 95% O_2 and 5% CO_2. Kanamycin sulphate antimicrobial was added to the Ames medium.

Electrophysiology

Patch pipettes were used to make small holes in the inner limiting membrane and ganglion cells were targeted under visual control. Spiking was recorded with a cell-attached patch electrode (5–6 MΩ), filled with superfusate. Light responses were used to confirm the viability of the recording setup as well as the general health of the cell.

Pharmacological agents were applied locally by micro-injection through a glass pipette using a PicoPump (World Precision Instruments, PV830). This method generated a quick wash in and out of the drug and our results were consistent with bath-applied methods. Spiking was blocked with 25 µM Tetrodotoxin (TTX, Sigma). For the synaptic blocker experiments we used a combination of 5 mM Curare, 1 mM CNQX and 10 mM AP-7 (Sigma). Data was analyzed in Matlab (MathWorks).

Light Stimulus and Data Acquisition

The stimulus presentation and data acquisition software (Presentinator) were written by E. Eizenman and G. Spor. Light stimuli were projected onto the retina from below through an LCD panel (CRL OPTO) and focused on to the photoreceptor outer segments. Background light levels were set to bleach rod responses [12] and retinas were adapted at this level for > 10 minutes prior to electrophysiology. Light stimuli consisted of 1 second stationary flashed squares (100, 200, 300, 500 or 1000 μm) centred at the soma. Contrast levels were set to 200% unless otherwise specified.

Electrical Stimulation

Electrical stimulation consisted of charge-balanced biphasic pulses: cathodic and anodic pulses were typically equal-and-opposite square wave pulses (Multi-Channel Systems STG2004 and MC Stimulus II). Cathodic pulses were delivered first and intervals between phases were large enough so that the neural response was completed before the onset of the anodic phase (typically 5 ms). At higher-frequency stimulation, the interphase interval was set to 2 ms. Electrical pulses were delivered using Platinum–Iridium electrodes (MicroProbes Inc., impedances 10–100 kΩ). Stimulating electrodes were placed (under visual control) approximately 15 μm from the somata of targeted ganglion cells.

Results

Separating Neuronal Activity from Electrical Artifact

A typical response to the cathodic phase of an electrical stimulus pulse is shown in Figure 19.1a (solid line). Large transient currents (horizontal arrows) were recorded which temporally correlate with the onset and terminations of the individual phases of the stimulus pulse. The response was modulated by TTX, a blocker of neuronal action potentials (Figure 19.1a, dotted line, the difference is most prominent in the gray box). This suggests that one or more action potentials are obscured by the large transient currents. Subtracting the TTX response from the control response unmasked a single spike (Figure 19.1b, solid line) which was similar in magnitude and kinetics to light evoked spikes (Figure 19.1b, dotted line). In response to the anodic phase of the stimulus pulse, there was no difference between the control and the TTX responses (data not shown).

Short Pulses Elicit a Single Spike

The spike was elicited immediately after the onset of the pulse (Figure 19.1b), suggesting that it was elicited at the leading edge of the pulse. To test this, we shortened the duration of the stimulus pulse. The responses to a 0.2 ms

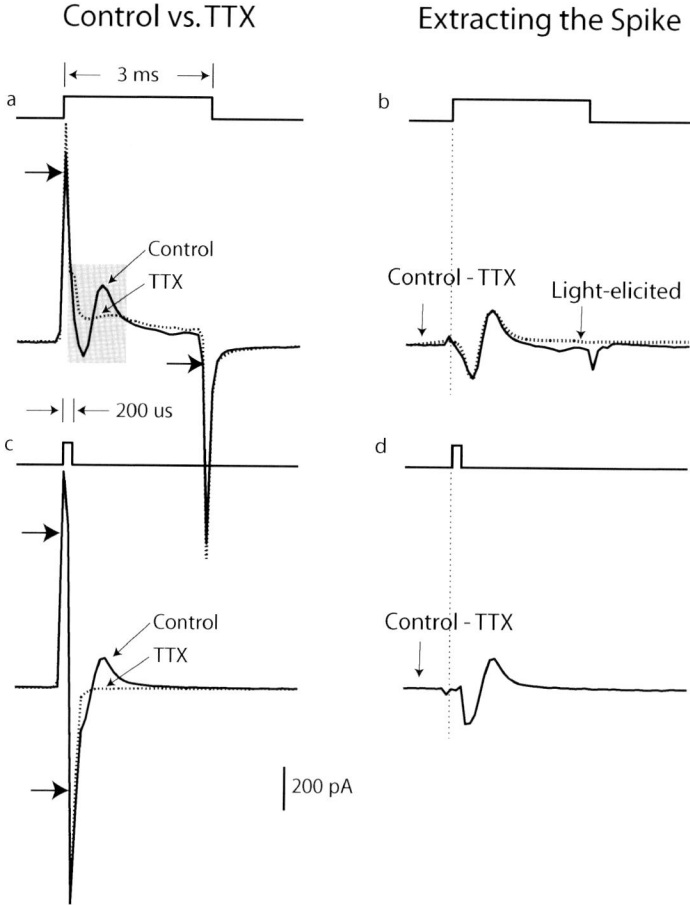

FIGURE 19.1. Single spikes are elicited by the leading edge of electrical pulses. (a) Voltage clamp response to a 3 ms 50 μA cathodic pulse (solid trace). Large current transients are seen (horizontal arrows) at pulse onset and offset (timing given by square wave above). The response was sensitive to TTX (dotted trace) but only in the time period immediately following the pulse onset (gray box). (b) Subtracting the TTX record from control extracts the spike which is similar to light-elicited spikes (dotted trace). (c) Voltage clamp response to a 200 μsec 50 μA electrical pulse (solid line). Large transient currents are again seen at pulse onset and offset (horizontal arrows). The response here was also TTX sensitive (dotted trace). (d) Subtracting the TTX record from control extracts the spike.

pulse under control conditions and in TTX (Figure 19.1c, solid and dotted lines respectively) were different, again suggesting the presence of neuronal activity. Subtraction again revealed the presence of a spike (Figure 19.1d). The response to a short pulse in control conditions was always triphasic, consisting of an upward, downward and then second upward deflection. In TTX, the response was always biphasic (no second upward deflection). In later experiments with short duration pulses the second upward deflection was a reliable marker to indicate that a spike had been elicited.

Longer duration pulses elicited longer latency spiking that occurred after completion of the stimulus pulse spikes (Figure 19.2a, left, asterisks). These

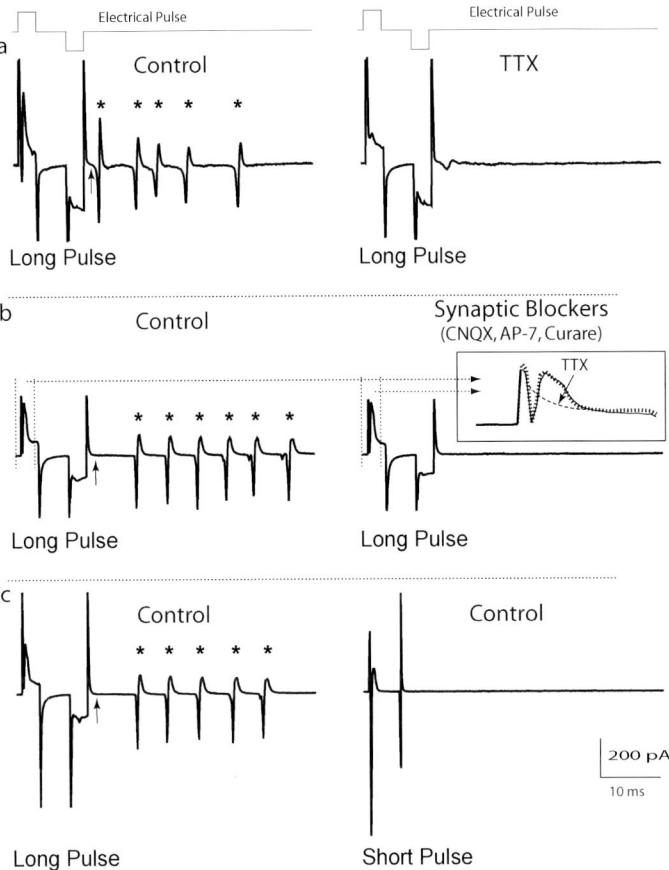

FIGURE 19.2. Early and late phases of spiking are activated by different mechanisms. (a) Left trace: A second phase of spiking (individual late-phase spikes indicated by asterisks) is elicited after completion of long duration pulses (vertical arrow indicates completion of pulse in a–c). Spikes are eliminated in TTX (right trace). Stimulus pulse waveform and timing is represented at top. (b) Late phase spiking in a different cell (left trace) is completely eliminated by 1 mM CNQX, 10 mM AP7 and 5 mM Curare (right trace). Inset: expanded time scale of the interval between the dotted lines from control (dotted) and excitatory synaptic blocker (solid) traces. The normalized TTX response from a different cell is shown for comparison. The early phase spike remains intact. (c) Late phase spiking initiated by long pulses (left, asterisks) is eliminated by reducing the duration of the pulse to 0.1 ms (right trace). The scale bar in (c) applies to all traces. Cathodic and anodic pulses were 3 ms for long pulses (a–c) and 0.1 ms for short pulse (c). Interval was 5 ms for all pulses.

spikes were also eliminated in the presence of TTX (Figure 19.2a, right). The number and timing of these spikes was variable and generally increased with increasing pulse amplitude or duration. Short duration pulses (~ 0.1 ms) did not elicit any of these spikes at all (Figure 19.2c, right). We refer to spiking responses that occur immediately following the onset of the cathodic pulse as early phase (Figure 19.1) and spikes that are elicited following completion of the entire pulse as late phase (Figs 19.2a–c, left).

Different Mechanisms Underlie Early and Late Phase Spikes

All of the late phase spiking was eliminated in the presence of a cocktail that blocked excitatory synaptic input to ganglion cells (Figure 19.2b, compare left and right panels). This indicates that excitatory synaptic input to ganglion cells is

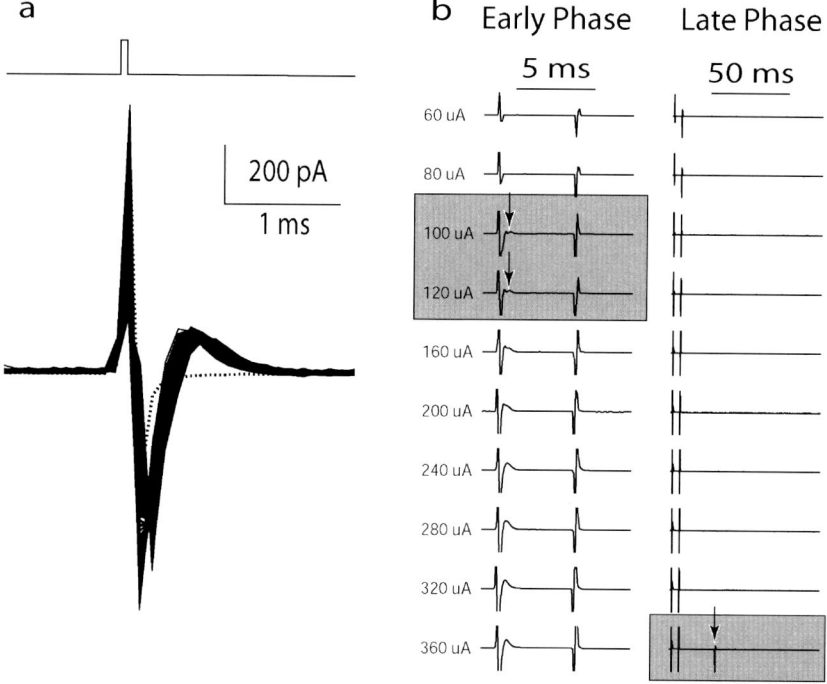

FIGURE 19.3. Short electrical pulses reliably elicit one spike per pulse. (a) Response to 200 consecutive 0.1 ms pulses delivered at 200 Hz (overlaid black lines). A second upward deflection is present in each response indicating a single spike is elicited by each pulse. Single TTX record (dotted line) is presented for comparison. Cathodic–anodic interval: 2 ms, pulse amplitude: 20 μA. (b) Response to increasing amplitude 0.1 ms pulses at two different time scales; left traces shows early phase spiking, right traces show late phase spiking. Threshold for early phase spiking is approximately 100–120 μA (shaded region, vertical arrow) and for late phase spiking is 360 μA (shaded region, vertical arrow). Stimulation frequency is 100 Hz.

required for the generation of these spikes, and suggests that long-duration pulses activate presynaptic neurons that release excitatory neurotransmitter. The early phase spike was not eliminated in the presence of these blockers (Figure 19.2b, inset), indicating that it arises from direct activation of the ganglion cell.

Reliable Generation of One Spike per Pulse

At higher stimulation frequencies, short duration pulses continued to elicit one spike per pulse (Figure 19.3a). The responses to 200 individual pulses, applied at 200 Hz (Figure 19.3a, solid lines), are all tri-phasic, indicating the presence of a neuronal spike in each. Comparison with the TTX response (dotted line) confirms that every stimulus pulse elicited a spike.

A single spike per pulse could be reliably elicited over a wide range of pulse amplitudes as shown in Figure 19.3b. For this cell, pulse amplitudes above 120 μA consistently elicited a single spike. Above 360 μA, late phase spikes were elicited, likely due to activation of presynaptic neurons.

Discussion

We have developed a stimulus paradigm that elicits one neuronal spike for each stimulus pulse and is effective over a wide range of stimulation frequencies and amplitudes. This is important because it will allow electrical prosthetic devices to generate prescribed patterns of spiking in different ganglion cell types, including those patterns that are normally generated by the retina in response to light stimulation.

Short pulses are likely to simplify the generation of spatially complex patterns of activity because they activate only ganglion cells. Longer pulses activate bipolar cells in addition to ganglion cells and are also likely to activate amacrine cells, either directly or indirectly from activated bipolar cell input. Nearly all activated amacrine cells will generate feedforward inhibitory input to ganglion cells and/or feedback inhibition to bipolar cells [16]. This suggests that the net effect from a single long pulse will be to raise the threshold for activation when a subsequent pulse (via direct stimulation) is delivered. In response to light, amacrine cells generate inhibitory signals with time constants on the order of 100 ms which further suggests that prosthetic stimulation which activates amacrine cells may have a limited temporal response. The use of short pulses avoids the long latency and broad spatial activity.

Short pulses elicit spiking in ganglion cells using less total charge than long pulses. Currently, charge density provides the limiting factor in reducing the size of stimulating electrodes [17]. If we can elicit ganglion cell responses with less total charge, we can reduce the size of the stimulating electrode and possibly generate a more focal response, a strong advantage when designing high-density microelectrode arrays for prosthetic stimulation.

Acknowledgments. We wish to thank Matt McMahon and David Zhou from Second Sight, LLC for technical advice, Tommy Chong for help with data processing. This work was supported by grants from DARPA, NIH, and funding from Second Sight, LLC.

References

1. Friedman, D. S., B. J O'Colmain et al.(2004) The prevalence of age-related macular degeneration in the United States. "Arch Ophthal" **122**; 564–572.
2. Foundation Fighting Blindness, Pamphlet on Retinitis Pigmentosa, 2004. http://www.blindness.org/RetinitisPigmentosa/
3. Bunker, C. H., E. L. Berson et al. (1984). "Prevalence of retinitis pigmentosa in Maine." *Am J Ophthalmol* **97**(3): 357–365.
4. Curcio, C. A., N. E. Medeiros et al. (1996). "Photoreceptor loss in age-related macular degeneration." *Invest Ophthalmol Vis Sci* **37**(7): 1236–1249.
5. Stone, J. L., W. E. Barlow et al. (1992). "Morphometric analysis of macular photoreceptors and ganglion cells in retinas with retinitis pigmentosa." *Arch Ophthalmol* **110**(11): 1634–1639.
6. Humayun, M. S., M. Prince et al. (1999). "Morphometric analysis of the extramacular retina from postmortem eyes with retinitis pigmentosa." *Invest Ophthalmol Vis Sci* **40**(1): 143–148.
7. Santos, A., M. S. Humayun et al. (1997). "Preservation of the inner retina in retinitis pigmentosa. A morphometric analysis." *Arch Ophthalmol* **115**(4): 511–515.
8. Humayun, M., R. Propst et al. (1994). "Bipolar surface electrical stimulation of the vertebrate retina." *Arch Ophthalmol* **112**(1): 110–116.
9. Rizzo, J. F., 3rd, J. Wyatt et al. (2003). "Perceptual efficacy of electrical stimulation of human retina with a microelectrode array during short-term surgical trials." *Invest Ophthalmol Vis Sci* **44**(12): 5362–5369.
10. Chow, A. Y., V. Y. Chow et al. (2004). "The artificial silicon retina microchip for the treatment of vision loss from retinitis pigmentosa." *Arch Ophthalmol* **122**(4):460–469
11. Zrenner, E., F. Gekeler et al. (2001). "[Subretinal microphotodiode array as replacement for degenerated photoreceptors?]." *Ophthalmologe* **98**(4): 357–363.
12. Roska, B. and F. Werblin (2001). "Vertical interactions across ten parallel, stacked representations in the mammalian retina." *Nature* **410**(6828): 583–587.
13. O'Brien, B. J., T. Isayama et al. (2002). "Intrinsic physiological properties of cat retinal ganglion cells." *J Physiol* **538**(Pt 3): 787–802.
14. Berry, M. J., D. K. Warland et al. (1997). "The structure and precision of retinal spike trains." *Proc Natl Acad Sci USA* **94**(10): 5411–5416.
15. Berry, M. J., 2nd and M. Meister (1998). "Refractoriness and neural precision." *J Neurosci* **18**(6): 2200–2211.
16. Wassle, H. (2004). "Parallel processing in the mammalian retina." *Nat Rev Neurosci* **5**(10): 747–757.
17. Margalit, E., M. Maia et al. (2002). "Retinal prosthesis for the blind." *Surv Ophthalmol* **47**(4): 335–356.

20
Electrophysiology of Natural and Artificial Vision

John R. Hetling
Department of Bioengineering
University of Illinois at Chicago

Introduction

The retina is a sensory structure, an interface between the brain and the environment. It is a critical gateway for the majority of the information a normally sighted individual receives about his or her surroundings. A photon reflected or emitted from an object enters the eye and is captured by a photoreceptor cell; the geographical location of the photoreceptor on the retina tells us about the direction from which the photon arrived. The energy of the captured photon is transformed into a change in molecular structure of a photopigment molecule, the first step in the phototransduction cascade. The end result of this cascade is a decrease in membrane potential of the photoreceptor, and thus a reduction in transmitter release at the synaptic terminal that communicates the capture of this photon to the post-synaptic cells. From here, the influence of this photon, along with the influences of many others, is propagated through the neural network of the retina, traveling by electrotonic (passive) conduction and, more rarely, action potentials, communicated between neurons primarily via chemical synapses. The information contained in these parallel streams of photons, captured by the ~ 100 million photoreceptor cells in each eye, is integrated, compared, filtered in time and space, and used as feedback in a feat of image processing that has yet to be replicated in a silicon device. The end result is that all of the information we gather from our eyes (space, color, brightness, contrast, and movement) is contained in a binary code of spikes on the roughly one million information channels (axons) of the optic nerve. These spike trains are the precursors of the coherent images of the mind's eye.

The goal of a prosthetic device for vision is to create a new interface with the world of light when the retina becomes dysfunctional due to disease. To reach this goal, a *retinal* prosthesis must sense key information in a visual scene, and then communicate this information to the mind's eye by way of the dysfunctional retina. The prosthesis becomes responsible for transduction, signal processing, information coding, and stimulation of the remaining retinal neurons. Figure 20.1 illustrates,

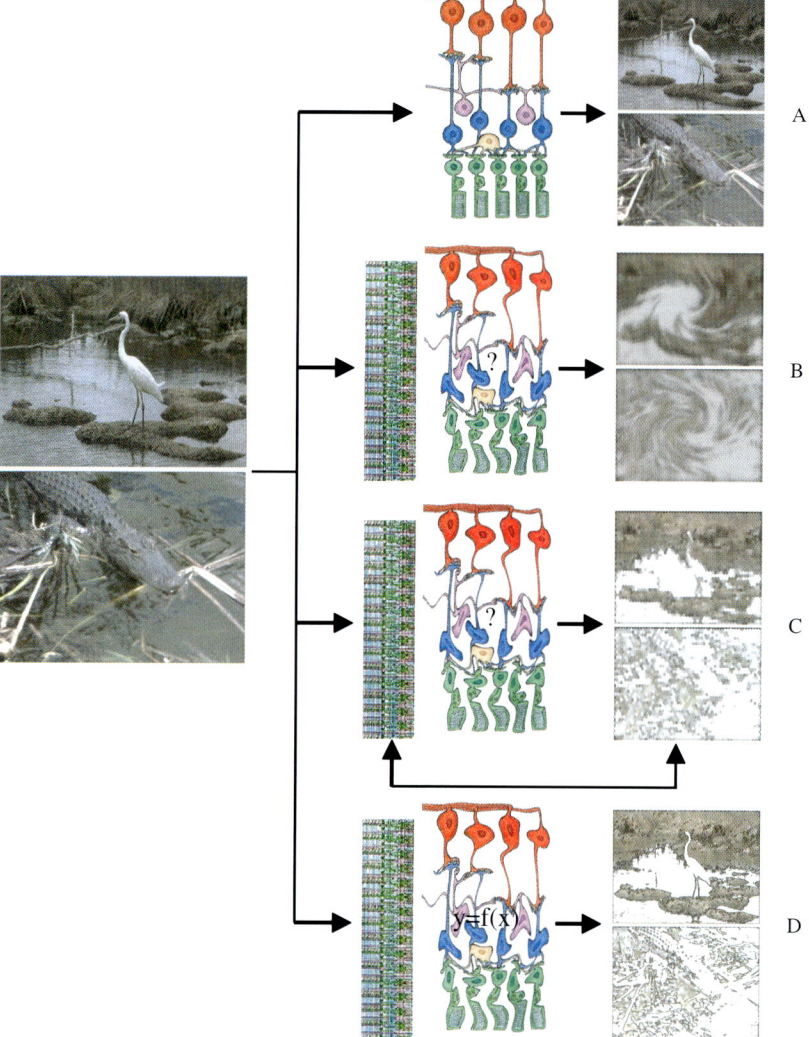

FIGURE 20.1. Knowledge of the input–output relationship of the diseased retina will aid prosthesis design, resulting in a higher-quality percept. On the left are two distinct visual scenes, on the right are the visual percepts under four different conditions (a–d, conceptual images only, not based on actual patient reports). (a) A healthy retina creates a high-quality percept. (b) A dysfunctional retina receives input from an electronic prosthesis. Initial clinical results with prototypes are encouraging, reporting percepts that are robust, but unpredictable and variable in size, shape, brightness, color, threshold, and retinotopic origin. (c) By using patient feedback to optimize the image coding strategy of the prosthesis, with larger numbers of electrodes, and through patient behavioral adaptation to the prosthesis (i.e. scanning the visual scene), it is anticipated that a percept containing a useful amount of information will be achieved. Note that a "useful" percept will be defined by individual patients, and may be very fundamental compared to normal vision. (d) With knowledge of the input–output relationship of the diseased retina (absent in (b) and (c)), it may be possible to further improve the quality of the percept by optimizing electrode design and stimulus parameters.

conceptually, the general development of retinal prostheses from a functional point of view. Two primary challenges are to understand the relationship between neural activity in the diseased retina and the formation of images in the mind's eye, and the relationship between microelectrode-associated currents and neural activity in the diseased retina. We know almost nothing about the former, and very little about the latter.

The *objectives* of this chapter are to summarize techniques that have been used extensively during the past several decades to learn about the response of the retina to natural, photic stimuli, and to then describe how these techniques can be modified to understand the response of the retina to electrical stimulation. Results from two of these modified techniques, applied to understand the response of the retina to subretinal stimulating electrodes, are presented. The scope of this chapter is limited to in vivo electrophysiological techniques, or those methods which can be used to measure the responses of the retina when it is in as natural a state as possible, in the intact subject (animal or human). Psychophysical tests, in which a patient reports his or her perception of a stimulus, are also excluded due to the complexity of applying these approaches in animal studies (in which case the perception must be inferred from a behavioral response).

With knowledge of how the retina responds to artificial electrical stimulation gained from animal studies, and knowledge of the perceptions of humans for given electrical stimuli, we may be able to infer the activity in the human retina, understand the relationship between activity in the diseased retina and perceptions, and then make effective decisions to improve the design of the prosthesis.

Electrophysiology of Natural Vision

The light response of the retina has been measured by a number of electro-physiological methods, each involving one or more recording electrodes which transduce the ion-based potential differences within the body to electron-based potential differences in metal conductors. The measured potentials are either membrane potentials or field potentials. Membrane potentials are measured by placing one electrode inside a neuron, and measuring the potential difference relative to a second electrode outside the neuron. Due to the high mechanical precision and stability required to place the intracellular electrode without damaging the target neuron, these approaches are rarely used in vivo, and will not be covered here. It should be noted that membrane potential can also be measured using a membrane-bound dye that changes its fluorescent emission properties in a manner proportional to the potential across the membrane to which it is bound; for a number of reasons, these potentiometric probes, or voltage-sensitive dyes, are best suited to in vitro applications when single-cell resolution is the goal.

Local field potentials (LFPs) result from the combined contribution of all nearby neurons, each of which is associated with an electric field due to its membrane potential and shape. LFPs are recorded only from nearby neurons for two reasons: First, the magnitude of an electric field falls off with the square of the distance from the source. This is described by Poisson's equation, and for the very weak electric fields generated by single neurons, means that the field is too weak to be detected above noise levels with typical electrophysiological recording technology. Second, the dipole moment, indicating the magnitude and direction of the electric field generated by an individual neuron, tends to be randomly oriented from neuron to neuron. Therefore, the contributions of a number of neurons sum to zero over any moderate distance from the dipole sources. LFP recordings tend to be low-frequency signals, where the recorded potentials reflect the summed electric fields generated by many action potentials traveling along nearby neurons.

Somewhere between intracellular recording (one neuron per electrode) and what is typically referred to as local field potential recording (roughly tens to hundreds of neurons per electrode) is the technique of extracellular recording. In this approach, a very small electrode (on the order of a neuron soma diameter, $\sim 10\,\mu$m) is placed in close proximity to one or more neurons, essentially in direct physical contact, and records the changes in membrane potential that accurately reflect the dynamics of single action potentials. Due to the small distance between recording electrode and target neuron, these signals are much larger in amplitude than the contributions of more distant neurons. Central nervous system neurons are properly referred to as units, and extracellular recordings are categorized as single-unit or multi-unit recordings. While single-unit recordings are generally preferred, the activity of individual neurons in a multi-unit recording can be separated using any number of available spike-sorting techniques (for reviews of available techniques, see Refs. [1, 2]). A straightforward approach would be a simple threshold, where the spikes attributed to two different neurons were distinguished by peak amplitude. Amplitude of the signal recorded from each neuron is strongly influenced by distance to the electrode and orientation of the effective dipole generated by each neuron's activity. The potential recorded by an extracellular electrode does not directly reflect the membrane potential, V_m, but generally resembles a first or second time derivative of the membrane potential (dV_m/dt or d^2V_m/dt^2), due to the capacitive nature of the coupling between the cell membrane and the metal conductor.

If large populations of neurons are aligned in space, and demonstrate activity coordinated in time, as in the highly organized and stratified retina, the dipole sources and associated field potentials of many millions of neurons sum and can be recorded some distance from the source. Measuring these field potential recordings is perhaps the simplest electrophysiological technique, and is the most common type of clinical measurement, because it can be performed non-invasively. For example, the electroencephalogram (EEG) and electrocardiogram (ECG or EKG) use electrodes applied to the surface of the skin. However, the simplicity of this approach comes at a cost of spatial resolution. That is, it is difficult to know the location of the dipole source within the body that is associated with a potential difference recorded at the skin. Further, recall that the magnitude of an electric field falls of with the square of the distance from the source; this means that the contribution of an individual cortical neuron to the potential recorded at the scalp (a few centimeters away) is extremely small.

Field potential recordings made from the body surface are often referred to as surface potential recordings. These signals are potential differences measured between two points on the body surface. If you picture the dipole source of the potential difference within the conductive body, the associated electric field can be visualized as a series of increasing shells surrounding the source (like layers of an onion), where each shell represents all of the locations in space that are at the same potential at a given snapshot in time. These isopotential surfaces eventually intersect the boundary of the body, where the electrodes are located. The potential difference recorded on the body surface results from the two electrode locations intercepting two different isopotential surfaces. Attempts to mathematically infer the location of the dipole source from knowledge of the surface potentials is solving the inverse problem (as opposed to solving for the field potential a known distance from the dipole, the forward problem) and is called functional mapping; this is a highly developed area of biomedical modeling. Much of the challenge in functional mapping arises from the complex shapes of the isopotential surfaces due to anatomy and differences in conductivity of the various tissues between the dipole source and the electrodes.

Recent trends in electrophysiology are driven by attempts to overcome the limitations of each of the methods just described. The relative attributes of each method are summarized in Table 20.1.

New technologies and analysis strategies are aimed at the ultimate goal of recording from many individual neurons without sacrificing the ability to resolve individual action potentials. The development of multielectrode arrays attempts to scale up the extracellular recording approach, but is ultimately limited by the displacement of tissue required to introduce the many electrodes into the target structure. The spatial resolution of functional mapping from surface

TABLE 20.1. Summary of four basic strategies for recording neural activity. Existing methods are a compromise between resolution (temporal and spatial) and the number of individual units that can be resolved in the recorded signal.

Method	Temporal resolution	Spatial resolution	Number of units
Intracellular Recording	Excellent (direct measure of V_m)	Excellent	Very Limited (1 unit)
Extracellular Recording	Excellent (spikes resolved; signal is a derivative of V_m)	Very Good	Limited ($\sim$ 1–3 units)
Local Field Potential Recording	Moderate (spikes not resolved)	Moderate ($\sim$ 1 mm^3)	Moderate (100's–1000's of units)
Field (Surface) Potential Recording	Moderate (spikes not resolved)	Poor (resolve voxels of a few mm^3)	Many (e.g. entire cortex accessible)

potential recordings is improved by the use of a large number of electrodes, but is ultimately limited by the small contribution to the measured potentials by individual neurons. Electrophysiological functional mapping has recently been combined with structural imaging techniques that have high spatial resolution but poor temporal resolution (e.g. MRI, CT); these multi-modal imaging strategies exploit the strengths of each approach. The technology that allows us to monitor thousands of individual neurons, in real time, in vivo, has yet to be developed. In the meantime, we have several techniques that provide a great deal of information about the nervous system. The following section provides an overview of approaches used specifically to measure the response of the retina to stimulation with light (i.e. photic stimulation).

Non-Invasive Techniques

The earliest technique used to record from the retina was the corneal electroretinogram (ERG) [3]. In its simplest form, this is a straightforward surface potential measurement, where two electrodes are placed on the body surface near the eye, and a potential difference that varies with time is recorded following the delivery of a light stimulus. Typically, one electrode is placed in contact with the cornea of the eye itself, and a second reference electrode is placed on the skin a few inches away (on the temple, forehead, or earlobe).

Almost all techniques for electrophysiological recording, including ERG recording, employ a differential amplifier, in which the potential difference between a reference electrode and ground electrode is effectively subtracted from the potential difference between the recording electrode and the same ground electrode. Ideally, the potential on the reference electrode would reflect only contributions from noise sources that are also picked up by the recording electrode. Thus, the signal downstream of the differential amplifier consists of only the desired signal:

$$(\text{Signal} + \text{Noise}) \quad - \quad (\text{Noise}) \quad = \quad (\text{Signal})$$

Recording Electrode	Reference Electrode	Amplifier Output

The degree to which this strategy works can be quantified by the common mode rejection ratio (CMRR), which is the ratio of the differential mode gain (DMG, the gain applied to the potentials not appearing on both electrodes) to the common mode gain (CMG, the gain applied to the potentials common to both electrodes).

$$\text{CMRR} = \text{DMG}/\text{CMG}$$

The CMG is ideally zero, but is typically close to one in real amplifiers; this is much smaller than the DMG of 100–1000 typically applied to field potential signals. A CMRR of 100 or better is considered quite good.

When recording the corneal ERG in human subjects, the recording (or active) electrode and the reference electrode are often both contained in a specially fabricated contact lens. As long as the two electrodes intercept different isopotential surfaces (where the retina is the dipole source), a differential signal will be recorded. The contact lens serves the important function of stabilizing the contact between the cornea and the electrodes, and it allows light (the stimulus) to enter the eye.

There are many variations of corneal ERG recording, primarily distinguished by the type of stimulus used to evoke the response. The simplest variation is the response to a single, brief (less than about 1 ms) flash of light delivered to a dark-adapted subject. The major components of this response are illustrated in Figure 20.2, and are dominated by the rod pathway in the retina. By exploiting known differences in temporal and spectral responsivity of different cell types within the retina, the stimulus can be designed, and the recorded signal can be analyzed, in order to study the response of a single class of retinal neurons or the response of a single neural pathway. There are more possible variations than can be adequately described in a single chapter. A few variations commonly used in the clinic, and which might be adapted to study responses of the retina

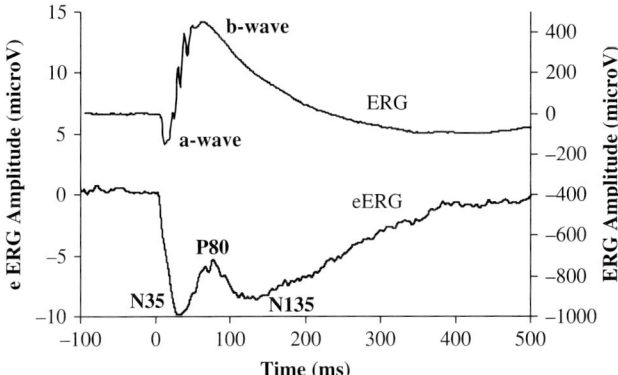

FIGURE 20.2. Electroretinogram responses to light and electrical stimuli (presented at time zero) recorded in rat. **Upper trace** plots a representative ERG response to a brief, moderately bright flash of light presented to a dark-adapted animal. The major components are the negative a-wave (leading edge reflects photoreceptor activity), the positive b-wave (dominated by contributions from ON-type bipolar cells), and the oscillatory potentials superimposed on the leading edge of the b-wave. **Lower trace** plots the eERG response recorded under one specific set of conditions (see text). The major components of the response are labeled according to the convention used to describe pERG responses, where **N** and **P** indicate negative or positive peaks, respectively, and the numbers refer to the peak latency, in msec. Pharmacological dissection of the eERG response (summarized in Figure 20.8) suggest that the origins of the each component are as follows: N35, OFF-type bipolar cells; P80, ON-type bipolar cells; N135, ganglion cells in the OFF pathway. These findings are consistent with the sign of the contribution of these cell types to the ERG in a healthy retina.

to artificial electrical stimulation, are described here. These are the focal ERG (fERG), the paired-flash ERG (pfERG), the multifocal ERG (mfERG), and the pattern ERG (pERG). The interested reader is referred to the International Society for Clinical Electrophysiology of Vision (ISCEV) standards for applying each of these techniques. These standards facilitate comparison of results collected in the many clinical laboratories around the world, and are available from the ISCEV website (updates are published on a regular basis; e.g. Refs. [4, 5, 6]). Several available texts provide guidance for applying the various techniques to assess functional loss in the retina [7, 8].

The *focal ERG* is used to elicit a response from a local area of the retina by using a focal stimulus (small spot of light). The fERG is typically used to assess macular function, but can be applied to any portion of the retina. Since a small portion of the retina is activated by the stimulus, there are a smaller number of dipole sources contributing to the field potential recorded at the cornea; therefore the fERG signal is quite small, on the order of 10 microvolts. In practice, the spot is turned on and off at a fixed rate and the responses from several (typically 100–300) responses are averaged to improve the signal to noise ratio (SNR). Also, in order to reduce the influence of areas of the retina outside the desired spot that may be stimulated by scattered light, the spot is surrounded by a constant-luminance annulus typically brighter than the spot (in order to saturate the area of the retina surrounding the region of interest). This is perhaps the most straightforward approach to adapt to electrical stimulation, where the electrodes in contact with the retina are intuitively focal stimuli. However, as discussed below, there are special considerations when substituting an electrode for a focal light stimulus.

The *paired-flash ERG*, an extension of the single-flash technique, can be used to isolate the contribution of the photoreceptors to the ERG response recorded at the cornea [9, 10]. In the response to a single flash, the leading edge of the first major component (the negative a-wave) reflects the activity of photoreceptor cells alone, before the other cell types in the retina begin their response to the stimulus (Figure 20.2). The pure photoreceptor response can be recorded at the cornea from the time of the stimulus onset to just before the a-wave peak, beyond which the recorded signal reflects the summed contributions of several cell types. However, it is known from in vitro single cell recording studies that the photoreceptor response lasts several hundred milliseconds. Most of the photoreceptor response is masked in the recorded ERG by the strong contributions of other cell types to the field potential at the cornea.

The paired-flash method relies on the fact that a very bright flash will rapidly drive all of the photoreceptors to saturation. That is, a very bright flash delivered to a dark-adapted eye will elicit the largest a-wave possible (A_{sat}). The first step in this approach is to deliver a brief flash of any arbitrary intensity; this is the flash you want to measure the response to, and is commonly referred to as the *test flash*. At a known time, t, after delivery of the test flash, but during the time-course of the photoreceptor response, a second very bright, saturating flash is delivered. The excursion from the prevailing baseline to the peak of the a-wave induced by the bright flash, $A_{sat}(t)$, is a measure of how far the photoreceptors were from saturation at the time the bright flash was delivered. Performing the subtraction $A_{sat} - A_{sat}(t)$ reveals how far the photoreceptors were from saturation due to the test flash at the time the bright flash was delivered, $A(t) = A_{sat} - A_{sat}(t)$. By delivering the bright flash at several different times after the test flash, in several paired-flash trials, $A(t)$ can be calculated for the full time course of the photoreceptor response. An important feature of this approach is that it is used to isolate the response of the first-order neurons contributing to the corneal ERG, and this is the property of the pfERG technique that may be exploited in studying electrical stimulation.

The following two techniques are common in clinical electrophysiology of the eye. Because they exploit network properties of the retina, their application to studying the response of a diseased retina to electrical stimulation is much less straightforward than the fERG and the pfERG. They are described briefly here as an illustration of the variety of ERG techniques available, and with the hope that once we know more about the physiology of the degenerate retina, correlates may be developed for use in studying electrical stimulation.

The *pattern ERG* is primarily used to measure the activity of ganglion cells. The third-order neurons of the retina are most sensitive to spatial differences in retinal illumination. The stimulus of the pERG is therefore a high-contrast checkerboard pattern subtending the central visual field that inverts (black becomes white, white becomes black) at a regular frequency (1–10 Hz). The mean luminance of the checkerboard remains constant, which minimizes the influence of the photoreceptors and second-order neurons to the recorded response. Like the pfERG, the pERG can provide information about a specific class of cells in the retina, but does not convey information about the location of a focal retinal anomaly. That is, the recorded response is still the summed response of contributions from all areas of the retina subtended by the stimulus pattern (typically the macular region).

The *multifocal ERG* is a technique used to perform functional mapping of the retina, and can provide information related to cell type and spatial location [11]. However, in contrast to most functional mapping methods that use multiple recording electrodes, the mfERG uses a single corneal electrode. The goal is to produce a map of the retina that indicates the strength of the local response to a given stimulus. This is accomplished by stimulating different areas of the retina, and then correlating the area in visual space associated with the stimulus to an area on the retina. Imagine viewing a large screen divided into two halves, upper and lower. If the upper half is bright and the lower half is dark, the inverted image of the screen will illuminate the inferior half of the retina, while the superior half of the retina remains in the dark. The signal recorded at the cornea would be the response originating in the stimulated inferior region of the retina; the superior region would not contribute. If the screen was then divided into, say, a 10×10 array of pixels, the response to each box could be mapped to one of one hundred regions in the retina. In practice, this method must account for the mean illuminance of the retina, which determines the state of light adaptation, and therefore has a large influence on the recorded response. Commercially available mfERG systems use hexagonal pixels arranged in rings of increasing diameter, and subsets of the entire group of pixels are illuminated in a pseudo-random sequence of flashes. A single corneal response is measured following each flash. Sophisticated computational routines then assign relative contributions from each of the illuminated hexagons to each recorded response, resulting in a map of activity. The mfERG response is thus a mathematical derivation, not a direct measure of activity. The mfERG response can be calculated in a number of different ways, with the first-order and second-order kernels being most common (kernel order refers to the relative timing of the stimulus and response). It should be kept in mind that the contributions of the various cell types in the retina to the mfERG response are not yet completely understood. This is proving to be a powerful technique, but due to limitations on our understanding of the mfERG response of healthy retina, and the complex relationship between response features and cellular activity, its application to electrical stimulation of degenerate retina is likely far off.

Invasive Techniques

Fine wire microelectrodes can be introduced into the eye through the lumen of a needle that pierces the sclera. These electrodes are typically made of small-diameter insulated tungsten wire, and are either cut off square, exposing the metal at the cut end, or are drawn or etched to a fine tip ($\sim 1\,\mu$m diameter), and insulated everywhere except the final few microns. If such an electrode is inserted into the retina, a local field potential can be recorded which is influenced most strongly by a small volume of retinal tissue, the intra-retinal ERG. Because the retina is organized in distinct lamina, where each layer is occupied by a limited number of cell types, a local response recorded at a known depth in the retina can be attributed to a single cell type (or at least a small number of cell types). However, it is difficult to know with great certainty how deep an electrode has penetrated the retina, especially in vivo. If a microelectrode is placed on the vitreal surface of the retina, in the nerve fiber layer or just below, it is possible to record the spiking activity of ganglion cells (both single-unit and multi-unit). Single-unit recording has been used successfully for decades in cat and primate to study ganglion cell properties [12, 13]. For example, a single ganglion cell would be targeted for recording, and then focal stimuli (small spots of light, on the order of 10–100 μm in diameter) delivered to the retina in a "search pattern" to learn which areas of the photoreceptor layer the ganglion cell received input from. The retinal area which produced a change in firing rate, up or down, in the ganglion cell would define the receptive field for that cell, which could be related to an area in the visual space. Though possible to perform with no permanent damage to the eye, intraocular microelectrode approaches are most suitable for animal studies, and are most amenable to animals with large eyes. Extracellular recordings have been made in vivo in both rats and mice, but the technical challenges are significant [14].

Identifying Response Origins –
Pharmacological Dissection

In the moments after light reaches the photoreceptors, a pattern of activity propagates through the neural network of the retina, from cell to cell, primarily by chemical synapses. The pharmacology of synaptic transmission in the retina is orderly yet complex (recently reviewed by Yang [15]). There are just two principle transmitters found here: glutamate and γ-aminobutyric acid (GABA). Glutamate is excitatory, and GABA is (mostly) inhibitory. Glutamate mediates information flow primarily in a vertical direction, from photoreceptors to bipolar cells to ganglion cells. GABA mediates information primarily in lateral directions, or as upstream feedback, and is released by horizontal and amacrine cells. Different cell types express variations of the glutamate and GABA receptors, with the first distinction being ionotropic (ion channels, e.g. iGluRs and $GABA_A$) and metabotropic (coupled to G-proteins, e.g. mGluRs and $GABA_B$) types. There are further subtypes of iGluRs and mGluRs, and a distinct $GABA_C$ receptor. The various subtypes of glutamate receptors are generally named for the agonist they are sensitive to. An agonist mimics the action of glutamate, whereas an antagonist binds to the receptor but does not elicit the response associated with glutamate. So, for example, the glutamate agonist N-methyl-$_d$-aspartate (NMDA) binds to a subset of iGluRs known as NMDA channels. Not all agonists are highly selective, and L-aspartic acid (aspartate), a glutamate agonist, binds readily to all glutamate receptors in the retina. In the presence of aspartate, the rods and cones respond to light, but none of the post-synaptic neurons are able to respond to the light-induced change in transmitter release from the photoreceptors. Thus, any ERG response recorded in the presence of aspartate can be attributed to activity of the photoreceptors alone, without contributions from other cell types.

A great deal of work has been done to identify the various receptor subtypes expressed by the different types of retinal neurons. Indeed, the functional subclasses of retinal neurons are defined by the receptor types they express (though subclasses can also usually be distinguished based on morphological differences). The point to make here is that the various agonists and antagonists for receptor subtypes can be used to suppress the activity of subpopulations of retinal neurons. Thus the complex response of the retina can be "dissected" by judicious removal of cell types from the retinal network [16, 17]. The typical protocol is to introduce the drug into the eye at an appropriate concentration such that it binds to the appropriate receptor at a saturating level (i.e. binding to all available receptors). In the case of an agonist, the cells so bound become maximally depolarized (or hyperpolarized), and do not change from this state following a light stimulus. Thus the contribution of this subpopulation of neurons is removed from the recorded response (recall that physiological signals are generally recorded with AC amplifiers, and the static field potential generated by a constant membrane potential is filtered out; only *changes* in membrane potential contribute to the recorded response).

If the response of the retina is recorded before introduction of an agonist or antagonist, and then again after introduction of the drug, the latter response waveform can be subtracted from the former to reveal the component of the response that was sensitive to the drug. If the cell type which binds that particular agonist or antagonist is known, then the response component suppressed can be correlated with that cell type. This strategy has been used extensively in the retina, and it is possible that new agonists and antagonists will be discovered, which may reveal new subtypes of receptors, and therefore new physiological classifications of retinal neurons.

Introduction of the drug to the retina in vivo is accomplished via an injection into the vitreal cavity. A small diameter (24 to 30 gage needle) is used to pierce the sclera just below the outer margin of the iris, and the tip is advanced at an angle that avoids the lens until it is within a millimeter or two of the retinal surface. Then a small volume of solution containing the drug is injected. The volume must be small relative to the vitreal volume so that the intraocular pressure is not significantly raised. It is common to mix a small quantity of biocompatible, non-reactive dye into the drug solution to provide visual confirmation that the injection was successful. The drug diffuses within the vitreal volume, and within 20–60 minutes is evenly distributed throughout the retina. Some drugs will be cleared by the circulatory system in a few hours, but generally it is not practical to wait for the drug to be "washed out" in these experiments. A strategy used to provide a control for effects of the injection procedure itself is to inject one eye with the drug, and the other eye with physiological saline, and then measure the response to the same stimulus presented to both eyes. Pharmacological dissection can be employed in conjunction with both invasive and non-invasive recording techniques.

Electrophysiology of Artificial Vision

For a retinal prosthesis to be successful, it must stimulate the retina in such a way that a useful percept is elicited in the patient. This goal might be reached by treating the retina as a black box, and then modifying implant parameters during a training period until the patient percept is optimized. However, only the amplitude and kinetics of the electrical stimulus can be modified after a prosthesis has been implanted in the eye (via a radio frequency emitter–receiver, as is commonly used to communicate with cardiac pacemakers after surgery). Physical design features such as electrode shape, size, spacing, and location relative to the retina cannot be altered without further surgery. Each new surgery entails patient risk. Therefore, there is strong motivation to optimize physical design features before a prosthesis is implanted. This can be achieved in part by studying the response of the retina, at a cellular level, in experimental animals [18]. The response of the retina to electrical stimulation had not been reliably recorded until very recently, and there is little known about the origins of its components. We now hope to study this response using adaptations of the techniques originally developed to study the response of the retina to light [14]. The remainder of this section describes the approaches developed to study the response of the retina to electrical stimulation.

The methods described below were developed within certain criteria. First, the response of the retina had to be measured in vivo. This criterion was important because maintaining the natural physiological support system (circulatory system, temperature control, osmolarity, chemical environment, physical support, juxtaposition to supporting tissues, etc.) would help ensure that the response recorded during the experiment would be similar to the response of the retina in a patient. Also, the interface between chronic stimulating electrodes and target neural tissues is known to change over time following implantation due to immune and other tissue reactions. Thus, it is of interest to study the efficacy of prosthesis design over time scales of weeks, months, and eventually years. This is not possible with an in vitro preparation.

A second criterion was that the technique should be available for use in appropriate animal models of eye disease. As will be addressed below, the response of a degenerate retina, to either light or electrical stimulation, is not likely to resemble the response of a healthy retina. Mouse is the most common species for developing genetic models of human eye disease, but the eye size of mouse is prohibitively small for introducing electrode arrays adjacent to the retina. Rabbit, cat, and pig all have larger eyes, but they are more expensive to maintain and there are fewer appropriate models available. Rat is second only to mouse in terms of available models, and the eye is just large enough to accommodate small electrode arrays. Thus, rats with appropriate retinal defects were chosen as a compromise between eye size and expense. Some of the appropriate rat models of degenerative retinal disease are the P23H, RCS, and s334ter (see Refs. [19, 20], for reviews of animal models of retinal degeneration). The ERG and electrically elicited ERG (eERG) data in this chapter were recorded from a novel pigmented P23H transgenic rat, where the transgene has a proline substituted for histidine at position 23 in the gene that encodes the photopigment rhodopsin [21]. This is the most prevalent mutation in human patients with autosomal dominant retinitis pigmentosa (adRP).

Working with Degenerate Retina

The response of the healthy retina to light stimulation has been studied for many decades, and there is yet some controversy over the cellular origins of prominent components of the corneal ERG. During the course of retinal degenerative disease, the changes that occur in the retina can be severe, including cell loss and significant remodeling of the retinal network [22]. It is therefore critical to evaluate the response of the retina to artificial stimulation with respect to the degree of degeneration present.

Motivated by the prospect of delivering artificial stimulation to a retina that no longer responds to light, it is now of great interest to study in detail the biology of severely degenerate retina. We need to understand not only what cell types remain and in what numbers, but how the morphology of these cells has changed following loss of neighboring and connecting cells. If the goal of artificial stimulation is to produce meaningful retinal output (i.e. spatial and temporal ganglion cell spiking patterns), it will be of great benefit to understand the properties of the network we are stimulating. These network properties are of importance for both subretinal stimulation, which targets distal neurons, and epiretinal electrodes, which are capable of stimulating cells presynaptic to the ganglion cells.

The work described below used P23H rats that showed moderate degeneration of the retina. That is, cell loss was 30–40%, and the retina maintained a reduced, yet still easily recorded, ERG response to light stimuli. Figure 20.3 shows a micrograph of a typical histological cross section of retina from one of these experimental animals, along with a cartoon drawn to represent the relative thickness of each layer at this stage of degeneration. This age was a compromise between demonstrating that a response to electrical stimulation could be elicited in a diseased retina and choosing a retina that retained enough normal structure so that we could begin to interpret the response.

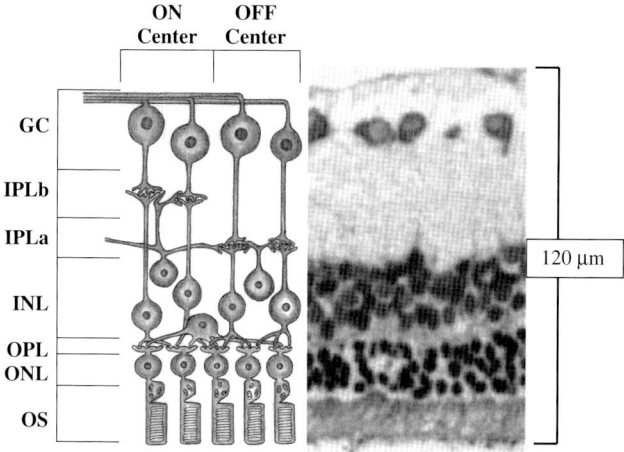

FIGURE 20.3. Retina of a 16-week-old pigmented P23H transgenic rat, showing moderate degeneration. The schematic on the left is drawn to represent the retina pictured in the micrograph on the right. The major lamina are labeled: OS, photoreceptor outer segments; ONL, outer nuclear layer, containing photoreceptor cell bodies; OPL, outer plexiform layer, containing synaptic connections between photoreceptors and bipolar cells; INL, inner nuclear layer, containing cell bodies of bipolar, horizontal and amacrine cells; IPLa & IPLb, inner plexiform layer, containing synaptic connections between bipolar cells and ganglion cells, the sublaminae containing connections of the ON and OFF pathways are distinguished; GC, ganglion cell layer, containing cell bodies of the ganglion cells.

Delivering the Stimulus

The electrical stimuli delivered to the retina by existing prototype prostheses are generated by microelectrodes. We have designed and fabricated small electrode arrays that can be placed in and around the eye, allowing the investigation of several electrode configurations. Shown in Figure 20.4 is one example of our retinal stimulating array (RSA). The design constraints included the size of the rat eye and biocompatibility of materials. The final device is just 12-microns thick, and can include any arbitrary electrode design. Feature size is limited to about 10 microns (minimum) using standard photolithography processing, and the connector that interfaces the RSA with the external cable supports 27 conductors (which could support 27 individual electrodes). The substrate is polyimide; conductors and electrodes are gold, although recently an activated iridium oxide–film has been incorporated on the electrode surfaces to increase the charge transfer characteristics [23]. While polyimide is not suitable for chronic applications, it serves well for acute experiments lasting a few hours.

The RSA dimensions were chosen so that it could be inserted in the subretinal space. This is accomplished by making a small incision through the sclera, choroid, and retinal pigment epithelium, but not the retina, and then inserting the RSA through this opening and advancing it until the electrodes are in the desired location [14]. The thin device is flexible enough to follow the curvature of the eye beneath the retina without causing retinal detachment or tears. The electrode configuration is determined by the relative positions of the active and reference electrodes. A number of configurations are under investigation for retinal prostheses, including placing the reference adjacent to the active electrode, on the backside of the prosthesis, on the opposite side of the retina, in the vitreous, or outside of the eye. Each of these configurations can be achieved by

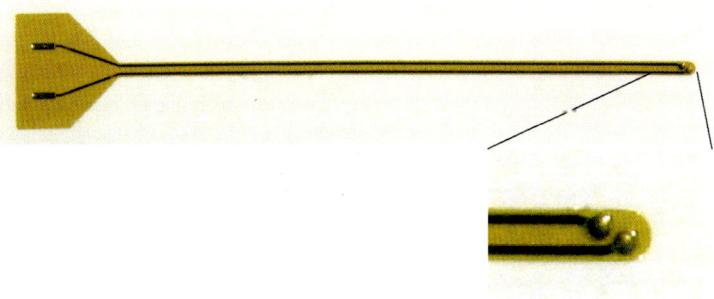

FIGURE 20.4. Photograph of the retinal stimulating array (RSA) developed for use in rat. Inset shows a magnified view of the terminal end containing the electrodes. Any arbitrary design of electrodes, with feature size $\geq 10\,\mu m$, can be made on site in a few days using standard photolithographic microfabrication. This example contains two 500 micron diameter round electrodes; up to 27 electrodes can be accommodated (limited by the 27-contact connector, not shown). Total length = 50 mm, width of terminal end = 1 mm.

using two RSA devices positioned relative to the retina; e.g. one in the subretinal space, one in the vitreous.

The stimuli used were constant current, biphasic square-wave pulses, although any arbitrary stimulus kinetics can be achieved. Square-wave pulses are convenient to generate, and are commonly employed. The surface potential of the stimulating electrode is often orders of magnitude greater than the response amplitude sensed by the recording electrode. Because the stimulating and recording electrodes are relatively close together (both in contact with ocular tissue), a substantial stimulus artifact is recorded. This artifact can be either computationally subtracted, or minimized by temporarily disconnecting the amplifier during stimulus delivery.

Non-Invasive Recording

As described above, non-invasive recording has the advantages of simplicity and the potential for clinical application. The response of the retina to electrical stimuli can be measured at the cornea using the same recording technology as is employed with standard ERG techniques. For clarity, we will refer to the response of the retina to electrical stimulation as the **eERG**. Both the fERG and the pfERG have straightforward correlates when measuring the eERG; each will be described below.

Adaptating the Focal ERG

Adaptation of the fERG to the eERG is essentially unavoidable when using small electrodes to stimulate the retina. By virtue of design, a small area of the retina is activated, and the resulting signal can be recorded using a corneal recording electrode identical to that used to measure the ERG. However, there are important differences. A single micron-scale electrode in contact with the retina will stimulate only a tiny fraction of the retinal neurons, even less than typically targeted by the fERG. Consider that the maximum ERG amplitude recorded in a dark-adapted wild-type rat in response to a single full field flash of light is approximately 1 mV. The surface area of a rat retina can be conservatively approximated by assuming that it lines the posterior hemisphere of an eye of 7 mm diameter. This yields an area of 77 mm^2. A 100-μm-diameter electrode has a surface area of 0.0079 mm^2, and thus subtends 0.01% of the retina. If the electrode only stimulated the neurons directly above it, the recorded response would be expected to be no larger than 0.01% of the response to the full-field light stimulus, or 0.1 μV. To record this response with a minimum acceptable SNR of 5, the noise level would need to be 0.02 μV. This low noise level is rarely achieved in real-time recording, and necessitates the use of averaging (the root mean square (RMS) noise level is inversely proportional to the number of responses averaged).

Figure 20.5 shows representative eERG responses elicited by relatively large (500 μm diameter) bipolar electrodes placed in the subretinal space; the good

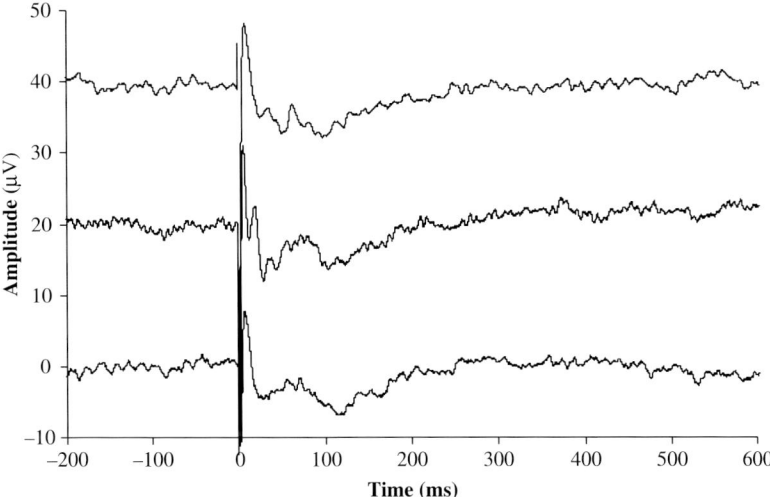

FIGURE 20.5. Representative eERG waveforms recorded in three animals. Each waveform is the average of 100–300 responses to a biphasic square pulse stimulus ($100\,\mu A$ each phase) delivered at approximately 3-second intervals. Animals were pigmented P23H rats, 6–12 weeks of age. Retinal stimulating array electrodes were $500\,\mu m$ diameter gold electrodes (cf. Figure 20.4). Major response components (see Figure 20.2) are consistently present across animals.

SNR was achieved by averaging several hundred responses to the same stimulus. The peak amplitude of about $\sim 10\,\mu V$ is consistent with the area of the retina subtended by the electrode pair. The lower trace in Figure 20.2 plots the mean eERG response recorded with these electrodes in six animals (different from those in Figure 20.5), and represents the typical waveform. Comparison with the light-induced ERG above reveals obvious differences between the responses to light and the electrical stimulation. The cellular basis for these differences is explored below, but first a number of technical considerations should be noted.

When delivering a large number of repetitive stimuli, several questions must be answered. At what frequency should the stimuli be delivered? Does the retina adapt or fatigue in response to the frequent stimuli? A typical interval between light flashes when performing a typical single-flash ERG experiment is 1–2 minutes, to allow the animal to dark-adapt between stimuli. Light adaptation occurs at several levels in the visual pathway, and so it may be assumed that adaptation to electrical stimulation may also occur. Are there any other physiological changes that might occur during the tens of minutes required to deliver hundreds of repeated stimuli? All of these factors must be considered within the constraint of the practical duration for an in vivo experiment, during which an animal must be kept anesthetized, or a patient able to maintain concentration and avoid excessive discomfort.

Two other issues require special consideration when recording the response to electrical stimulation. These are the determination of *threshold* and *saturation*, two fundamental characteristics of any response. In a stochastic, binary system, such as a spiking neuron, threshold is typically defined as the stimulus strength required to elicit a just-measurable response (i.e. a single action potential) 50% of the time. For a signal that is continuous, such as the eERG amplitude, threshold can be defined as a signal that is some criterion amount above the baseline noise level (e.g. $3\times$ the RMS noise, or 3 standard deviations above the mean noise level). However, for small signals, the noise level is inversely proportional to the square root of the number of responses averaged, and the threshold is thus a function of the recording protocol, and becomes difficult to assess.

Saturation is defined as the maximum response of the system being measured. A plot of response amplitude vs. stimulus intensity for neural systems often takes the form of a saturating exponential function. As the stimulus strength increases, eventually a maximum response is elicited, and further increases in stimulus strength result in no further increase in response amplitude. In the healthy retina, the saturated ERG response occurs when the number of photons delivered to the eye is sufficient to result in the closing of all of the cGMP-gated cation channels that mediate the photoreceptor dark current. This happens at about 1000 photons absorbed per photoreceptor, and doubling the number of photons delivered to the eye does not further increase the response.

The special consideration when dealing with electrical stimuli is that of stimulus field containment and recruitment of cells. When the potential applied to a stimulating electrode increases, the field potentials around the electrode also increase. If you assume that there is a threshold value of field potential that results in activation of a retinal neuron, then as the field potential at a given point in space increases, it becomes more likely that a neuron in that location will reach threshold and contribute to the recorded response. The region of the retina that is brought above threshold by direct effect of the delivered electrical stimulus (i.e. not via synaptic connections) is termed the *stimulus field*. The number of neurons contributing to the response (due to both direct stimulation and synaptic connections) then becomes a function of the stimulus strength, which is only limited by the potentials that cause electrode or tissue damage. The plot of eERG amplitude vs. stimulus strength is not well fit by a simple saturating exponential, because as neurons near the electrode become saturated at high stimulus strengths, neurons farther away from the electrode are just reaching threshold. This issue of cell recruitment also illustrates the need for electrode designs that create consistent stimulus fields over a range of stimulus strengths, to avoid recruitment of neurons outside of the desired stimulus field when using stronger stimuli.

Adaptating the Paired-flash ERG

Adapting the pfERG technique to electrical stimulation depends on the ability to establish two critical stimulus amplitudes. The lower stimulus amplitude is defined as that which elicits a response just above threshold (analogous to the first flash in the pfERG protocol). The upper stimulus amplitude is one in which the stimulus field is the same as that for the low stimulus (attained through judicious electrode design), and which drives the first-order neurons in the stimulus field to saturation (analogous to the second, bright flash in the pfERG protocol). In practice, electrical stimuli would be delivered as paired pulses, and the response of the first-order neurons to the first pulse would be titrated at arbitrary times by the second pulse. Just as the pfERG allows reconstruction of the full time course of the photoreceptor response, the paired-pulse eERG allows the reconstruction of the full-time course of the first-order neurons stimulated by electrodes. The utility of this approach lies in its ability to measure the kinetics of the response. This information can then be used to identify the type(s) of "first-order" cells in the stimulus field and gain insight into their physiology and response characteristics relative to stimuli of arbitrary design.

Adapting Invasive Recording Techniques

Invasive recording, using microelectrodes targeting single cells or small populations of cells, provides spatial information that is inaccessible with the non-invasive techniques. The extent of the electrical stimulus field cannot be visualized directly, so the population of neurons targeted is uncertain. Further, due to the complex and atypical network connectivity in a degenerate retina, the extent of the *response field* (the population of neurons responding to the stimulus due to both direct *and* synaptically mediated stimulation) is also unknown when recording at the cornea. If, however, a microelectrode is brought into contact with the ganglion cell layer of the retina, making single-unit recordings, the response field at this level can be determined by scanning the electrode across the retina during repeated delivery of the stimulus. A map of the lateral extent of neurons whose activity is correlated with stimulus delivery is created. The technical considerations are identical to those used to measure the response to light stimuli, with the added minor challenges of moving the recording electrode in carefully defined increments within the eye, and stimulus artifact suppression (discussed above). The former challenge is easily met using a commercially available robotic micromanipulator combined with both visual feedback through the pupil and by inferring the location of the electrode by the recorded response.

A second protocol which can be employed with invasive recording is that of modifying the parameters of the electrical stimulus until the resultant spike train recorded from a single ganglion cell mimics that elicited by a defined light stimulus. Here, a light stimulus is delivered to an area of the retina that is within the stimulus field of an electrode, and the ganglion cell response is recorded. Then the same area of retina is stimulated with the electrode, and the stimulus parameters altered until the ganglion cell response is similar to that obtained in response to the photic input. Key features of the stochastic spike train to be mimicked are latency to first spike, peak (or minimum for an OFF-type cell) firing rate, and latency to peak (or minimum) firing rate, each of which are stereotypical for individual types of ganglion cells. Obviously, this protocol can only be carried out in a retina that still responds to light, but even so it will likely provide many insights into the relationship between the input–output functions of the degenerate retina for light and the electrical stimuli.

Pharmacological Dissection of the eERG

The strategy of pharmacological dissection has been one of the most powerful in learning to interpret the ERG, and so it will prove very useful in identifying the cellular origins of the various components of the eERG. Figure 20.6 illustrates the relationship between the retina and a non-uniform electric field generated by an active-reference pair of subretinal electrodes. Also indicated are the cell types in the retina that are incapacitated by three common glutamate agonists. The properties of these agonists are summarized in Table 20.2.

   The procedure for studying the response to electrical stimulation is identical to that for studying response to light stimulation, but there is an important distinction that must be made when analyzing the results. In a healthy retina responding to light stimuli, the photoreceptors become active first, and the visual signal propagates through the retina in a reasonably well-understood manner. The relative order of neurons in the retinal network, i.e. presynaptic vs. postsynaptic for each cell type, determines the order in which activity propagates. This is somewhat complicated by lateral pathways in the retina and feedback, but the principle flow of information is from photoreceptors to bipolar cells to ganglion cells.

   However, an electrical stimulus might directly stimulate neurons at several levels in the retina. That is, the stimulus field may subtend the cells closest to the electrode as well as those post-synaptic to these neurons. (In the case of epiretinal electrodes, the stimulus field may subtend ganglion cells as well as cells further from the electrode that are *pre*-synaptic to the ganglion cells). If the stimulus field extends across synaptic layers in the retina (the inner and outer plexiform layers), the effects of agonists and antagonists targeting the same receptor type become quite different.

   An agonist will fully depolarize or hyperpolarize the target cell. This will positively block synaptic transmission to this cell type. It may or may not prevent the membrane potential of the target cell from changing in response to an applied electric field, and so this cell may or may not contribute to the recorded response. This is an important question that has yet to be approached experimentally or via computational modeling. Unpublished work by our group suggests that the passive properties of the neural membrane, the morphology of the cell, and the intersection of the cell with an inhomogeneous electric field largely determine the response to applied electric fields, and so a cell may still respond even if the receptor-mediated biophysics are inoperative. A further question is then whether this response (i.e. change in membrane potential) is within a range that will modulate transmitter release and thus elicit additional response contributions from downstream neurons.

   An antagonist blocks synaptic transmission, but does not alter the membrane potential of the target neuron significantly from resting level. Therefore, if this

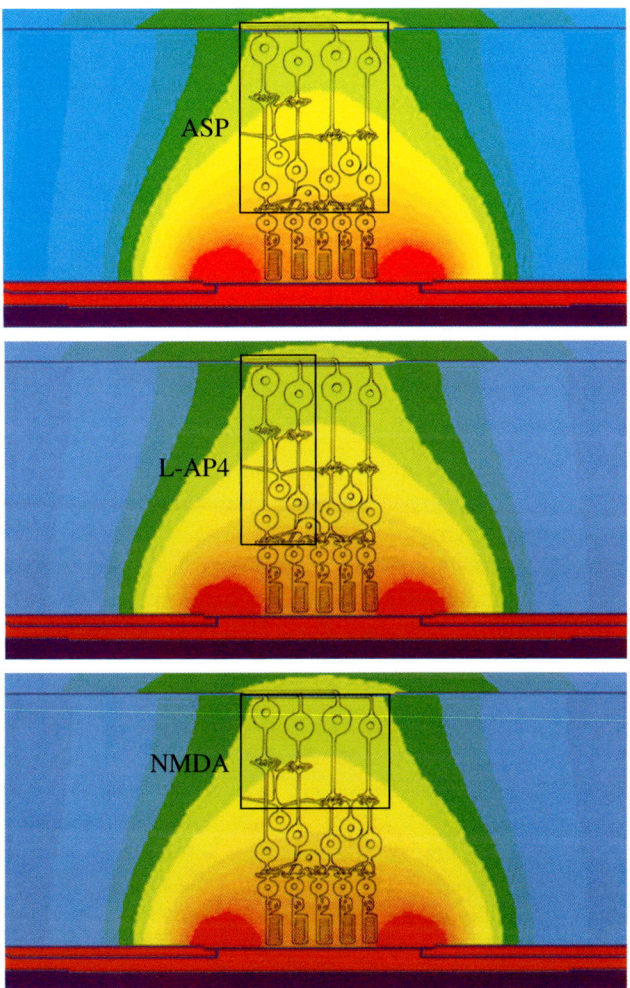

FIGURE 20.6. Interaction between a complex electric field and the retina, and the targets of common glutamate agonists. A cartoon of a moderately degenerated retina is superimposed on a color map of potential gradient (second spatial derivative of the field potential) generated by a finite element model of an eye containing a pair of subretinal electrodes. For the non-uniform electric field created by these bipolar electrodes (lower left and lower right in each panel), the super-threshold stimulus field subtends only subpopulations of retinal neurons, which may be identified by pharmacological dissection of the response. Boxes indicate the cells that are saturated by the indicated glutamate agonist (see Table 20.2). Aspartate (ASP) binds to receptors of all post-receptor neurons. 2-amino-4-phosphonobutyrate (L-AP4) binds selectively to the bipolar cells of the ON-pathway. N-methyl-$_D$-aspartic acid (NMDA) selectively binds to third-order neurons.

TABLE 20.2. Glutamate agonists useful in pharmacological dissection of the summed response of the retina to photic or electrical stimulation.

Compound	Receptor	Effect	Target cell types	References
1) APB (L-AP4) 2-amino-4-phosphonobutyrate	mGluR4, 6–8	Agonist	Photoreceptors (?), Amacrine & On Bipolar Cells	[24, 25]
2) ASP Aspartate	All GluR	Agonist	All Retinal Cells	[26]
3) PDA cis-2,3-piperidinedicarboxylate	All iGluR	Weak NMDA Agonist, KA & AMPA Antagonist	Off Bipolar, Horizontal & Third-Order Cells	[27–29]
4) NMDA N-methyl-D-aspartate	NMDA	Agonist	Amacrine & Ganglion Cells	[30]

cell is within the stimulus field of the electrode, it can readily respond to the applied electric field, and presumably pass the visual signal to post-synaptic neurons in a natural manner.

Pharmacological dissection of the eERG response is performed with two objectives: First, the cellular origins of each component in the response need to be identified. Once this first objective is met, this knowledge can be used to identify the direct cellular targets of the stimulus (i.e. the cell types within the stimulus field).

The use of agonists, which are historically more frequently employed in vision science, have been used to meet the first objective. The compounds chosen for this series of experiments – aspartate, L-AP4, and NMDA – are described in Figure 20.6 and Table 20.2. Figure 20.7 illustrates the experimental results obtained in one animal using NMDA to block activity in third-order neurons (amacrine and ganglion cells). The protocol involves using ERG responses to light stimuli to verify the effect of the agonist, which aids in interpreting the eERG responses.

The results of the experiments using all three glutamate agonists are summarized in Figure 20.8. The interpretation of these results reveals the cellular origin of all three main components of the eERG response. In the presence of L-AP4, which blocks activity in the ON-pathway bipolar cells, the positive phase of the eERG is abolished; therefore, P80 is attributed to this population of neurons. This is consistent with the contribution of this cell type to the ERG, the corneal-positive b-wave (see Figure 20.2). The early negative component of the eERG, N35, is preserved in the presence of both L-AP4 and NMDA. The primary population of retinal cells not affected by either of those compounds is the OFF-pathway bipolar cells, so N35 is attributed to this cell type. In the presence of NMDA, and in the presence of L-AP4, the late negative phase of the eERG is suppressed. Third-order neurons in the ON-pathway are blocked by both of these compounds, and therefore N135 is attributed to this group of cells. With knowledge of the cellular origin of each major response component, changes in the eERG response due to progressive retinal degen-

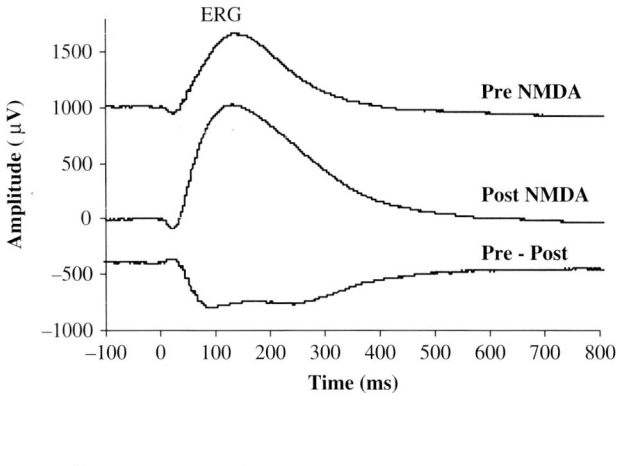

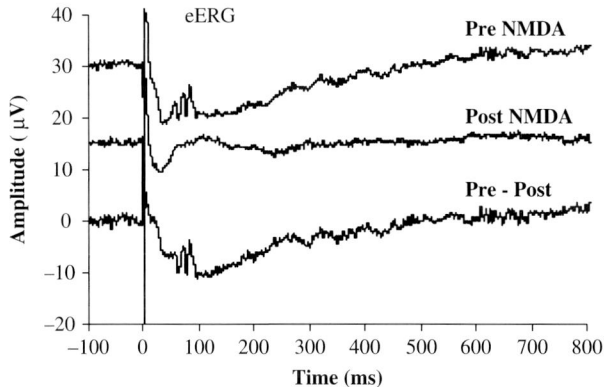

FIGURE 20.7. ERG and eERG responses recorded in one experiment employing pharmacological dissection. ERG responses to light stimuli are plotted on the upper axes; eERG responses to electrical stimulation are plotted on the lower axes. Waveforms are shifted vertically for clarity; all pre-stimulus baselines have average values of zero. The **top trace** in each panel was recorded under baseline conditions, and are typical ERG and eERG response waveforms. The **middle trace** in each panel was recorded approximately 20 minutes following intravitreal injection of NMDA, which binds to third-order neurons. Removing the corneal-negative contribution of third-order neurons enhances the b-wave in the ERG (consistent with previous reports), and eliminates the N135 component of the eERG. This strongly suggests that third-order neurons are responsible for the N135 eERG response component. The bottom trace in each panel plots the difference waveform (pre-NMDA minus post-NMDA), and isolates the portion of the baseline response removed by the presence of the drug.

eration or to changes in electrode design or stimulus parameters can now be studied.

The second objective of using pharmacological dissection described above (identifying the direct targets of electrical stimulation) requires particular care

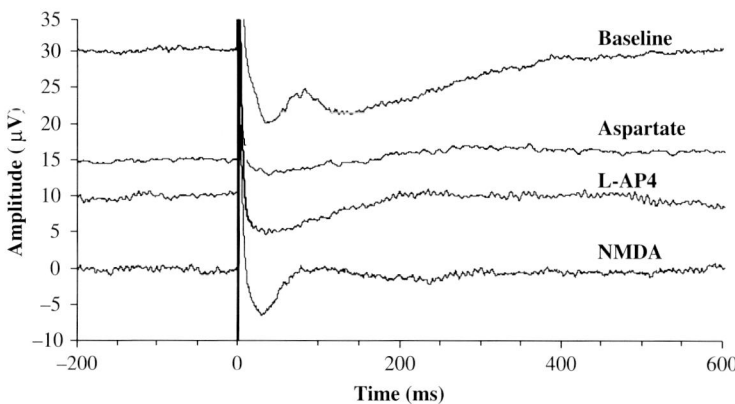

FIGURE 20.8. Summary of pharmacological dissection results. Each waveform is the average response recorded in two animals, except the Baseline response, which is the average across all six animals. The lower three waveforms were recorded in the presence of the glutamate agonist indicated. See text for interpretation.

in interpreting results. For example, our results show that in the presence of aspartate (analog/agonist which binds to all post-receptor glutamate receptors), the eERG response to subretinal stimulation using bipolar electrodes is abolished. One interpretation of this result is that the stimulus field subtends only the photoreceptor layer, and the photoreceptors themselves do not contribute significantly to the recorded response. Thus, the recorded response is comprised of contributions of cells post-synaptic to the photoreceptors. An alternate interpretation is that the stimulus field extends deep into the retina and subtends all of the retinal cell types above the electrode. We still conclude that the photoreceptors do not contribute strongly to the eERG, but we cannot rule out the possibility that the second- and third-order neurons, which are all maximally hyperpolarized or depolarized, are unable to respond to the stimulus in the presence of aspartate. Agonists can be useful in identifying cellular origins of eERG components, but antagonists are required to determine the depth of the stimulus field.

Summary

Invasive and non-invasive recording, and pharmacological dissection, have been used for decades to teach us a great deal about the response of the retina, both healthy and diseased, to natural light stimuli. We have described a number of these historical techniques and suggested ways that they may be adapted to teach us about the response of the retina to electrical stimulation. Early results demonstrating eERG recording, and the use of pharmacological dissection to understand this response, have been presented. While the example of eERG waveforms provided represent typical responses, these were obtained under a very specific set of conditions: one electrode geometry, size, material, configura-tionand placement relative to the retina, delivering one amplitude and one pulse shape of current to one disease state in one animal model. If any of these param-eters were changed, the response would also likely change, suggesting a veritable universe of experiments to be performed. We are directing our experimental efforts by paying attention to the electrode designs and stimulus parameters employed by groups developing prototype prostheses, with the goals of aiding in the interpretation of the response of the retina, at the cellular level, to electrical stimulation, and to the optimization of electrode and stimulus design. A major objective is to learn enough about the input–output relationship of the prosthesis-diseased retina system to support detailed computational models, which will ease the experimental burden, and facilitate rapid prosthesis design. If we do our job correctly, it may be possible some day to determine the physiological state of the patient's retina, enter this information into a simulation environment, and then create a prosthesis design optimized for the individual. However, I suspect that readers of this book will play a larger role in attaining that goal than the authors.

Acknowledgments. The author would like to thank Dr. Monica Baig-Silva, Dr. Casey Hathcock, Safa Rahmani, and Patrick Axtell for providing many of the figures in this chapter. Funding provided by a Bioengineering Research Grant from The Whitaker Foundation.

References

1. Lewicki MS (1998). A review of methods for spike sorting: the detection and classification of neural action potentials. *Network-Computation in Neural Systems.* **9**(4):R53–78.
2. Buzsaki G (2004). Large-scale recording of neuronal ensembles. *Nature Neuro-science* **7**(5):446–51.
3. Granit R (1933). The components of the retinal action potential in mammals and their relation to the discharge in the optic nerve. *J Physiology* **77**:207–239.
4. Marmor MF 1989). An international standard for electroretinography. *Documenta Ophthalmologica.* **73**(4):299–302.
5. Marmor MF (1995). An updated standard for clinical electroretinography. *Archives of Ophthalmology.* **113**(11):1375–6.
6. Marmor MF & Zrenner E (1999). Standard for clinical electroretinography (1999 update). International Society for Clinical Electrophysiology of Vision. *Documenta Ophthalmologica* **97**(2):143–56.
7. Fishman GA & Sokol S (1990). *Electrophysiologic Testing in Disorders of the Retina, Optic Nerve, and Visual Pathway.* American Academy of Ophthalmology (Pub).
8. Lam BL (2005). *Electrophysiology of Vision: Clinical Testing and Applications.* Taylor and Francis Group, Boca Raton (Pub)
9. Pepperberg DR, Birch DG & Hood DC (1997). Photoresponses of human rods in vivo derived from paired-flash electroretinograms. *Visual Neuroscience* **14**:73–82.
10. Hetling JR & Pepperberg DR (1999). Sensitivity and kinetics of mouse rod flash responses determined in vivo from paired-flash electroretinograms. *Journal of Physi-ology (London)* **516**:593–609.

11. Bearse MA Jr. & Sutter EE (1996). Imaging localized retinal dysfunction with the multifocal electroretinogram. *Journal of the Optical Society of America, A, Optics, Image Science, & Vision.* **13**(3):634–40.

12. Enroth-Cugell C & Robson JG (1966). The contrast sensitivity of retinal ganglion cells of the cat. *Journal of Physiology* **187**:517–552.

13. Passaglia CL, Troy JB, Ruttiger L & Lee BB (2002). Orientation sensitivity of ganglion cells in primate retina. *Vision Research.* **42**(6):683–94.

14. Baig-Silva MS, Hathcock CD & Hetling JR (2005). A preparation for studying electrical stimulation of the retina in vivo in rat. *Journal of Neural Engineering* **2**:S29–S38.

15. Yang, X-L (2004). Characterization of receptors for glutamate and GABA in retinal neurons. *Progress in Neurobiology* **73**:127–150.

16. Robson JG & Frishman LJ (1999). Dissecting the dark-adapted electroretinogram. *Documentia Ophthalmologica* **95**:187–215.

17. Xu L, Ball SL, Alexander KR & Peachey NS (2003). Pharmacological analysis of the rat cone electroretinogram. *Visual Neuroscience* **20**:297–306.

18. Hetling JR & Baig-Silva MS (2004). Neural prostheses for vision: Designing a functional interface with retinal neurons. *Neurological Research,* **26**:21–34.

19. Hafezi F, Grimm C, Simmen BC, et al. (2000). Molecular ophthalmology: an update on animal models for retinal degeneration and dystrophies. *Br J Ophthalmol.* **84**:922–927.

20. Jones BW & Marc RE (2005). Retinal remodeling during retinal degeneration. *Experimental Eye Research* **81**:123–137.

21. Machida S, Kondo M, Jamison JA, et al. (2000). P23H rhodopsin transgenic rat: correlation of retinal function with histopathology. *Invest Ophthalmol Vis Sci.* **41**:3200–3209.

22. Marc RE, Jones BW, Watt CB, Strettoi E (2003). Neural remodeling in retinal degeneration. *Prog Retin Eye Res* **22**:607–655.

23. Meyer RD, Cogan SF, Nguyen TH, Rauh RD (2001). Electrodeposited iridium oxide for neural stimulation and recording electrodes. *IEEE Trans Neural Syst Rehabil Eng* **9**:2–11.

24. Slaughter MM & Miller RF (1981). 2-amino-4-phosphonobutyric acid: A new pharmacological tool for retina research. *Science* **211**:182–185.

25. Brandstätter JH, Koulen P, Kuhn R, van der Putten H & Wässle H (1996). Compartmental localization of a metabotropic glutamate receptor (mGluR7): two different active sites at a retinal synapse. *Journal of Neuroscience* **16**:4749–4756.

26. Murakami M, Otsuka T & Shimazaki H (1975). Effects of aspartate and glutamate on the bipolar cells in the carp retina. *Vision Research* **15**:456–458.

27. Slaughter MM & Miller RF (1983). An excitatory amino acid antagonist blocks cone input to sign-conserving second-order retinal neurons. *Science* **219**:1230–1232.

28. Davies J (1982). Conformational aspects of the actions of some piperidine dicarboxylic acids at excitatory amino acid receptors in the mammalian and amphibian spinal cord. *Neurochemical Research* **7**(9):1119–33.

29. Hashimoto K & Kan M (1998). Presynaptic origin of paired-pulse depression at climbing fibre Purkinje cell synapses in the rat cerebellum. *Journal of Physiology* **506.2**:391–405.

30. Massey SC & Miller RF (1990). N-methyl-D-aspartate receptors of ganglion cells in rabbit retina. *Journal of Neurophysiology* **63**:16–30.

Index

BIOLOGICAL AND MEDICAL PHYSICS
BIOMEDICAL ENGINEERING

The fields of biological and medical physics and biomedical engineering are broad. multidisciplinary and dynamic. They lie at the crossroads of frontier research in physics, biology, chemistry, and medicine. The Biological & Medical Physics/Biomedical Engineering Series is intended to be comprehensive, covering a broad range of topics important to the study of the physical, chemical and biological sciences. Its goal is to provide scientists and engineers with textbooks, monographs, and reference works to address the growing need for information.

Editor-in-Chief:

Volumes Published in This Series:

Printed in Singapore